Commendations for Ma  nt? ([1990

In conducting biblical Christian Leadership Seminars for Christian educators and leaders I have challenged participants by example and admonition to reinterpret their disciplines, professions, or areas of endeavor. James Nickel, responding to this challenge, determined to reinterpret one of the most difficult disciplines – mathematics – from a biblical framework. His resulting work is of trailblazing significance.

The Christian community is indebted to James Nickel for the prodigious research and protracted labor this study represents. It is my hope that his scholarly reinterpretation of mathematics will produce a multiplication of biblical Christian education and leadership.
Dr. Glenn R. Martin (1935-2004), Professor of History and Political Science, Indiana Wesleyan University

Having had the privilege of working alongside James Nickel for about three years, I absorbed his scholarly and spiritually insightful approach to the study of mathematics. This book is groundbreaking in that it presents a clearly understandable Christian worldview explanation of the meaning and purpose of mathematics.

Peter Cain, LLB (Hon), BMath (Hon), DipEd, Senior Manager (A/g), Australian Capital Territory Civil and Administrative Tribunal

James Nickel took a huge step of faith to leave the USA in 1982 to travel with a new wife and young son to a village of 300 people in the north of South Australia to teach at a fledgling Christian school of a dozen students. It was there in Booleroo Centre that the Lord led James to find a richness in mathematics that few dream of and even fewer understand.

Peter Frogley, BA, DipEd (Art), Director, Light Educational Ministries, Australia

Mathematics: Is God Silent? is a book that I wish I had written—a sweeping look at the history of mathematics and its Christian influences. Its 400 pages are greatly expanded over the first edition, giving more opportunity for more quotations from primary sources and more opportunity for Nickel to offer support for his position. That position is a strong Reformed tradition, even Kuyperian. Yet not narrowly so. His debt to Jesuit philosopher of science Stanley Jaki comes through frequently.

An extensive bibliography, index, Scripture index, and time-line make the book a useful reference. End-of-chapter questions and two extensive pedagogical sections make it a useful resource in the training of mathematics teachers or in teaching mathematics itself. Its inexpensive … price is further discounted for classroom use to make it more affordable for a college course or ambitious high school course in history of mathematics.

Nickel avoids commonplace oversimplifications of the history of mathematics. For example, he properly credits Newton's first law of motion to Philoponus and Buridan. He properly distinguishes between Cartesian mechanism and Hobbesian materialism. He also avoids the kind of ephemeral mathematics-theology connections that Leonhard Euler, the greatest 18th C. mathematician, mocked when he said to Diderot, "Monsieur, $(a+b^n)/n = x$; donc Dieu existe, réponse! " – "Sir, *a* plus *b* to the *n*th divided by *n* equal *x*, therefore God exists. Reply!"

Nickel intersperses inspiring examples of the beauty of mathematics.

Every Christian teacher of mathematics should read this book.

Dr. Gene B. Chase, Professor Emeritus of Mathematics and Computer Science, Messiah College

James Nickel's book *Mathematics: Is God Silent?* explores the riches of mathematics fully. There is no other book length work of its kind on the market, and it fills a real need for people who are interested in a Christian approach to knowledge.

Dr. Vern S. Poythress, Professor of New Testament Interpretation, Westminster Theological Seminary, former Assistant Professor of Mathematics, Fresno State College

A very helpful resource for beginning the task of integration of academic areas and the knowledge content of Christian teaching is James Nickel, *Mathematics: Is God Silent?*.

Dr. Dallas Willard (1935-2013), Professor of Philosophy, University of Southern California

In this book, James Nickel clearly demonstrates for us that the Creator of the universe has not forgotten any aspect of His creation, nor has He disappointed us by hiding Himself so that He cannot be found in the systems of numbers He is allowing us to uncover. Truly, Jim has lived out the proverbial exhortation, "It is the glory of God to conceal a matter, but the glory of kings to search out a matter."

In Proverbs 12:11, we read, "When a man plows his field he has plenty of bread, but when he pursues vain things he lacks heart." Have you and your students lost the "heart" to pursue mathematics diligently? *Mathematics: Is God Silent?* will rekindle those little first that enthused you about mathematics when you first learned the wonder of counting 1, 2, 3, … on your fingers.

Paul Hawkins, MA, DD, Associate Dean, University of the Nations, Youth With A Mission

A very easy read, this handsome book surveys many of the great mathematicians and illustrates how their Christian assumptions about God and God's creation led to their systematic work in the field of numbers. If God speaks to all of life – as the Bible insists! – then God surely has something to say to math.

Byron Borger, Hearts and Minds

Any Christian that is a student or teacher of mathematics ought to read this excellent volume. Teachers of science and religion will also find it quite fascinating. Citing well-documented historical sources and using cogent argumentation, it overwhelmingly convinces the reader that mathematics ought to be studied and taught from a distinctly Biblical perspective.

Douglas Osborn, President of Biblical Integration Resources

Commendations for *The Dance of Number*

I came across James Nickel's book *Mathematics: Is God Silent?* as a Christian Mathematics teacher. His book was like rain in a parched land. I do not think that I overstate the case when I say that reading *Mathematics: Is God Silent?* was a watershed moment for me as a teacher.

When I was afforded the privilege to work through an advance copy of *The Dance of Number*, I did so with great anticipation. I was not disappointed. I only wish I had been taught with such a textbook – it would have made the study of Mathematics meaningful and much more interesting. In addition to the excellent pedagogy evident throughout the book, Nickel firmly establishes Mathematics as a field worthy of study by any person interested in thinking God's thoughts after Him.

James Nickel's work will benefit schools, teachers and children across the globe. He has done a service to the body of Christ which will, I believe, prove to be of inestimable value in preparing Christian children to take their rightful place in the world, and in setting aright so much that has gone wrong in education.

Hanna Pretorius, HOD Academics, Domino Servite School, South Africa

Work of a lifetime. People are known for their contributions in fields of inquiry, participation in great events, experiments which create discoveries, and ideas which break down walls of unknowing. Few are they whose life work assists in all four.

With James Nickel's latest masterpiece he has given mathematics, publishing, discovery, and think-tank worlds a contribution to which all teachers, students, thinkers, and generations will point. No one has both the breadth of knowledge combined with the educational practice to match Nickel's curriculum. No one ever needs try again.

The Church, the school, the academe, the world should sit up and take notice. Mathematics reflects the language The Father used to create the universe; its principles and truths permeate all of life, every problem, and solution. Jesus, second person of The Trinity, the God-Man, is the source, unification, sustainer, and will be the culmination of all truth. The Spirit now amplifies creation's message to every one of God's creatures: "Look! See! Know! Math is God's witness to earth from Heaven!" All mathematical knowledge throughout creation is intended for Trinitarian glory. James Nickel has held a mirror up to mathematics to announce The Personal Eternal Triune Creator's lordship over all.

Dr. Mark Eckel, President, The Comenius Institute, Indianapolis, Indiana

No field has more frequently been offered as proof of the a-religious nature of vast areas of human knowledge than mathematics, and no scholar has so incisively refuted this notion than James Nickel. In this comprehensive work, Nickel exhibits the inescapably Christian character of mathematics – logical, beautiful, and useful precisely because it is God's creation in the Son. There is no work in history quite like this one – and likely none in the foreseeable future.

P. Andrew Sandlin, Founder & President, Center for Cultural Leadership

James Nickel may well had just written a book which will change the Christian community for the good. For most of Christian history, the people of the Book were people of both the word and people of the number. Modern thought severed "humanities" from "sciences" and associated mathematics with the latter. But letting no man sunder what God has joined, Nickel reconciles the two traditions. If you think that math is not for theologians, or if you think that theology is not for mathematicians, prepare to be surprised.

If you got off the math track somewhere around High School because you got tired of memorizing formulas or because you were convinced that you just aren't a "math person", it's not too late. Nickel's book allows you to go back and pick up where you left off. It also allows you to give the young people in your life a fair chance at the beauty and value of higher mathematics. And most important of all, it affords you a chance to worship God not just with words, but with numbers. He counts the number of the stars; He names them, each one. He is the God of name and number. His glory fills the whole world.

Jerry Bowyer, S.T.L., Economist, writer, and founder of Bowyer Research

THE DANCE OF NUMBER

ELEMENTS OF THE DANCE
BECOMING PROFICIENT IN ARITHMETIC

THE DANCE OF NUMBER

ELEMENTS OF THE DANCE
BECOMING PROFICIENT IN ARITHMETIC

VOLUME 1

BY JAMES D. NICKEL

Nickel, James D.
The Dance of Number: Elements of the Dance, Volume 1
Text /James D. Nickel—1st Edition
Index

Published in the United States by Sound Mind Press, Wenatchee, Washington
www.biblicalchristianworldview.net/donCurriculum.html
First printing by CreateSpace: 2018

Printed in the United States of America
Cover design by James D. Nickel.
Image source: iStockPhoto

SOUND MIND PRESS

Content summary:
1. Arithmetic. 2. PreAlgebra. 3. Geometry. 4. Theology–Trinitarian.

ISBN: 978-0-9991054-0-5 (paperback)

v.1.1 printing, August 2018
v.1.2 printing, Septemeber 2018
v.1.3 printing, May 2019

According to Gregory of Nyssa (ca. 335-ca. 395), creation is a wonderfully wrought hymn to the power and love of the Father. The Father's Word, His Son, brought forth and currently sustains the rational order of the form and matter of the universe as a kind of musical harmony, richly and multifariously toned, guided by the Spirit with an inward rhythm and accord, a joyful dance of giving and receiving, indeed, the symphony of the pleasurable joy of the Triune dance. The intermingling of the melody and cadence of the cosmic elements, of their motion and rest, sing the glory of their Triune Creator. It is this orchestration of all things one with the other which is music in its truest, most perfect, and most glorious form.

Amplification of the summary of the thoughts of Gregory of Nyssa, *In Inscriptiones Psalmorum* (1.3:19-22), by David Bentley Hart, *The Beauty of the Infinite*, p. 275.

Dedicated to my three grandsons, Isaac, Aaron, and Levi. Above all, this book was written with you in mind. May you learn to dance to the tune of the Triune God's orchestrated universe.

PREFACE

As a high school teacher of mathematics, I have observed far too many students who can't do algebraic work because they have not mastered Arithmetic. When it comes to dividing by a fraction, many cannot remember what to invert, a symptom of a deeper conceptual issue: "What does it mean to divide a number by a fraction, and what should my answer approximately be?" I have also noted that very few students possess what I call the joy of number sense, i.e., a *feel* for numbers, the pulse of their patterns – how they work, how they interact, and how they reveal rational order.

Why? It is our digital world of calculators and computers that may be at fault. (Note: I am not an opponent of technological progress.) Too often students succumb to pushing buttons instead of doing an addition or multiplication problem in writing. In our world of digital cash registers, when was the last time that you saw a store clerk return change to you by counting backward, one of the lost arts of mental Arithmetic? Maybe our educational system, by experimenting with all kinds of pedagogical fads, has lost its way and lost far too many students to math illiteracy as a result.

As you can see by Figure 2 (next page), the mastery of arithmetic operations is the foundation for success in Algebra, Geometry, and Trigonometry. We can understand Algebra, as Sir Isaac Newton (1642-1727) observed, as generalized Arithmetic. In Algebra, instead of working with numerals – symbols of the number concept – you are working with symbols of numerals, which are symbols of symbols. These symbols consist of letters that mathematicians combine with numerals. Because mathematics is consistent, you add, subtract, multiply, and divide algebraic symbols in the same way you add, subtract, multiply, and divide numerals in Arithmetic. Algebra, combined with principles of Geometry and Trigonometry, forms the backbone for understanding Calculus, the mathematics of motion at an instant, a method that Newton invented, along with Gottfried Wilhelm Leibniz (1646-1716), when trying to solve scientific problems. The historian Arnold Toynbee (1889-1975) notes the significance of Calculus:

> Looking back, I feel sure that I ought not to have been offered the choice [whether to study Greek or Calculus – JN] … calculus ought to have been compulsory for me. One ought, after all, to be initiated into the life of the world in which one is going to live. I was going to live in the Western World … and the calculus, like the full-rigged sailing ship, is … one of the characteristic expressions of the modern Western genius.[1]

Calculus is the rigging of the ship of modern science and technology. It is also the foundation of higher branches of mathematical study, subjects you will study in University. From the mountaintop of higher mathematics, you can behold a multitude of rational wonders. To climb to this height is a worthy and achievable goal, but you must master the rudiments of counting and algebraic operations as your first steps in the journey upward.

The pages that follow examine the dance of number, explorations that will enable you to discern the rhythm of numerical and spatial patterns, both within mathematics and in the physical creation. These investigations begin with the premise that the Triune God of Scripture, the interpenetrating dance of His being,[2] is the ground for those rational patterns. The more you are at home with these patterns, the better you can understand and use the good gifts that God has granted to you in the physical world. If you want to understand the nature and structure of the physical creation, you must master arithmetic and algebraic processes. May the living God bless you with delight and understanding as you endeavor to learn the dance of these patterns.

James D. Nickel, BA, BTh, BMiss, MA, Wenatchee, WA, 2018

---

[1] Arnold Toynbee, *Experiences* (New York: Oxford University Press, 1969), pp. 12-13.

[2] Theologians have coined a special Greek word for this reality, *perichoresis*. It means "to encompass" or "to mutually indwell."

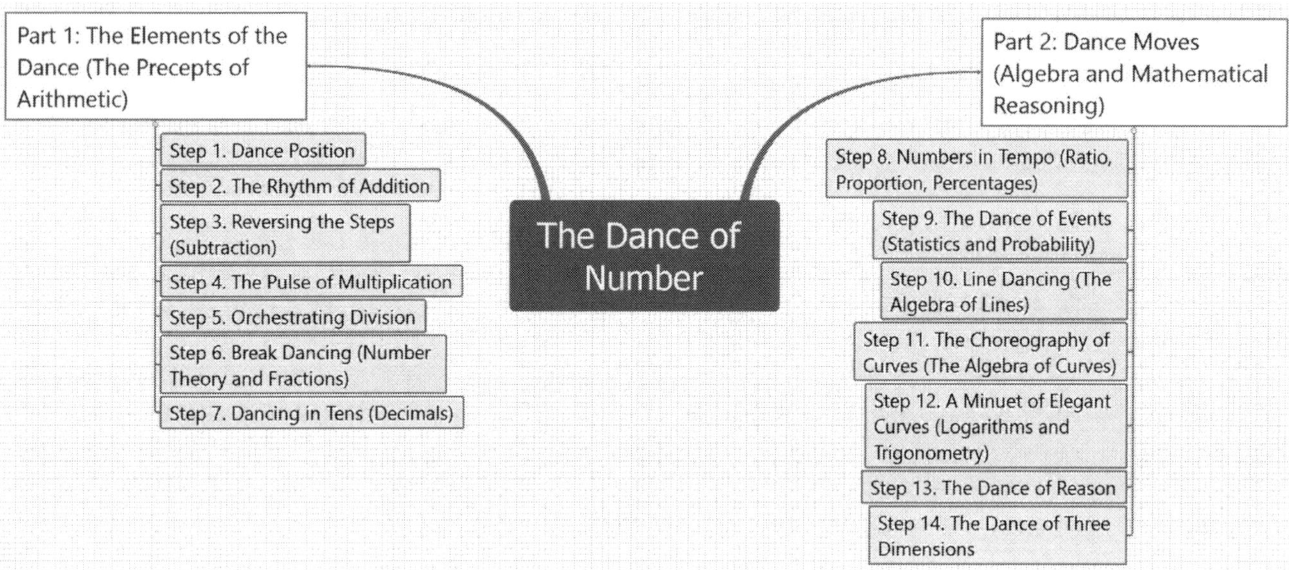

Figure 1.

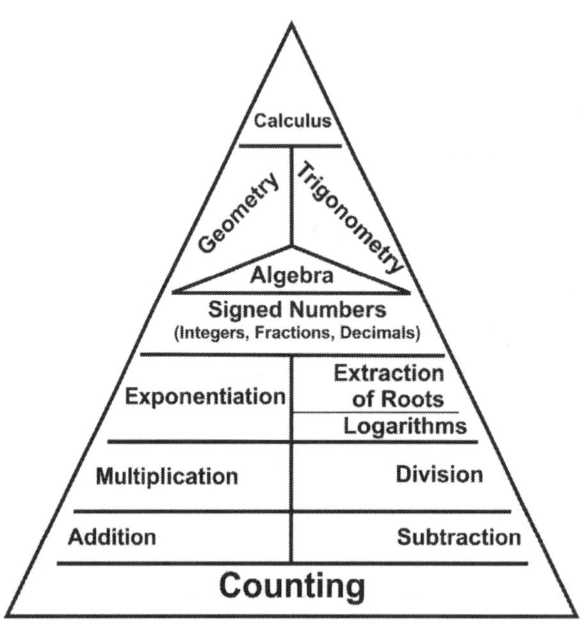

Figure 2.

> It is impossible to understand algebra if you have not mastered arithmetic: it is impossible to understand calculus if you have not mastered algebra. If you attempt the impossible, without realizing what you are doing, your morale will suffer.
>
> Walter Warwick Sawyer, *Mathematician's Delight* (1944), p. 44.

CONTENTS

INTRODUCTION

TO THE READER OF DILIGENCE AND HONOR

You have in your hands a book about Arithmetic, but this book is different. It starts from different assumptions about the nature and character of the physical world outside of you and the nature and character of you as a person, the observer of this world.

Years ago, Athanasius (296-373 AD[1]), bishop of Alexandria, Egypt, penned these words:

> ... there is nothing that is and takes place but has been made and stands by Him and through Him, as also the Divine John 1:1 says, "In the beginning was the Word, and the Word was with God, and the Word was God; all things were made by Him, and without Him was not anything made." For just as though some musician, having tuned a lyre, and by his art adjusted the high notes to the low, and the intermediate notes to the rest, were to produce a single tune as the result, so also the Wisdom of God, handling the Universe as a lyre, and adjusting things in the air to things on the earth, and things in the heaven to things in the air, and combining parts into wholes and moving them all by His beck and will, produces well and fittingly, as the result, the unity of the universe and of its order, ... He produces ... a marvellous and truly divine harmony.[2]

In its history, mathematics as a language has resolutely pointed to a given order, a specificity of order, in the universe. Naturalistic[3] mathematicians and scientists have done their best to avoid the implications of such order, the revelation of consistent rationality in the universe that points to an ultimate ground of rationality beyond it, the dance of the Triune God – Father, Son, and Spirit. The beauty, symmetry, and wonder revealed in the harmonious dance of number are a portal through which you can see a faint glimpse, an infinitesimal glimmer, of the beauty and wonder of the Author and Sustainer of this dance.

Mathematics: from the Greek μαθητης (mathetes) meaning "learner" or "disciple." It is the root of the Old English *mathein,* "to be aware," and the Old German *munthen,* "to awaken." Mathematics, therefore, is a body of knowledge that is teachable and learnable.

The discipline of mathematics, with counting and arithmetic operations as its foundation, is one way in which to order numerical and spatial structures. Since man, male and female, is made in the Triune God's image, our mind has a "sixth sense" in relation to numbers and space. In other words, we can think numeri-

God has endowed his creation with a rationality and beauty of its own in created correspondence to his transcendent rationality and beauty ... the more the created universe unfolds its marvelous symmetries and harmonies to our scientific inquiries, the more it is bound to fulfill its role as a theater which reflects the glory of the Creator and resounds to his praise.
Thomas F. Torrance, *Reality and Evangelical Theology* (1982), pp. 10-11.

---

[1] AD is an abbreviation for the Lain *anno domini,* in the "year of our Lord."

[2] Athanasius, *Contra Gentes* (ca. 320), par. 42.

[3] Naturalism is the belief that everything in the universe can be adequately explained from within the universe. There is no transcendent truth that speaks to, or is the ground of, the reason and purpose of the universe.

cally and spatially because we are made in the image of God.[4] In Genesis 1, we read of the Triune God speaking all things, the space of the heavens and the place of the earth, into existence. Space and number are in unity; one interpenetrates the other. The spatial and numerical patterns revealed in creation have been made in, by, through, and for Christ, the *logos,* the voice, the Word of the Father.

Figure 1. Source: iStockPhoto.

Figure 2. Source: iStockPhoto.

There is a dance, a harmony of order in creation revealed by law-like patterns, all reflecting God's creational covenants (Genesis 8:22). Mathematics, which is the search for patterns in the created order, or mind investigating matter, is reflective of God's covenantal order founded in the Address and Agency of Creation, Jesus Christ the Word (John 1:1-3; Colossians 1:15-17). Since we have been gifted with an orderly mind, our thinking, if it is in submission to the Triune God, will point us to the Ultimacy in rationality, order, harmony, symmetry, proportion, balance, and interconnectedness, the ground and center of the dance of the universe, Jesus Christ. As mathematician William Dembski (1960-) states, "My thesis is that all disciplines find their completion in Christ and cannot be properly understood apart from Christ."[5]

Knowledge of the dance of number is an essential prerequisite for a proper understanding and use of God's created order, which includes the study of higher branches of mathematics. The more you understand God's universe, the more you will discover mathematical connections. God's universe is replete with meaningful patterns that you can understand using mathematics. His creation reveals many numerical secrets. For example, flowers have spirals that line up with a special arrangement of numbers, the Fibonacci numbers, and you will learn more about them in this book. Seashells form perfect mathematical curves called equiangular or logarithmic spirals. A well-rounded study of mathematics should *awaken* you to detect these wonderful patterns.

In addition to observing these creational connections and if you want to be faithful to God's dominion-related calling for your life, whatever that calling entails as a businessman, astronaut, carpenter, homeworker, etc., you need to handle arithmetical problems with ease and assurance. Who knows? God might use you to further explore the wonderful world of mathematics as a vocational calling.[6]

Think, for a moment, about how numbers are a part of your life:

- What time is it?
- How many hours ago did you eat breakfast?

[4] In Latin, *imago Dei.*

[5] William Dembski, *Intelligent Design: The Bridge between Science and Theology* (Downers Grove: IVP, 1999), p. 20.

[6] The Triune God calls everyone to a mission in life, doing something unique based on a grace-based relationship with the Father through Jesus Christ in the Holy Spirit. God has gifted everyone first with existence, the knowledge of being eternally loved of the Father, and finally with abilities and talents as a vessel in His kingdom. Many people think the goal of education is about getting jobs and earning money to pay bills. For the follower of Christ, discerning the Triune God's calling to fellowship in vocation is of the highest priority. As you study and learn about the nature of God's world, He will direct you toward His calling and purpose for you. Be faithful in your studies, and God will engineer your circumstances to thrust you into kingdom energies, no matter what it might involve.

- How many days until Christmas?
- When were you born?
- How many children are in your family?
- How long ago was World War I, World War II?
- How old was the great scientist, Isaac Newton, when he died?
- How tall are you?
- How much do you weigh?
- How much weight have you gained in the last three years?
- How much does that drink cost?
- What is the cost of that drink per ounce?
- How do you balance a checking account?
- What is your street address?
- What is your phone number?
- In an emergency, what numbers do you dial?
- If I'm going to feed twenty people and I have a recipe for five people, how much of each ingredient do I need to add to the mix?
- What does it mean that you can earn 12% per year on a certain investment?
- How many gallons of paint should I buy to paint my room?
- How much wood do I need to build a dog house?
- Is there enough space on my iPod[7] to hold all my music files?

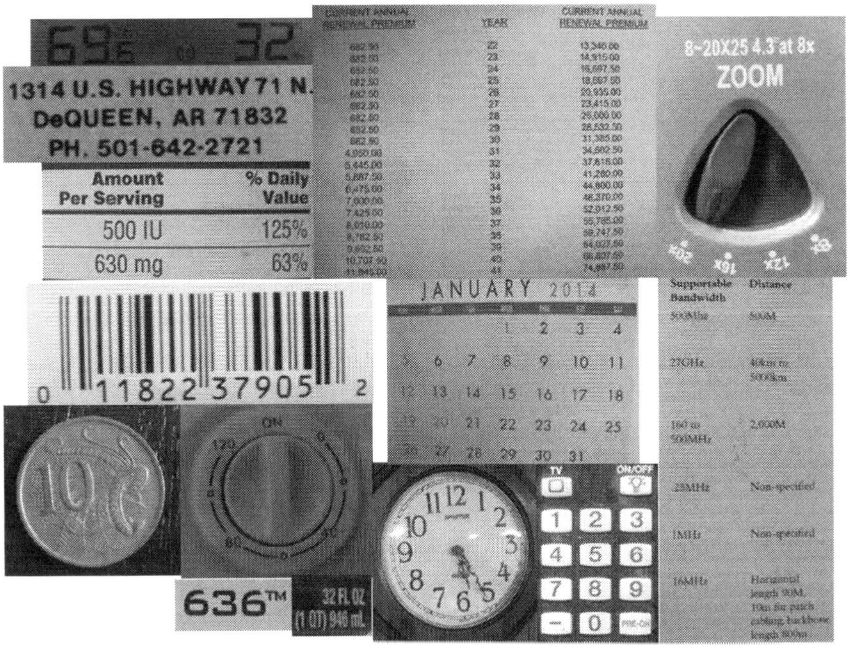

Figure 3. Source: James D. Nickel.

The questions can go on and on.

After Arithmetic, branches of mathematics like Algebra, Geometry, and Trigonometry open fascinating doors of understanding. In the words of Denning Miller:

> The foolish act of jay walking is related to a Euclidean theorem [geometry–JN]. Mathematics will tell you, in moments of enforced contemplation, why bathroom floors can be laid with square or six-sided tiles, but not with eight-sided ones. The stars in their nightly courses reside close to that infinity where parallel lines were conceived by the Greeks to meet. Light, sound, and electricity are easy to comprehend in mathematical terms, but it is the simpler things as well, whose laws mathe-

[7] An iPod is a late 20th and early 21st century marvel, created by Apple; it is a small (handheld) digital computer that can store data, music, and movies. The original iPod has evolved into the iPhone and iPad, devices that are portable computers.

matics will give you. A great suspension bridge silhouetted against a twilight sky, the path of a falling bomb, a sprung diving board vibrating to rest, such everyday things as these you will understand more clearly and enjoy more deeply if you know something of the inwardness of the laws that govern them.[8]

> ... mathematics is a language which can ... vitalize the simplest aspects of everyday life.
> Denning Miller, *Popular Mathematics* (1942), p. ix.

I hope you realize that numbers are a part of your everyday life and without knowledge of numbers and how they work, you could not function efficiently in the life that God has granted to you. In the life of the triune God, you live, and move, and have your being (Acts 17:25-28). He has created not only all things visible, but all things invisible, matter and the structure of that matter. Structure includes the principles that undergird mathematics. The physical creation, what goes on outside of you, has a patterned structure. You, made in God's image, have a mind, what goes on inside of you. Creational patterns reveal numerical relationships, and you can understand this patterned law order, this marvelous dance.

The world outside of you and the world inside of you cohere because of a common Creator. In sync with your diligent study of the harmony of mathematics, my prayer for you is for an increase in the depth of your participation in the internal fellowship of the Triune Creator, the Son's loving worship of the Father in the Holy Spirit.

As you begin this study, pause for a moment and answer the following questions:

- Is math[9] hard for you?
- Do you hate math?
- Do you ever ask yourself, "Why must I learn math?"

> The language of mathematics ... is solely an aid to using your mind – as a matter of fact, a tremendous aid .... In learning any language you have to have a vocabulary first, and there is no vocabulary that is easier to acquire than the mathematical one.
> Denning Miller, *Popular Mathematics* (1942), pp. viii-ix.

If you answered "yes" to any of these questions, consider *developing* the following habits for the responsible and enjoyable learning of arithmetic principles:

1. Renew your thinking about math, from negative talk to positive talk. Since arithmetic operations help you understand more about God's creation, set your mind to think about math in this context.
2. Ask questions of the text. Probe the text for understanding.
3. Learn the language of math. To do so takes practice and plenty of it. Diligent attention to detail and a thorough completion of homework exercises will pay the rich dividends of the joy of experiencing "a job well done."

[8] Denning Miller, *Popular Mathematics: The Understanding and Enjoyment of Mathematics* (New York: Coward-McCann, Inc., 1942), pp. ix-x.

[9] I will use *math* as an abbreviation of mathematics. In Britain and countries that are part of the British Commonwealth of Nations (for example, Australia, New Zealand, and South Africa), *maths* is the abbreviation of mathematics.

4. Understand as well as memorize! Do not solely rely on memorization. Learn the why, not just the how.
5. Read the text carefully! Most people know how to read for information, but only a rare few know how to read for understanding. Seek to wrestle with the text; be diligent, rise early at the task, ask questions, and reread if you do not understand. Work hard at thinking and ruminating over the arithmetical principles taught. If you meet a word that you do not understand, try to understand its meaning by the context. Then, look up its meaning in a dictionary. By practicing this type of mind work, you will gain a most valuable skill. As the saying goes, "Those who read, lead." As you learn to read for understanding, God will work in you the character traits of a rare breed – a servant leader.

> A very considerable part of the reputation mathematics has for being difficult is due to its being taught from the outside, as a memory task, rather than an exercise in understanding and insight.
>
> Walter Warwick Sawyer (1911-2008), *Vision in Elementary Mathematics* (1964), p. 4.

> When reading a book on mathematics, you cannot read it quickly. Acquire the habit of reading with a paper and pencil in hand; this is the only way to own mathematical principles.
>
> James D. Nickel

6. Study math in a way that fits your learning style. Find assistants, a tutor, a parent, other students, or older members of your family to help you if you run into trouble.
7. Don't wait to get help if you do not understand something.
8. Relax as you learn and develop your arithmetic skills. It is an enjoyable skill.
9. Learn to talk math as you observe the dance of number and arithmetical operations around you.
10. Realize that you are responsible for your success or failure in mathematics. Yet, this responsibility does not rest in you *alone*. As you participate in the redemptive reality of Christ for you, and in you, then Christ generates this responsibility (Galatians 2:20). God has gifted you with a mind that can "sense numbers" and how they work. You have the mind of Christ, the Christ who created all things, visible and invisible, and holds all things together and the "all things" include how we understand mathematical laws (Colossians 1:15-17). Participate in this gifting in prayerful dependence upon the Holy Spirit, the gift of Jesus the Word. The Spirit is your Master Teacher. Since the Triune God is the ultimate Author of all law-like patterns, including the dance of number, God by His Spirit can instruct you. He can teach you as you seek to participate, to apply yourself to the task at hand. Remember, fire, oxygen and fuel are the necessary ingredients for starting a fire. The Holy Spirit will light a fire of joyous energy in you as you provide the oxygen of your mind and the fuel of your diligent study.

> Mathematics is a subject where you must read the same sentence several times over, and as with poetry, you must read at an appropriate pace.
>
> Luke Heaton, *A Brief History of Mathematical Thought* (2015), p. 7.

> Mathematics is the loom upon which God weaves the fabric of the universe.
>
> Clifford A. Pickover, *The Loom of God: Mathematical Tapestries at the Edge of Time* (1997), p. 16.

> The pattern of learning mathematics follows this order: First, the beautiful (how wonderful!), then the good (I want to learn this!), and finally, the true (now I want to understand!). We expose the student, no matter what the age, to some of the wonders of mathematics (e.g., Pascal's Triangle, The Fibonacci sequence, casting out Nines, multiplying with fingers, the hexagonal structure of the honeycomb and of snow flakes, etc.). This exposure awakens a profound desire to investigate, experiment, play. Then the actual participation in wonder teaches the student, from the inside, the rules and rhythms, the dance of mathematics. A completely inadequate way of drawing a student into the world of mathematics would be to start with a clarification of the rules or with a set of drills. Rather, show the student the beauty of mathematics, and he will want to learn it, and having participated in it, he will know.
>
> James D. Nickel, application of some thoughts of Thomas Aquinas (1225-1274).

How to Use this Curriculum

Perspective Awareness

The reason for the popularity of some math curricula is its ease of use for both students and teachers/parents. These curricula, especially those designed specifically for homeschoolers, are primarily formula- or skill-based. There is little serious interpenetration with history, philosophy, theology, science, or learning to read and write real mathematics, i.e., its syntax.

> The formulation of a problem is often more essential than its solution, which may be merely a matter of mathematical or experimental skill.
>
> Albert Einstein, *The Evolution of Physics* (1938), p. 95.

The only way most math textbooks can achieve popularity is by stooping to the lowest level, i.e., they give the student math formulas or techniques that help the student to get answers for specific types of problems. This cookbook approach limits the development of conceptual thinking by trumping the why question with the how question, i.e., methods eclipse reasoned justification. Understanding and doing math requires the interplay of two thought components: (1) inductive reasoning, the search for patterns or the derivation of a way to solve a problem and (2) deductive reasoning, the establishment of the absolute truth of certain propositions or the derivation of the correct steps to get the correct answer to a problem. Instead of dictating methods, the student should learn how to discover them via induction and justify them via deduction. Many math curricula sacrifice what is essential in mathematics, problem-solving and critical thinking skills, under the facade of making math easier to learn.

The goal of this math curricula is to develop mathematical utility, the how, in the context of mathematical exploration, how the how is obtained, and mathematical justification, the why.

GUIDELINES

This curriculum is divided into two parts, Part 1 (The Elements of the Dance: Becoming Proficient in Arithmetic) and Part 2 (Dance Moves: Algebra and Mathematical Reasoning).

- Each part is divided into two volumes. Each part is divided into seven chapters, where each chapter is called a Step, a step of the dance.

Part	Volume	Content	Steps
1	1	The Elements of the Dance: Becoming Proficient in Arithmetic	1 – 4
	2		5 – 7
2	1	Dance Moves: Algebra and Mathematical Reasoning	8 – 11
	2		12 – 14

- This material is suitable for ages 12 to 16 (grades 6 to 10). Some of the material in Steps 11 and 14 is somewhat advanced so it would be useful as a reference for students aged 17 to 18. If students age 12 to 16 master the material in Steps 1 to 10, they should be able to follow the reasoning and methods taught in the last four Steps.
- If a student age 11 to 13 needs to master arithmetic operations, I recommend working only through Part 1, or Steps 1 to 7, doing it in one or two years.
- In Part 2, beginning with Step 8, we start exploring algebraic ideas, i.e., solutions to linear, quadratic, trigonometric, and logarithmic equations. Although Steps 8 to 14 are algebraic in their thrust, the algebraic work is linked with geometric, statistic, logarithmic, and trigonometric concepts. The arithmetic employed in these steps depends upon the instruction of the previous steps. Therefore, I do not recommend that students pick up this text and start with Step 8. Steps 1 to 7 need to be done first.
- I have designed this two-part text for three to four years of study. There is no need to rush math matters, especially when mastery of concept is so important at this stage.[10] Part 1, Steps 1 to 7, can be done in one year accelerated or at a slower pace in two years. Two years should be set aside to complete Part 2, Steps 8 to 14.[11] This schedule can vary depending on the maturity of the student.
- Many traditional day schools recycle books from year to year. Therefore, very rarely do students get to keep their math textbooks. There are three reasons why a student should retain this textbook series:

---

[10] In recent decades, there has been an unfortunate tendency in math education, public and private, to rush a student to take Algebra as early as possible, sometimes as early as Grade 7. From 30 years of involvement in math education, I know that a rigorous study of Algebra requires a maturity level that most students age 12-13 do not have. There are exceptions to this rule, but I advise educators to make sure that arithmetic fundamentals are thoroughly mastered before students enter the world of Algebra. If mastery of arithmetic does not exist, studying Algebra is not only a waste of time, it will destroy morale by positioning students in the "I hate math" mode permanently. And remember, there is a significant transition in the mental process as one goes from Arithmetic to Algebra. Professor Warren W. Esty, in "Algebraic Thinking, Language and Word Problems" (*Communication in Mathematics, K-12 and Beyond*, p. 52), advises caution because "students who are at a numerical level of thought will not necessarily be prepared to work with algebraic objects of thought at the conceptual level assumed by algebra texts and instructors." It appears that some schools, public and private, elevate themselves by advertising that their students are doing Algebra in Grade 7. Don't let this drive for elite status fool you. *It is better to go slow and master the material, than to go fast and get lost.* In this context, Part 1 of this curriculum focuses on the mastery of arithmetic, where algebraic thinking is introduced to prepare the student for Part 2.

[11] The scope of concepts taught in Part 2 include material included in the traditional Algebra 1 and Algebra 2 syllabi. That is why, if necessary, a student should take two years to complete Part 2.

1. As a record of their learning experience.
2. As a quality reference for future math work.
3. As an addition to their personal library, a library that all educators should encourage students to build.

Also, because of the interconnected nature of this curriculum, the student needs to have access to previous volumes as they work through the four-year program. The cost these textbooks in the paperback version is considerably less than the typical hardcover book, so I encourage schools to purchase them for student ownership. Or, better yet, have the parents or students do the purchasing. Personal ownership of a book encourages the discipline of self-government, the handling of care of one's property.

- Mastering the material in this two-part curriculum would be more than adequate preparation for the study of a formal, rigorous course in either Euclidean Geometry or PreCalculus. The added and unique bonus of this text is its introduction and demonstration of the mathematical methods of proof and concepts foundational to the infinitesimal Calculus.
- Any parent or teacher of K-5 levels would benefit by studying Steps 1 to 7. The methods taught, along with supporting justifications, can be employed no matter what textbook is used for these levels, but be aware that most the textbooks on the market do not teach left to right methods taught in these steps. What these textbooks teach and what these steps teach, especially in the employment of these streamline methods and concept explanations, differ.
- Any parent or teacher of any level, or any interested individual, will reap rich benefits by studying the methods and concepts taught in this curriculum.
- I include quotations throughout the text for three reasons. First, some of these quotes entertain. Believe it or not, mathematics has a humorous side to it, so enjoy some of this joviality. Second, some edify. By that, I mean that these quotes, as truth statements, are meant not only to inform, but to elevate, to guide, and to nurture. Encourage students to ponder them carefully. Third, some challenge. They are meant to stretch the mind, engage thinking, and generate this question, "What is meant by this?" If students wrestle with them, they will develop a rewarding skill that will stay with them for their lifetime.
- All Scripture references should be read slowly and carefully. I believe that the source of Scripture is (1) the breathing who is the Holy Spirit, the *pneuma*, who proceeds from the Father (II Timothy 3:16) and (2) Christo-centric in nature (Luke 24, John 5:39), glorifying the Father's Son (John 17). The purpose of the words of Scripture is to lead the reader to know Christ the Word, the creator and sustainer of all things, the ground of every facet of rationality, the exact representation of the Father, and the One through whom the Father has reconciled an evil, crooked, sick, and antagonistic world to Himself (John 1:1-14, Colossians 1:10-23, Hebrews 1:1-3, II Corinthians 5:9-6:3).
- The material contained in every footnote should be read.
- Every homework problem should be attempted and completed. These problems are not designed so that they can be skipped. In many cases, the homework exercises of subsequent lessons depend upon successful completion of the homework exercises of the current lesson.
- Detailed solutions manuals, including suggested scheduling, are available; these solutions do not simply state the answer but show how one reasons to the answer.
- Refer to http://www.biblicalchristianworldview.net/donCurriculum.html for updates and more information on this curriculum.

ACKNOWLEDGEMENTS

This text has been written as the fruit of nearly 40 years of combined study in and teaching of mathematics. In 2001, I wrote a rough draft of this material in six months during a full-time work schedule. I have spent the subsequent years teaching the material and revising it via feedback and further research.

I have used a host of resources. See the bibliography for a good number of them. The list is given in Volume 2 of both parts.

The fundamental paradigm that I use for teaching addition, subtraction, multiplication, and division is based upon Edward Stoddard, *Speed Mathematics Simplified* (Dover Publications, [1962, 1965] 1994), a book I discovered in 2005. Unlike most speed math books, rather than providing a series of disjointed tricks, Stoddard's book teaches a streamlined and consistent method, based upon the Japanese Soroban and its use of complements. The connections and techniques he explores are sound, sensible, speedy, interesting, fascinating, and enjoyable. I have taken the liberty of changing Stoddard's methods of teaching multiplication and division to ensure additional efficiencies. In doing so, I have incorporated some of the methods taught in the schools of India, commonly known as Vedic mathematics.

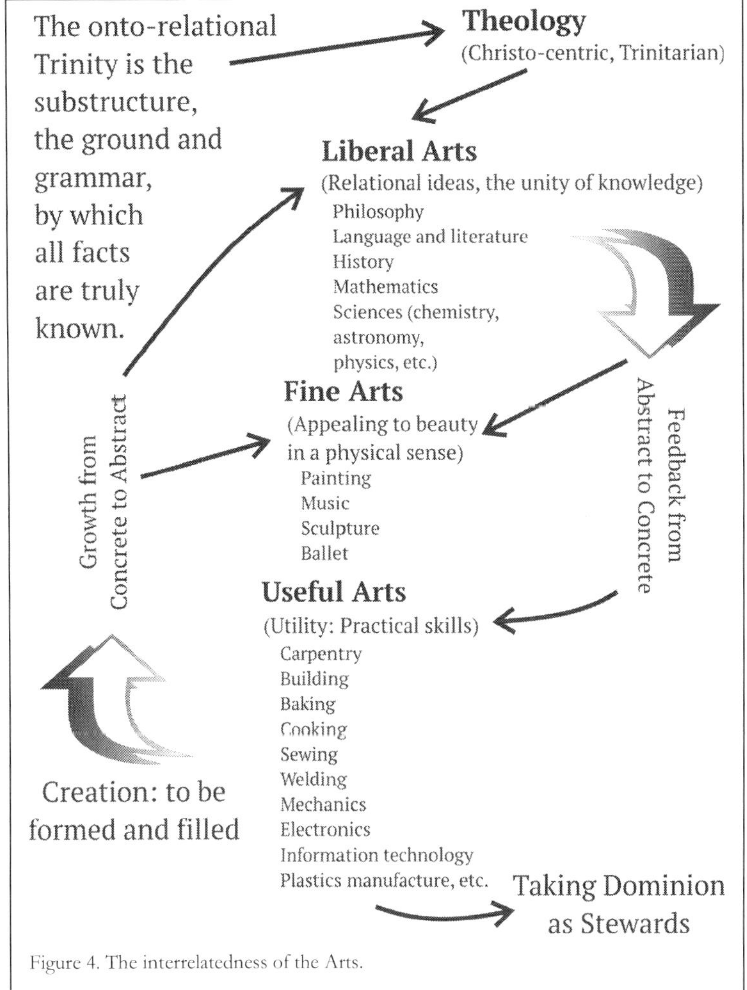

Figure 4. The interrelatedness of the Arts.

Contrary to popular opinion, teaching Arithmetic is not easy. A teacher must understand fine nuances and instruct by building layer upon layer. Israeli university professor Ron Aharoni (1952-) discovered how difficult and challenging teaching Arithmetic is when he had to teach it to young children in a kibbutz. The lessons he learned, along with many insightful observations and teaching methodologies, are found in his excellent book *Arithmetic for Parents* (Sumizdat, 2007). I have gleaned a goodly portion of wise insights by studying this short but powerful book.

I am thankful for the textbooks authored by Harold R. Jacobs (1939-). In 1982, they served to apprentice me as a novice mathematics teacher.

Without peer in the exposition of the interpenetration of mathematics in the physical sciences, I am grateful to longtime New York University Professor Morris Kline (1908-1992), especially for his seminal works, *Mathematics in Western Culture* (Oxford University Press, 1953) and *Mathematics and the Physical World* (Thomas Y. Crowell Company, 1959).

For his profound insights into the language and syntax of Algebra, I am indebted to Warren W. Esty, Emeritus Full Professor of Mathematics at Montana State University.

For an understanding of the Trinitarian ground of the rationality that is pervasive in the universe, I am indebted to the writings of Thomas F. Torrance (1913-2007), professor of Christian Dogmatics, University of Edinburgh, and a 1978 Templeton Prize winner.

> The future seems to hold, for mathematics instruction, an increased dependence on technology in the form of computers, hand-held calculators, and video presentation. Surely, I cannot be the only person to notice the clear correlation between the declining mathematical abilities of American students and the aggressive introduction of these technologies into the mathematics classroom. The decline began as the technology came in.
>
> Jerry P. King, *The Art of Mathematics* (1992 [2006]), p. 176.

RECOMMENDED RESOURCES

TOOLS

- Kitchen measuring aids
- Analog Dial Caliper (precision: 0.001 inch)
- Metric and British Imperial ruler
- Spring scale (grams/Newtons)
- Compass for geometric constructions (preferably one made by Staedtler)
- Protractor or a Carpenter's Angle Finder
- Japanese Soroban
- Buret, Graduated Cylinder, and Beaker

SOFTWARE

- Spreadsheet (Microsoft Excel is preferable)
- The Geometer's Sketchpad (http://www.dynamicgeometry.com/index.html)
- Ultra Fractal (http://www.ultrafractal.com)
- Graphmatica (http://www.graphmatica.com)
- Desmos (https://www.desmos.com/calculator)
- Great Stella Polyhedron Exploration (http://www.software3d.com/)
- Graphing Calculator 3D (http://www.runiter.com/graphing-calculator)

Part 1: Elements of The Dance, Volume 1

Becoming Proficient in Arithmetic

> Behind the intelligibility of the universe, its openness to the investigation of science, there lies, I believe, the fact of the Word of God. The Word is God's agent in creation, impressing his rationality on the world. That same Word is also the light of men, giving us thereby access to the rationality that is in the world.
>
> John Polkinghorne, *The Way the World Is: The Christian Perspective of a Scientist* ([1983, 1992] 2007), p. 109.

STEP 1. DANCE POSITION

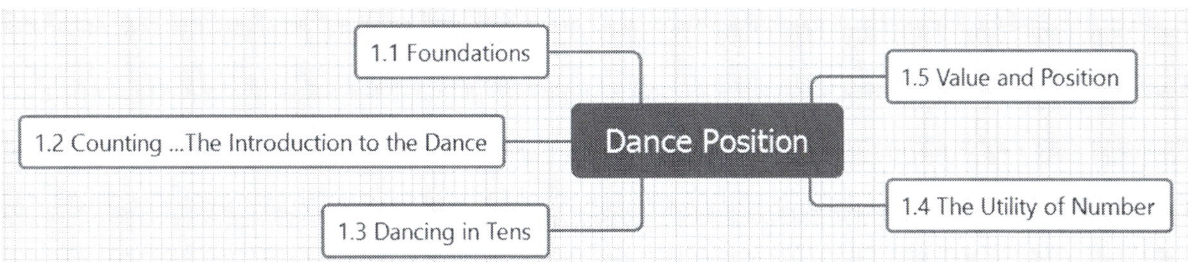

> Every point of view is
> a view from a point.

1.1 Foundations

Charles Lutwidge Dodgson (1832-1898), or Lewis Carroll of *Alice in Wonderland* fame, was not only a writer, he was a mathematician. In his book, *Through the Looking Glass*, he engages a narrative between the White Rabbit and the King:

> The White Rabbit put on his spectacles. "Where shall I begin, please your Majesty?" he asked. "Begin at the beginning," the King said very gravely, "and go on till you come to the end: then stop."

Terms, Symbols & Concepts Introduced
1. Algebra
2. *al-jabr*
3. *Anno Domini*
4. Arithmetic
5. *Arithmos*
6. cf.
7. Elohim
8. ff.
9. Geometry
10. i.e.
11. *imago Dei*
12. *Logistica*
13. *Logos*
14. Mathematics
15. Naming as a tool of dominion
16. Number
17. Ontology
18. Onto-relational
19. Perichoresis
20. Trinity

The Bible tells us about the beginning of time and space as we know it;[1] and it is at this beginning, a beginning that tells us about the Creator and the creation, that we commence our journey into the world of mathematics.

Creation

> In the beginning God created the heavens and the earth (Genesis 1:1).

The Hebrew for God is ***Elohim***, a plural noun.[2] This plurality is exemplified in the phrase "Let Us make" (Genesis 1:26). In the next verse, we see the Spirit[3] of God hovering over the face of the waters (**cf.**[4] Genesis 1:2). Then, Elohim created all things by the speaking forth of His Word, by the breath of His mouth (Genesis 1:3**ff.**[5]; Psalm 33:6).[6]

In the first three verses of the Bible, the act of creation is circumscribed in plurality: God, Spirit, and Word. In the New Testament, we discover Elohim in the fullness of plurality in unity, Elohim is **Trinity**, One in Three and Three in One: Father, Son, and Spirit.

> Creation is "poetry in motion" – the great dance being played out through the scenes of our lives.
>
> C. Baxter Kruger, *The Great Dance: The Christian Vision Revisited* (2000), p. 88.

Figure 1. Source: United States Postal Service.

> In the beginning was the Word [*logos*], and the Word was with

---

[1] The why of this beginning, the creation, is the extravagantly loving purpose of the Triune God for humanity, a plan secured before the creation of time and space. See Ephesians 1:1-14 and II Timothy 1:7-10.

[2] The masculine plural form of this word may be regarded (1) as intensive to indicate God's fullness of power, (2) as majestic to indicate God as one being above all, or (3) as an allusion to the plurality in the being of God Himself.

[3] Spirit, in the Hebrew, is transliterated into the English as *ruach*. It is a feminine noun that means spirit, wind, or breath. In the Greek, spirit is the neuter πνεῦμα (*pneuma*). In John 16:13, πνεῦμα is linked to the neuter ἐκεῖνος (that one), but used in connection with masculine personal "he will guide" (ὁδηγήσει), "he will speak" (λαλήσει), "himself" (ἑαυτοῦ), "he may hear" (ἀκούσει), and "he will declare" (ἀναγγελεῖ). Irrespective of grammatical gender, she, it or he, by the Spirit being deeply relational, beautifully ordering, and fellowship-oriented, a communion that runs side to side with power, we know what it means to be an authentic person.

[4] cf. is the Latin abbreviation of *confer,* meaning "compare with."

[5] f. is the Latin abbreviation of *folio,* meaning "next page." In this instance, ff. means "the following verses."

[6] Thus, the Spirit and the Word, albeit distinct, are tied together since they are both linked to the voice of Elohim. The image given is that the Word is the articulate sound of Elohim, the Spirit is the wind, the breath of that voice.

> God, and the Word was God. He was in the beginning with God. All things came into being through Him, and apart from Him nothing came into being. [What] has come into being in Him was life, and the life was the light of men … And the Word became flesh, and dwelt among [in] us, and we saw His glory, glory as of the only begotten from the Father, full of grace and truth (John 1:1-3, 14).[7]

> For by Him [Christ, the ***logos*** of God – JN] all things were created, both in the heavens and on earth, visible and invisible, whether thrones or dominions or rulers or authorities – all things have been created through Him and for Him. He is before all things, and in Him all things hold together (Colossians 1:16-17).

THE CREATOR

God's speech in Genesis 1 is the Word, the *logos* (in Greek, λογος) of God, the Lord Jesus Christ. In the beginning, this Word already existed, and this Word was in face-to-face fellowship with God the Father. This Word has always been of the same essence, the same being as the Father. All things, the heavens and the earth, came into existence through this Word. Without this Word, not one thing, not one iota of creation, came into existence. What has come into existence is in this Word and this Word is life, the life who is the light of everyone.

Logos means reason, intelligence, wisdom, the living voice, eternal self-communication.[8] The Apostle John connects the person of Christ to the reason of creation, a reason grounded in relationship. Christ, the Word of God, is the source of the rationality that pervades the universe in its many forms. Since *logos* is centered in the person of Christ, all reasoning, to be valid reasoning, must also be grounded in Christ. Since Christ reveals the Father (John 14:9), and the Spirit of God reveals Christ (John 16:14), to say that Christ is the ground of rationality is also to say that the Father and the Spirit are the ground of rationality.

> Christ the Word envelops both Scripture and Creation.
> James D. Nickel

God is three persons in one being. God's triune nature is a joyful and self-giving mutual indwelling and interpenetration, union with distinction, indeed, a dance of three persons who share the same being, or the same essence. Using Greek words developed by the Nicene[9] church fathers of Cappadocia (4th century AD), God is three *hypostaseis* in one *ousia.* (In the Greek, *treis hypostaseis en mia ousia.*) To grasp the nature of this reality (John 10:38, John 14:10-11, 20; John 17:23), later theologians used the Greek word **perichoresis**, from *perichoreo,* meaning "to encircle," "to encompass," "reciprocal containing," "mutual inexistence," or "mutual indwelling."[10] The dance metaphor comes from the Greek word *perichoreuo*, meaning "to dance around." *Choreuo* is the root word used in choreography, the science of the sequence of steps and movements in a dance.

---

[7] Concerning the [What] addition, the New American Standard Version of the Bible does not translate the third verse in this way. The flow of the Greek text allows for it, though, since it has no punctuation marks. The New Revised Standard Version changes the punctuation based on that movement of thought. [Note: the Greek word translated "among" in verse 14 is εν (en). This preposition means "in." Note how John uses this preposition in the rest of the first chapter, specifically verses 2, 4, 5, 10, 23, 28, 31, and 33.]

[8] In contrast to ancient Greeks who understood *logos* as the impersonal principle of the universe. *Logos*, understand in its Hebraic framework, means "God's eternally self-communicating being." In Harold Wells, *The Christic Center: Life-Giving and Liberating* (Maryknoll: Orbis, 2004), p. 113.

[9] By Nicene, we mean the date of the writing of the Nicene-Constantinopolitan Creed (ca. 325, 381 AD). AD is an abbreviation of the Latin ***Anno Domini***, which means "in the year of our Lord.

[10] G. L. Prestige, *God in Patristic Thought* (London: SPCK, 1952), pp. 289-291.

God is Trinity. Father, Son, and Spirit are involved in one another, an involvement of intimate fellowship, deep and penetrating union. God is being in communion. **Ontology** is the study of the nature of being or existence. With this explanation in mind, we can say that the God of Scripture is **onto-relational**, being in relationship.

A post-Nicene church father, Aurelius Augustine (354-430) said that "we think God's thoughts after Him" when we study Scripture. Since the words of Scripture bears witness to Christ, the Word or Voice of God, studying Scripture is a way to relate to Him who is life eternal (John 5:39). In Genesis 1, we see God speaking creation into existence. His Word, Jesus (John 1:1-3), is an expression of His thoughts. Creation is the consequence of God's thoughts, or His plans, plans that came to fruition by His Word.

The astronomer Johannes Kepler (1572-1630) encountered the Living God as he explored the mathematical relations of God's harmonious creation. After developing the law of the elliptical[11] motion of the planets around the Sun, he fell to his knees and exclaimed, "My God! I am thinking Thy thoughts after Thee!"[12]

The scientist Albert Einstein (1879-1955) believed that the deeper one penetrates into nature's rational order, the greater one's respect for God. Although not a believer in the onto-relational Trinity of Christian orthodoxy, Einstein followed the thought of Scottish physcist James Clerk Maxwell (1831-1879), a believer, who said that in science, relation is the most important thing to know.[13] With this input from Maxwell, Einstein echoed Kepler's thoughts by saying, "What we [physicists] strive for ... is just to draw His lines after Him."[14]

There is a perichoretic structure, a dance, in creation. This dance is the unity behind the diversity of the cosmos,[15] the rational ground of all that exists. This unity exists because all things cohere in Jesus Christ, the Word of the Father, God the Son made flesh (John 1:14), who reveals to us the perichoretic nature, the dance of the dynamism of His relationship to His Father in the Holy Spirit.

NUMBER

> Human mind and culture have developed a formal system of thought for recognizing, classifying, and exploiting patterns. We call it mathematics.
> Ian Stewart, *Nature's Numbers* (1995), p. 1.

One aspect of the perichoretic structure of creation is **number**. Number is the rational order of things that makes the universe accessible to quantification (Lesson 2.1), **i.e.**[16], "to find out how much." We quantify objects by counting, by measuring, or by developing valid expressions using determinate, or fixed, quantities.[17] The dance of number is the dance of this rational order. **Mathematics** is the study and development of the rational order of the universe revealed by number.

---

[11] We will explore the ellipse in later lessons. It is an oval-shaped curve.

[12] Cited in Colin Humphreys, "Can Science and Christianity Both Be True?" in R. J. Berry, ed., *Real Science, Real Faith* (Eastbourne: Monarch, 1991), p. 116. Kepler exulted, with pleasure, in the works of God (Psalm 111:2). In 1604, Kepler wrote, "I may say with truth that whenever I consider in my thoughts the beautiful order, how one thing issues out of and is derived from another, then it is as though I had read a divine text, written into the world itself, not with letters but rather with essential objects, saying: Man, stretch they reason hither, so that thou mayest comprehend these things." Cited in Max Caspar, *Kepler* (Mineola: Dover Publications, [1959] 1993), p. 11.

[13] Maxwell also developed a relational way of understanding and doing mathematics. He named it embodied mathematics (Lesson 8.9). When he designed the famous Cavendish Laboratory in Cambridge University in 1874, he ordered this Latin inscription to be placed over the entryway: *Magna opera Domini exquisita in omnes voluntates ejus*, meaning "The works of the Lord are great, studied by all who delight in them" (Psalm 111:2).

[14] Cited in Thomas F. Torrance, *Theological and Natural Science* (Eugene: Wipf and Stock, 2002), p. 25.

[15] The cosmos is the physical universe. The word signifies "embroidery" giving the idea of the orderliness and harmony of its patterns. We get the word cosmetics from cosmos. Applying cosmetics to the face brings its elements into harmony and beauty.

[16] i.e. is the Latin abbreviation of *id est,* meaning "that is."

[17] For more about the meaning of determinate, see footnote 3 of Lesson 2.8, Lesson 9.1, and footnote 4 of Lesson 13.10.

Number, as a word, has an interesting derivation. The Old English or Anglo-Saxon, the English language from the 5th century to the 12th century[18] AD, used the word *rim* when referring to number. *Rim* also means "part." Middle English, the English language from the 12th century to the 16th century AD, used *nombre*, which came via Old French from the Latin *numerus*. We can trace this word to the ancient Greek word νεμο (*nemo*), meaning "distribute" or "share." It is of interest to note that we can recognize this word in νεμεσις (*nemesis*[19]), meaning "allotment" or "portion" of the anger of a god or goddess.

God's creation is imbedded with the rationality, the dance, of numerical and spatial patterns. Number forms the basis for **Arithmetic**[20] and **Algebra**.[21] Space, understood in the context of number, provides the basis for **Geometry**.[22] Number, the basis of the rational process of counting and measuring, reports on the nature of reality, a reality that is grounded in the person of Christ. Numerical and spatial patterns bring to light the rational order, the beauty, the perichoretic dance of creation.

The Triune God is the foundation, the ground, of all things, and our ability to count and measure is one of those things. God has ordered the things of creation by measure, number, and weight. Therefore, when we study mathematics and discern numerical and spatial patterns, we are in tune with the Triune God's ordering of things.

On the sixth day, the Triune God created man, male and female, in His image. The Latin for "image of God" is ***imago Dei***. Image in this sense means a reflection, to a limited and finite extent, of the person and character of the Triune God, the Creator. His dance is therefore in us. We are wired for perichoresis.

DOMINION

The Earth is the created gift of the Triune God to humanity, given to the *imago Dei* to form, to develop, to cultivate, to rule:

> The heavens are the heavens of the Lord, but the earth He has given to the sons of men. (Psalm 115:16)

The Father, Son, and Holy Spirit then told man to take dominion over the Earth with specific attention given to the animal creation (Genesis 1:26-27). An example of intelligent rule is Adam's naming of these animals. Note that God presented the animals before Adam in order "to see what he would call them" (Genesis 2:19). Although God named the stars (Psalm 147:4; Isaiah 40:26), He gave Adam the charge of naming the animals, i.e., the freedom to create names for the animals. The Triune God gave man the ability to observe the animals and the liberty to choose an appropriate name that aptly described their characteristics, i.e., the dance of their patterns. The Biblical understanding of naming is two-fold:

1. A name of an object reflects the characteristics, the essence, of that object, the meaning and value God gave to it, the discernment of its place in the cosmos, and
2. A person cannot name an object unless he has previously been given dominion over it; the object to be named is God's gift to mankind. Therefore, **naming is a tool of dominion**.

---

[18] Note how the century numbers are used, 5th and 12th. This usage speaks to the sequence of order. See Lesson 1.4.

[19] From the ancient Greek goddess of retribution, *Nemesis*.

[20] The ancient Greeks understood Arithmetic, ***arithmos***, as the "theory of numbers." They used another word to describe the skill of calculation with numbers: ***logistica***. We shall study the elements of both.

[21] Algebra is a shorthand way, using symbols, of generalizing Arithmetic to help solve specific problems involving number. The word itself is Arabic, ***al-jabr***, meaning "transposition" or "reunion." *Al-jabr*, used in a non-mathematical sense, was something a barber did in medieval times. A barber not only cut hair, he mended bones. Thus, an *algebrista* was a bonesetter. In Galatians 6:1, the Apostle Paul said, "Brethren, even if anyone is caught in any trespass, you who are spiritual, restore such a one in a spirit of gentleness; each one looking to yourself, so that you too will not be tempted." The Greek word for restore means "to join back together" as in setting a broken bone or "to mend" as in repairing torn nets. The Greek word for restore means the same as the Arabic word *algebrista*.

[22] Geometry is derived from two Greek words, *geo* and *metron*. It literally means "the measure of the earth."

Adam observed God's created order and, by naming it, classified it for use. What is the connection with math? The fixed patterns of God's creation are ordered by His lovingkindness, or covenant, a rulership grounded in the onto-relational Trinity.

> Thus says the Lord, "If My covenant for day and night stand not, and the fixed patterns of heaven and earth I have not established, then I would reject the descendants of Jacob and David My servant, not taking from his descendants rulers over the descendants of Abraham, Isaac and Jacob. But I will restore their fortunes and will have mercy on them." (Jeremiah 33:25-26)

For example, the rising and setting of the Sun constitutes a daily pattern. Seedtime and harvest, cold and heat, winter and summer, day and night (Genesis 8:22) manifest the faithfulness of God in the seasons of the year. The motion of the Moon, the planets, and the stars is patterned or cyclical. In the context of creation, *logos* means its law-like patterns that can be understood using reason. As we have noted, Christ is the foundation of the rational structures of creation and it is the duty of man, as a dominion bearer following in the footsteps of Adam, to observe this patterned reality and classify it for use.

Since humanity is created in the image of God, everyone has received the gift of reason and the gift of observational skills. These gifts include a sense for number and a feeling for space. Scripture reveals God thinking, God counting, and God measuring, albeit on a level far above ours (cf. Isaiah 55:8-9; Psalm 147:4; Isaiah 40:12). The *imago Dei*, therefore, thinks, counts, and measures "after God." As we have noted, our thinking, measuring, and counting are not exactly like God's. In our exploration of creation, we can see only, as it were, "through a glass darkly" (I Corinthians 13:12). Our seeing is never in the perfection that God sees, *but we can see!* This means that we can know truly but that we cannot know everything wholly.[23] We see only in part; only the Triune God can see everything perfectly. We see in part the observable patterns that originate in the Triune God's thoughts. We observe what the Father, Son, and Holy Spirit have created, i.e., the external world, and we are free to choose a name for the patterns we observe. Because we the *imago Dei*, the Trinity has gifted us (1) with the ability to see the patterns of the physical creation and (2) with the liberty to formulate relationships and consequences that both explain and predict.

What, then, is the ultimate reason to learn math? We learn math to catch a glimpse of the rhythm, the dance, of God's creational speech. We use our knowledge of that speech to:

1. Get "in step" with God's dance. Doing so will result in transformational change.
2. Steward God's creation under Him.

As we learn mathematics, we learn to discern the patterned order of creation. As we learn this patterned order, we are one step away from encountering the interpenetrating dance of the Living God, the source and sustainer of it.

EXERCISES

Define or explain the following terms and abbreviations:

1. *Elohim*
2. cf.
3. ff.
4. Trinity
5. *Logos*
6. Ontology
7. Onto-relational

---

[23] Although we can apprehend God the Father Almighty through the revelation of Jesus Christ by the Holy Spirit, we can never comprehend the Father in completeness.

8. *Anno Domini*
9. Perichoresis
10. Number
11. i.e.
12. Mathematics
13. Arithmetic
14. *Arithmos*
15. *Logistica*
16. Algebra
17. *Al-jabr*
18. Geometry
19. *imago Dei*
20. Naming is a tool of dominion

21. Find and define at least ten words that have the suffix of *logos* in the form -logy.

> … the Word of the Father is Himself divine, that all things that are owe their being to His will and power, and that it is through Him that the Father gives order to creation, by Him that all things are moved, and through Him that they receive their being.
>
> Athanasius, *On the Incarnation* (4th century AD; 1996 reprint), p. 25.

1.2 Counting ... The Introduction to the Dance

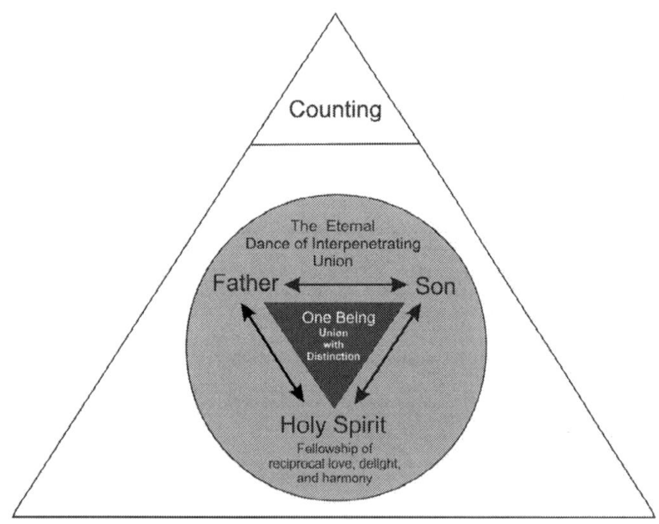

Figure 1. The ground of the rational order of counting.

Terms, Symbols & Concepts Introduced
1. Barred-gate system of counting
2. Contingence
3. Creation as a language fabric
4. e.g.
5. Providence
6. Theology

In **theology** (the *logos*, the word, concerning God, *theos*), perichoresis means mutual indwelling, the deep, interpenetrating union between the Father, Son, and Holy Spirit. In the **contingent**[1] realm of time and space, the physical creation reveals a multitude of instances where one object is contained in another in dynamic relationship. We can expect to discover, and we do find such connections embedded in creation, patterns of beauty and elegance. In this curriculum, we will be investigating many of these instances.

The discipline of mathematics is the search for perichoretic patterns; it is one of the peerless tools that we use to discover and investigate creational wonders, the beautiful rhythms of its multi-layered, interpenetrating structures. We see proximate revelations of perichoresis in the created order whenever we see relationship, link, bond, affiliation, correlation, interdependence, interaction, interpenetration, or interconnection. Perichoresis opens a whole world of understanding.

It Starts with 1, 2, 3

> As God created all things through his Word and Wisdom, and impressed upon them their rational form and order, so he maintains and rules over all things by the ordering activity of his Word and Wisdom.
> Thomas F. Torrance, *The Christian Doctrine of God* (1996), p. 224.

Counting is the foundation to all the operations of Arithmetic. The process of counting presupposes, or assumes, the existence of this God-given, **objective language fabric**. By objective, we mean that the created order is a reality that exists outside of our own existence. No matter what we think or feel, i.e., our existence, the patterns of creation are there. The Sun rises and the Sun sets. Winter follows autumn as summer follows spring.

God's creation patterns are objective to us; we are privileged to observe them, to classify them, and to make use of them. As part of the **providential** flow of history, man furthers his grasp and development of the number concept.[2]

---

[1] Contingence is the essential and rational order of the cosmos, an intelligibility grounded in an order beyond itself, in the wisdom of the Triune God, a wisdom centered in the *logos*, the Word of God who is Jesus Christ, the Son who is one with the Father in the communion of the Holy Spirit.

[2] *Providence* is the Triune God's creative and redemptive relationship with the physical world and the unfolding of history in which all that happens within these two realms comes under the Father, Son, and Spirit's rule, purposes, and provision as creator, sustainer, and redeemer of all things. The eternal purpose of the Father is a purpose of love and grace centered in His Son, the Word of God, revealed to us in the person and work of Jesus Christ by the Holy Spirit. In the words of David Bentley Hart, "God has willed His good in creation from eternity and will bring it to pass, despite their rebellion, by so ordering all things

Since God created all things through the Word of Christ, creation reveals an arrangement grounded in this language fabric. This language fabric is shown in law-like patterns, the sound or the dance of the cosmos. There is a perichoretic order, a deep rationality in the creation. This is a given that all men, Christian or not, must assume if they are to make sense of anything.

Mathematics reflects the pulse of the patterns of God's created order. The language of those patterns is number. Just as in any other language, we must understand the grammar of number if we want to obey God's command to take dominion over what He has made.

In this curriculum, we shall be exploring the ways in which man, *imago Dei,* has developed in his understanding and use of number. We will investigate the fundamentals, the dance steps, of the grammar of number. We shall explore how to work with numbers in a variety of contexts: whole numbers, fractions, and decimals. Along the way, we shall discover many interesting patterns. As we learn these patterns, we are training our minds to discern the numerical patterns in the physical creation. Let us engage this quest in earnest, realizing that the grammar of number is patterned after the creative and rhythmic voice of the Word of God, the Lord Jesus Christ. In the words of veteran high school mathematics teacher Larry Zimmerman:

> God commands His children to subdue and replenish the earth and take dominion over it.... mathematics is essential in subduing and replenishing tasks .... Without a working knowledge of the patterns of God's speech used in the creation, humans are powerless to replenish the earth and are in danger of being themselves subdued by it.[3]

> Mathematics, too, is a language, and as concerns its structure and content it is the most perfect language which exists, superior to any vernacular; indeed, since it is understood by every people, mathematics may be called the language of languages. Through it, as it were, nature herself speaks; through it the Creator of the world has spoken, and through it the Preserver of the world continues to speak."
>
> C. Dillmann, *Die Mathematik die Fackelträgerin einer neuen Zeit* (1889), p. 5.

In review, we read in Genesis 2 that the Triune God, the Father, Son, and Spirit, commanded Adam to name the animals. For each animal, Adam saw a pattern or characteristic and named that animal accordingly. In Scripture, naming is a tool of dominion and reflects ownership. The Triune God created the animals as a bequest to man, a gift to own and name as a steward, a steward in relationship to the Triune God.

Likewise, parents name their children, recognizing that children are a gift from the Creator. The Father, Son, and Spirit give stewardship of children to parents for a season. When children become responsible adults, parents commission them into the world to fulfill to represent God's name in the calling in their lives.

NAMING

To be able to name the rationality of the universe is a gift from God. Number is one way to name the many patterns of creation. Being imaged in God, the *imago Dei,* man owns number as a steward under God. As a steward, man names the objects of creation by numbering them. For example, we name "two days" using a symbol for two, 2, | |, etc.

---

toward his goodness that even evil (which he does not cause) becomes an occasion of the operations of grace." David Bentley Hart, *The Doors of the Sea* (Grand Rapids: Eerdmans, 2005), p. 82.

[3] Larry Zimmerman, *Truth and the Transcendent* (Florence: Answers in Genesis, 2000), p. 65.

> It is because things are amenable to rational treatment that we can apprehend them at all; we understand them or get light upon them, in so far as we can penetrate into their rationality and develop our grasp of it.
> Thomas F. Torrance, *Theological Science* ([1969] 1978), p. vii.

There is a dance to counting because counting follows a pattern. Numbers follow one after the other, marching as in single file. Given one number in this sequence, another will tag along after it. If we know that number, we can find the next number by adding one to it.

As we continue to count, numbers get larger and larger. One number follows another endlessly. There is a first number but not a last number. The order in this list reflects creational realities.

In history, different cultures have developed different ways of naming objects with number and structuring the outcomes of counting. This structuring of number is a revelation of man's calling to take numerical dominion over the created order.

ORGANIZATION

Imagine how you would develop an efficient number system. What symbols would you use to represent counting an object? Let's say that you are a pearl fisherman and that you wanted to keep track of the number of pearls that you have mined from the ocean floor. A good choice for a symbol would be something that looked like a pearl: **O**. On your first day, you inscribed your catch with circles on the sand:

OOOOOOO

On the second day, you caught **OOOO** more pearls. You now have this many:

OOOOOOOOOOO

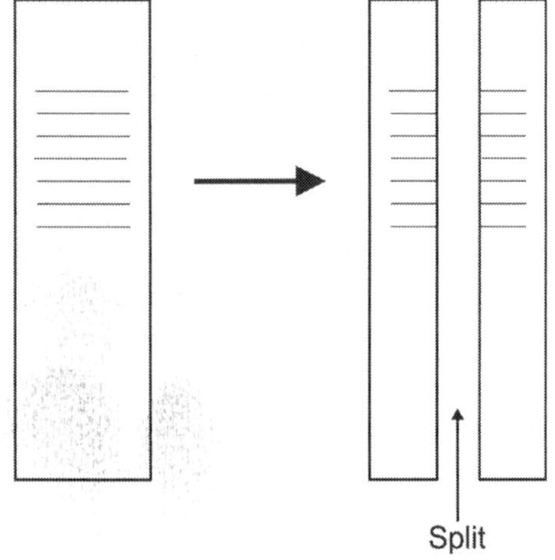

Figure 2. Splitting a tally stick.

Each day thereafter you add more pearls to your catch. At some point in this process, you might think of a better way to record the number of pearls that you have. Instead of representing each pearl, the unit pearl, with the symbol O, you might want to represent each pearl with the symbol |, as many civilizations have done. Then, you might want to group or bundle your symbols in some fashion, **e.g.**[4], ~~||||~~ . This grouping method is known as the **barred-gate system of counting** because it recognizes the inherent limit of recognizing groups of objects at four.

With a sharp instrument like a knife, you can notch a stick with lines. Notched sticks, or "tally[5] sticks," have been used for millennia and may have inspired the Roman numeral system, a system we shall investigate in Lesson 4.3. The benefit of using a stick is that it serves as a record of a transaction. A wooden tally stick could be marked with notches and then split lengthwise, each half duplicating the notches so that two parties to a transaction could record what happened. Do you see why

---

[4] e.g. comes from the Latin phrase *exempli gratia*, which means "for the sake of example."

[5] Tally comes from the old French *taille* meaning "cut or notch." *Tailler*, in modern French, means "to cut." In English, the word is *tailor*, the cutter of cloth.

this method ensures security and prevents fraud?[6] Can you see one-to-one correspondence, or matching, in this splitting of the tally stick?

The group of five may represent that you might have a bag to store five pearls. Your catch of OOOOOOOOOOO could now be represented by 𝍸 𝍸 |. These symbols tell you that your catch is now represented by two bundles of five, two bags, plus one; this representation is much easier to count. There are many ways in which one can name and structure the results of counting.

> A child kicks its legs rhythmically through excess, not absence, of life. Because children have abounding vitality, because they are in spirit fierce and free, therefore they want things repeated and unchanged. They always say, Do it again; and the grown-up person does it again until he is nearly dead. For grown-up people are not strong enough to exult in monotony. But perhaps God is strong enough ... It is possible that God says every morning, Do it again, to the sun; and every evening, Do it again, to the moon. It may not be automatic necessity that makes all daisies alike: it may be that God makes every daisy separately, but has never got tired of making them. It may be that He has the eternal appetite of infancy; for we have sinned and grown old, and our Father is younger than we.
>
> G. K. Chesterton (1874-1936), *Orthodoxy* (1908), p. 108.

EXERCISES

Explain the following words/phrases/symbol:

1. Theology
2. Contingence
3. Language fabric of creation
4. The objectivity of the language fabric of creation
5. Providence
6. Barred-gate system of counting
7. e.g.

> [There is] a subliminal hearing of a "speech" embedded in reality, which precedes, accompanies and shapes specific acts of scientific understanding.
>
> Thomas F. Torrance, *The Christian Frame of Mind* (1989), p. 79.

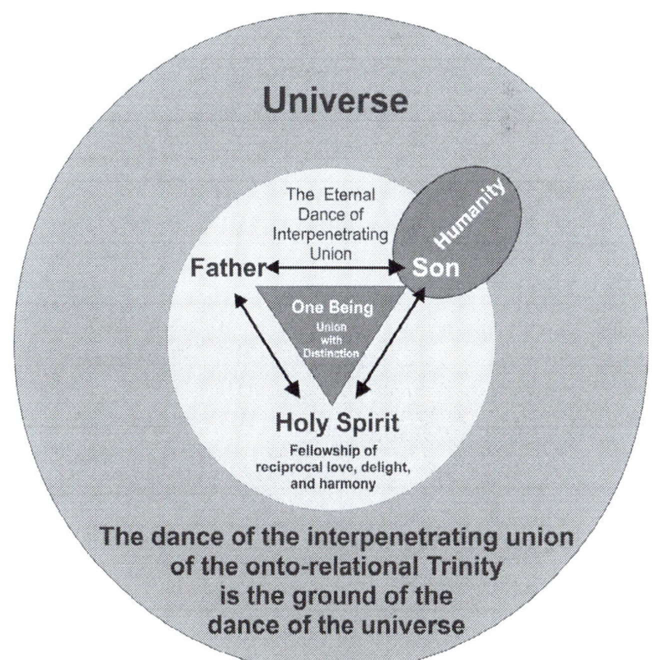

Figure 3. The ground of the rational order of the universe.

[6] These sticks were also named "stocks of wood." Each party to a transaction, in possession of half of the stock, was a "stock holder." Sound familiar? The phrase "bank stock" can also be traced to this ancient method of recording a business transaction.

1.3 Dancing in Tens

Terms, Symbols & Concepts Introduced
1. *Ad infinitum*
2. Base of a number system
3. ca.
4. Ciphers
5. Decimal system
6. Definition
7. Digit
8. Fixed-value system
9. Notation
10. Number system or scale
11. Numeral
12. Numeration
13. Orders of units
14. Periods
15. Place Value or Positional System
16. Uniform number system

A **number system**, or a **number scale**, provides a way of naming and structuring the results of counting. A number scale is a sequence of units that increase or decrease, reflecting a pattern or law. A number system is **uniform** if the increase and decrease is the same throughout the entire succession of units.

The Decimal System

A **digit**, alone or in combination, is a symbol used in a number system. Digit originally meant "finger" or "toe."[1] The number system that we are all familiar with is the **decimal system** of numeration in which we write numbers using ten symbols.[2] This system seems natural to us since God created us with ten fingers and ten toes.

We get the word decimate from decimal. If, in an ancient Roman army, some crime was committed and no one confessed, the commanding officer would line up the soldiers and kill every tenth soldier to enact a just punishment. Because of this procedure, decimate means to select by lot and kill every tenth person. In our decimal system numbers are written or symbolized using ten digits or combinations of these ten digits. As we shall see in our next lesson, the decimal system is one in which the increase or decrease is uniformly ten. Ten is also the **base** of our decimal number system.

Different cultures have used different digit symbols to name number. Archaeological excavations, dating about one hundred centuries before our time, have uncovered systems of counting in which markers, groups of pebbles, notches on sticks, or tokens corresponded directly with objects or goods they represented. In a digital system, a **numeral** is a combination of digits, which may be of differing lengths. A numeral is a symbol that represents the essence of number. Up until now, we have been writing numbers using number words. It is much easier and more efficient to represent numbers using symbols. The ten digits, or basic symbols, used in the decimal system are 0, 1, 2, 3, 4, 5, 6, 7, 8, 9. These symbols did not always look like this. They have a long history of development and refinement. They shape the basic toolkit of Arithmetic.

In review, a number system or a number scale is a way of naming and structuring the results of counting. A number scale is a sequence of units that increase or decrease revealing a pattern or law. A number system is uniform if the increase and decrease is the same throughout the entire succession of units. The decimal number system is uniform because this increase or decrease is uniformly ten. The base of the decimal system, like the number of digits it uses, is ten.

The key to the genius of the decimal number system is that in representing numbers, the digits are regarded as arranged in groups of ten. Therefore, we have first an indivisible unity, one, or a unit. Next, we have a group of ten units or tens. Then, we have a group containing ten tens or hundreds.[3] After that, we have a group containing ten hundreds or thousands, etc. This consistent grouping by ten is why the decimal system is uniform.

---

[1] In Latin, *digita* means "fingers." We also get the word digital from digit. As a unit of distance, digit represents the width of a man's index finger.

[2] *Decimal* is Latin for "ten."

[3] In essence, we are multiplying. One hundred is ten tens. See Step 4.

The French scholar Gerbert of Aurillac (945-1003), later to become Pope Sylvester II in 999, learned of the decimal system in his early life from the Arabs in Spain. He is generally credited as the one who introduced the Indian-Arab numerals to the rest of medieval Europe. This system, as far as historians know, was developed in India, by Hindu mathematicians, in the 6th or 7th centuries AD. When the Muslims conquered this region of the world in the 8th century AD, they not only captured people, they also captured the decimal system. Because of this history, the decimal system is also known as the Hindu-Arabic system of numeration.

Leonardo of Pisa or Leonardo Fibonacci (**ca.** 1170-1240),[4] Italian merchant and mathematician, was another person who helped spread the decimal system in Europe in his lifetime. His business travels led him to Islamic-controlled countries, and it was through this interaction that he noticed the way in which the Arabs wrote digits.

Figure 1. Leonardo Fibonacci. Source: Public Domain.

Grouping by Tens

Because of the use of ten digits, the decimal system and our bodies are connected. We have ready-made counters on our bodies, our fingers and our toes. This will help us count to 20 at a maximum. Teamwork is needed to go further. If you have a friend who is willing to be nudged each time you count to ten on your fingers, and to count the nudges on his or her fingers, it is possible to count to 100, ten tens. With three people, you can reach 1000, one thousand or ten tens of tens, … and with six people, you can reach 1,000,000, one million or ten tens of tens of tens of tens of tens, although the sixth person can take a long nap before he starts using his fingers.

Figure 2. Odometer. Source: iStockPhoto.

The same idea is used in devices such as the odometer, Greek for "journey measure," on your car. Behind the dashboard, or instrument panel, of older cars is a machine built of gears or wheels. Each wheel, when reaching ten, nudges the next.[5]

Old-style adding machines worked on the same principle. To find one of these machines, look in antique shops. A mechanical calculator, invented by the French mathematician Blaise Pascal (1623-1662) and named the *Pascaline* after him, uses this gear mechanism for addition and subtraction.[6]

Figure 3. Pascaline Calculator. Source: James D. Nickel.

[4] Fibonacci comes from an Italian phrase that means "son of Bonacci." Note, *ca.* stands for the Latin *curca* meaning "around" or "about." We are not sure of Leonardo's birth year so we estimate the approximate date as "ca. 1170."

[5] The odometer in new cars has been electronically digitized. The same principle is applied in electronic circuits.

[6] The Pascaline was 300 years ahead of its time. A few people of Pascal's day used them, but they were not popular. Production stopped in 1652 only to be resurrected in the 1950s to the 1980s, primarily as an educational tool. With the advent of the electronic hand-held calculator, these mechanical gems soon disappeared, but they are still excellent teaching devices because they demonstrate the mechanics of addition and subtraction.

Table 1: Decimal System

Hundred trillion	Ten trillion	One trillion	Hundred billion	Ten billion	One billion	Hundred million	Ten million	One million	Hundred thousand	Ten thousand	One thousand	Hundreds	Tens	Ones

We can use everyday objects to get a better feel for counting with the decimal system. Follow along with your imagination:

Step 1. Gather a pile of needles.
Step 2. With a rubber band, tie individual needles into bundles of ten needles.
Step 3. Take ten of these bundles of needles and put them into a pouch. This bag contains 100 needles.
Step 4. Take ten of these pouches and put them into a small box. This box contains 1000 needles.
Step 5. Take ten of these boxes and put them into the luggage compartment of your car. This car contains 10,000 needles.
Step 6. Take ten cars and put them into a truck transport. This vehicle contains 100,000 needles.
Step 7. Take ten trucks and put them into a ship. This ship contains 1,000,000 needles.

Seeing groups of ten in this manner, as bundles of something, gives us a feel for size. For example, we can picture the symbol 8159, eight thousand one hundred fifty-nine, by eight boxes, one pouch, five bundles of needles, and nine individual needles.

Notice again that, when you count in bundles like this, these packets establish an ordering sequence.

1. First order: needles or ones (the preferable word is ones).
2. Second order: bundles of ten needles tied with a rubber band, or tens.
3. Third order: pouches of ten bundles, or hundreds.
4. Fourth order: boxes of ten pouches, or thousands.
5. Fifth order: cars of ten boxes, or ten thousands.
6. Sixth order: trucks of ten cars, or hundred thousands.
7. Seventh order: ships of ten trucks, or millions.

The numerical base of a number system is the number of units in each bundle. In principle, any number can form the base of a number system. The decimal system is a base-10 system. In history, we see other number systems with these bases: 2, 5, 8, 12, 20, or 60.

Given a particular base, the bundles of each successive order all have the same number of units. Because of this same bundling arrangement, the decimal system is known as a **place value system**. In a place value system, the value of a digit depends on its place in its group; i.e., place counts! So, in a number like 333, each three has a different value. Going from left to right, the first three represents 300 or three bundles of 100, the second three represents 30 or three bundles of 10, and the last three represents three ones. Place value is also known as **positional notation**.

The ancient Egyptians and Romans (Lesson 4.3) used **fixed-value systems** in which a digit's value was always the same. No matter where you find it in the number, the Roman X is always 10, V is always five, and so on. Note that a uniform number system is a place value system while a varying number system is a fixed value system.

Let's go over some more basic **definitions**.[7] **Notation** is a method of writing or representing numbers by symbols or characters. **Numeration** is a way of reading numbers represented by symbols or characters. When we count the number of objects we may either use words or symbols, i.e., special characters. These special characters, as we have already noted, can be figures or letters. In the decimal system, ten figures, or ten symbols, are employed. In some older American textbooks, these character representations are named **ciphers**.[8] Cipher is a word derived from Dutch Arithmetic texts used in the 16th and 17th centuries.

> One for the money, two for the show, three to get ready, and four to go!

Here are the ten familiar symbols of the decimal system:

Table 2

Numeral	Names	Modern Greek	Latin	German	French	Russian	Spanish
0	Zero[9]	medén	Nulla	Null	Zéro	Nohl	Cero
1	One[10]	Hén	Unus	Eins	Un	Odyn	Uno
2	Two	Dúo	Duo	Zwei	Deux	Dva	Dos
3	Three[11]	Treis	Tres	Drei	Trois	Tri	Tres
4	Four	Téssares	Quattuor	Vier	Quatre	Chetyre	Cuatro
5	Five[12]	Pénte	Quinque	Fünf	Cinq	Piat	Cinco
6	Six	Héx	Sex	Sechs	Six	Shest	Seis
7	Seven	Heptá	Septem	Sieben	Sept	Sem	Siete
8	Eight	Októ	Octo	Acht	Huit	Vosem	Ocho
9	Nine	Ennéa	Novem	Neun	Neuf	Deviat	Nueve

[7] A definition is a statement that determines the exact nature of an idea or object. (Note: Every definition must rely on other undefined terms in its use of words.)

[8] *Cipher* comes from the Arabic word *sifr*, meaning "zero." *Syfer* in used in Africaans, the Dutch language of South Africa. Western scholars turned *sifr* into a Latin-sounding word, yielding *zephirum*, the root of the English word *zero*. Zero was so important to the new set of Hindu-Arabic numbers that people began to commonly call these numbers ciphers. We can see one instance of this popularity by observing that the French word for digit is *chiffre*. *Cipher* also means "secret code" because, when the Hindu-Arabic numbers were first introduced into Europe, businessmen immediately recognized their value and ease of use. Local governments hated these numbers because they could be easily changed or falsified; e.g., a 0 could be made a 6 quite easily. (Because of this problem, when writing checks, you must write the amount of the check in numerals and in words.) In 1299, the local government of Florence, Italy, banned the use of the Hindu-Arabic numbers. Despite this civil opposition, local merchants continued to use these numbers and some even resorted to using them to send encrypted messages, thus the connection between cipher and secret messages. See Lesson 13.12.

[9] Zero is also written or spoken as "naught."

[10] Nearly all the English names of numbers from one to nine are of Germanic origin and reflect finger counting. The number one, therefore, represents one finger.

[11] The Spanish word *tres*, meaning "three," and the French word *très*, meaning "very much," come from the Latin *trēs*, meaning "beyond." The word *three* in English is derived from the Anglo-Saxon *thria*, a word related to *throp*, meaning "pile" or "heap."

[12] In Indonesia, *lima* means "five" and it also means "hand."

> ... it took men about five thousand years, counting from the beginning of number symbols, to think of a symbol for nothing.
> Isaac Asimov (1920-1992), *Asimov on Numbers* (1982), p. 15.

The first numeral, zero, has a fascinating history.[13] It represents no value when standing alone.[14] It is the answer to the question, "How many living elephants do you have in your bedroom?" When not represented alone, as in the number 305, it tells us that the number 305 has zero tens.[15]

Classical Greek mathematicians had no symbol for zero nor any use for it because they did not understand its significance in terms of place value, i.e., as a number representing nothing, and, therefore, its importance in computation. According to Constance Reid, the Greeks were concerned "with learning the secrets of numbers, not with using them."[16] The Greek word for "nothing" or "no one" is *oudeís* (masculine gender). The Romans copied the Greeks in almost everything and likewise saw no need for the number zero. In Latin, *nullus*, also masculine gender, means "no one"; and *nil* means "nothing." The symbol itself may have come from the impression left on the sand when a pebble, used for counting, is removed. The first known use of the symbol for the digit zero, a small circle, appears on a stone inscription, dated AD 876, found at the Chaturbhuja Temple at Gwalior in India.

As we count with the decimal system, we represent the first nine numbers by a single digit, 1 to 9. These units are units of the *first* order, or ones. Note the ordinal usage with this and the following explanations (Lesson 1.4). By grouping ones of the first order, ten ones, we form a unit of the second order, tens.[17] Ten[18] is represented by writing 1 followed by 0 or 10, meaning one group of ten ones and zero groups of ones–i.e., zero ones.

Here are some ones that are good:
- Hole in one
- Number one
- At one
- One of a kind
- The chosen one
- The one and only
- One-off

Here are some ones that are bad:
- One-hit wonder
- One-horse town
- One-trick pony

Here are some other ones:
- One-man band
- One-way street

---

[13] See Robert Kaplan, *The Nothing that Is: A Natural History of Zero* (New York: Oxford University Press, 1999) and Charles Seife, *Zero: The Biography of a Dangerous Idea* (New York: Penguin Books, 2000).

[14] In Physics, absolute zero (-273.15°C) is the point at which the molecules of all substances have no energy and no motion, i.e. they freeze absolutely. See Lesson 7.2 and 7.12.

[15] *Love*, meaning zero, is used in tennis and is derived from the French *l'oeuf*, meaning "egg." (Note: An egg looks like 0.) The same thinking gave rise to the use in cricket of *duck* for a batsman's score of 0. Scratch is the golfing term for zero, as in *scratch out*, meaning to erase. A *scratch golfer* plays off a zero handicap. Nil is also used for zero since it is a simple abbreviation of *nihil*, the Latin word for nothing. In philosophy, nihilism argues that there is no meaning or purpose to life; i.e., life is zero.

[16] Constance Reid, *From Zero to Infinity* (Wellesley: A K Peters, Ltd., [1955, 1960, 1964, 1992] 2006), p. 5. Reid (1918-2010), a non-mathematician, made her literary mark in the world by writing about mathematics and mathematicians.

[17] The derivation of *ten* is probably from an old Indo-European word meaning "two hands," the literal meaning of the Gothic word *taihun*. Ten, in Old and Middle English, was written *tien*. For the Lengua people of Paraguay, the word for ten means "finished both hands."

[18] Ten is *déka* in Modern Greek, *decem* in Latin, *zehn* in German, *dix* in French, *desiat* in Russian, and *diez* in Spanish. The German *zehn* comes from the Old German *zehan* which resembles *zwei hand* meaning "two hands."

In the same manner are represented:

Table 3	
Two tens and zero ones or twenty[19]	20
Three tens and zero ones or thirty	30
Four tens and zero ones or forty	40
Five tens and zero ones or fifty	50
Six tens and zero ones or sixty	60
Seven tens and zero ones or seventy	70
Eight tens and zero ones or eighty	80
Nine tens and zero ones or ninety	90

The numbers between ten and twenty are represented by writing:

Table 4		
Eleven[20]	11	One ten and one one
Twelve[21]	12	One ten and two ones
Thirteen[22]	13	One ten and three ones
Fourteen	14	One ten and four ones
Fifteen[23]	15	One ten and five ones
Sixteen	16	One ten and six ones
Seventeen	17	One ten and seven ones
Eighteen	18	One ten and eight ones
Nineteen	19	One ten and nine ones

Following the same pattern, the numbers between 20 and 30 are represented as follows:

Table 5		
Twenty-one	21	Two tens and one one
Twenty-two	22	Two tens and two ones
Twenty-three	23	Two tens and three ones
Twenty-four	24	Two tens and four ones
Twenty-five	25	Two tens and five ones
Twenty-six	26	Two tens and six ones
Twenty-seven	27	Two tens and seven ones
Twenty-eight	28	Two tens and eight ones

[19] *Twenty* is derived from the Old and Middle English *twentig* or *twegentig*. *Twen* and *twegen* mean "two." The suffix *-tig* comes from the Gothic *tigjus* meaning "ten." Therefore, twenty means "two tens." In some languages, the word for whole person means "twenty." What do you think is the reason for this connection?

[20] *Eleven* is derived from the Old English *endleofan* where the suffix *-fan* may mean "to leave" or "left." Therefore, eleven means "one is left after counting ten," i.e., counting ten fingers.

[21] *Twelve* is derived from the Old English *twelfe* meaning "two left over" after counting ten. This derivation is more clearly seen based upon the Gothic *twailif*. *Douze* means twelve in French and is the basis of the English *dozen*.

[22] *Thirteen* means "three and ten." In the names thirteen through nineteen, the number of ones is mentioned first, then the suffix *-teen* meaning "tens."

[23] For the Lengua people of Paraguay, the word for fifteen means "finished foot." Do you see why? After counting ten fingers, counting five toes finishes the foot. Sixteen means "reached the other foot, one."

Table 5		
Twenty-nine	29	Two tens and nine ones

In our decimal system, what is the greatest number that can be represented by two digits? It is 99. By grouping ten units of the second, or ten tens, into a larger group, we form a unit of the third order, "hundreds."[24] One hundred is represented by writing 1 followed by 0 and 0, or 100, meaning one group of hundred, zero groups of ten, and zero groups of ones.

Table 6	
Two hundred	200
Three hundred	300
Four hundred	400
Five hundred	500
Six hundred	600
Seven hundred	700
Eight hundred	800
Nine hundred	900

What is the greatest number that can be represented by three digits? It is 999. By grouping ten units of the third order, or ten hundreds, into a larger group, we form a unit of the fourth order, thousands.[25] One thousand is represented by writing 1 followed by 0 and 0 and 0 or 1000, meaning one group of thousand, zero groups of hundred, zero groups of ten, and zero groups of ones.

> How close to zero is zero?
> John Prescott (1938-), former British Deputy Prime Minister.

As you can see, we can continue this orderly process forever, or, using a Latin equivalent, ***ad infinitum***.[26] The largest number that can be expressed by four figures is 9999. One more leads us to the fifth order, which will be ten thousand. The sixth order will be one hundred thousand, etc.

Orders of units are denoted by the position of the figures used in expressing a number. Thus, 875 represents eight units of the third order or eight hundreds, seven units of the second order or seven tens, and five units of the first order or five ones. 710 represents seven units of the third order or seven hundreds, one unit of the second order or one ten, and zero units of the first order or zero ones.

Here are two fundamental laws revealed in the order of the decimal system:

1. Ten units of any order in a number make one unit of the next higher order.
2. When any order of units in a number is vacant, the place is filled with the symbol for zero, or 0.

When numbers get colossal, grouping them in some manner makes for easier representation in writing and reading. It is generally accepted to group the successive orders into **periods** of three figures each. The periods are commonly separated by commas, but the convention in some countries is to separate by spaces, each period taking the name of its lowest order.[27] (Note: When writing numbers with four digits, it is not required to use commas.) Commas are never used with four-digit dates, and we say that Isaac Newton was born in "sixteen forty-two" (1642) instead of "one thousand six hundred forty-two."

---

[24] *Hundred* is derived from the Old English *hund.* Hund comes from the Gothic *taihun-taihund* meaning "ten tens" or "ten groups of ten." The suffix *-red* was added later meaning "to reckon." Therefore, hundred means "a hundred (ten tens) reckoning."

[25] *Thousand* is akin to the Old Norse *thushund* meaning "great hundred." The prefix *thus-* is the root of the English word *thumb,* which literally means "greater or stronger finger."

[26] *ad infinitum* means "endlessly" or "without limit."

[27] In scientific writing, commas are not used to separate periods because the comma is used as a decimal point in some countries. See Lesson 2.8 and 7.1.

Table 7: Place Value														
Trillions[28] (5th) period			Billions[29] (4th) period[30]			Millions[31] (3rd) period			Thousands (2nd) period			Ones (1st) period		
Hundreds	Tens	Ones	Hundreds	Tens	Ones	Hundreds	Tens	Ones	Hundreds	Tens	Ones	Hundreds	Tens	Ones

From Table 7, consider two questions: (1) As we move from right to left, what is happening to each order? (2) As we move from left to right, what is going on to each order?

When moving a figure one place to the left, we increase its representative value tenfold, the equivalent of multiplying by 10. (See Step 4.) When moving a figure one place to the right, we decrease its representative value tenfold, equivalent to dividing by 10. (See Step 5.) Now you know why the decimal system is a number system that is uniform. Remember, a number system is uniform if the increase and decrease is the same throughout the entire succession of units.

Table 8: Naming Differences	
American Name for Period	UK[32] Name for Period
Thousand	Thousand
Million	Million
Billion	Milliard
Trillion	Billion

EXERCISES

Define the following phrases, terms, abbreviations:

1. Number system or scale
2. Uniform number system
3. Digit
4. Decimal system
5. Base of a number system
6. Place Value or Positional System
7. Fixed value system
8. Definition
9. Notation
10. Numeration

---

[28] *Trillion* is named "billion" in countries other than the United States. See the footnote for *billion.*

[29] For clarification, the United States names numbers greater than one million based upon an old French system. In Europe, however, "a billion" is "a trillion" in the United States. So, a $18 trillion government debt in America would only be $18 billion in Great Britain. Some ignorant government official in America might say, "Let's move to England and cut the national debt!"

[30] The traditional British name for billion is "thousand million." In France, a billion is "milliard." Trillion is the traditional British and European name for billion. This way of naming can cause cross-cultural confusion, especially if you are an accountant.

[31] One thousand thousands is named a "million," from the Medieval Latin *millio* meaning "great thousand."

[32] UK stands for United Kingdom.

11. Numeral
12. ca.
13. Ciphers
14. *Ad infinitum*
15. Orders of units
16. Periods

> Blessed are the placemakers, for they shall be called the children of God.
> Matthew 5:9 (misprint from a Bible printed in 1562)

Developed in 1838 by the American artist and inventor Samuel B. Morse (1791-1872), combinations of dots and dashes are used to send messages (letters, punctuation marks, numerals) either by sound, as flashes of light, or on a telegraph.

Morse Code and the Decimal Digits	
Numeral	Code
1	• – – – –
2	• • – – –
3	• • • – –
4	• • • • –
5	• • • • •
6	– • • • •
7	– – • • •
8	– – – • •
9	– – – – •
0	– – – – –

198, in Morse Code, is: • – – – –   – – – – •   – – – • •

Write the following numbers in Morse Code:
17. 12
18. 29
19. 318
20. 1045

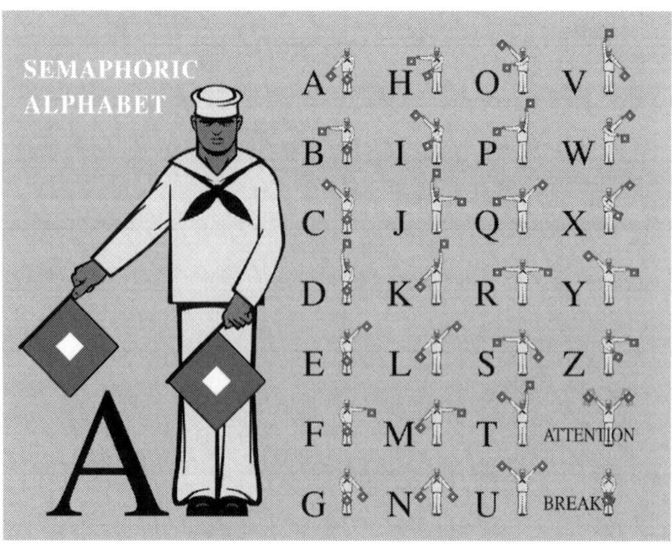

Figure 5. Source: iStockPhoto.

Translate the following Morse Code into numbers:

21. – – – • • – – – – – – – – – – – – – – –

22. – • • • • – – – – • • • • • •

23. • – – – – – – – – – • • • • – • • – – –

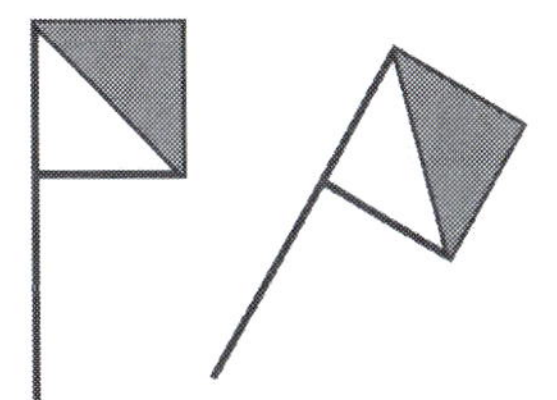

Figure 4. "Numbers follow" semaphore.

The semaphore[33] is a visual telegraph developed by the French engineer Claude Chappe (1763-1805). Historically, using semaphores to signal messages between ships has been of great importance. Signaling with mechanically or electrically controlled semaphore arms is still used by railroads to signal trains.

The letters in sequence A through I and K represent the 10 digits 1, 2, 3, 4, …, 9, and 0 when preceded by the sign meaning "numbers follow."

24. Translate the semaphores in Figure 6 into a three-digit numeral if the "numbers follow" sign has been given.

Letter	Numeral
A	1
B	2
C	3
D	4
E	5
F	6
G	7
H	8
I	9
K	0

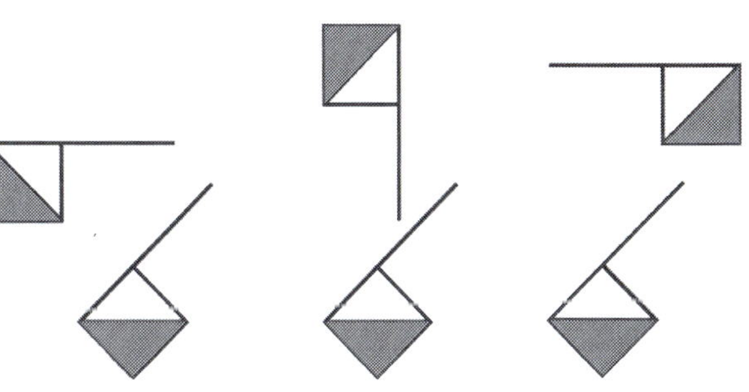

Figure 6.

Sign language for the deaf was developed by a Roman Catholic priest, father Charles-Michel de l'Épée (1712-1789). Thomas Gallaudet (1787-1851) developed sign language in America.

To express numbers like hundred and thousand, the C and M are used. To make 400, you sign four (See the table of the next page.) and then the hand signal for C: 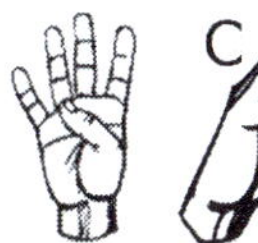

[33] *Semaphore* is derived from two Greek words, *sema* meaning "sign" and *pherein* meaning "to bear."

To make 9000, you sign nine and then the hand signal for M:

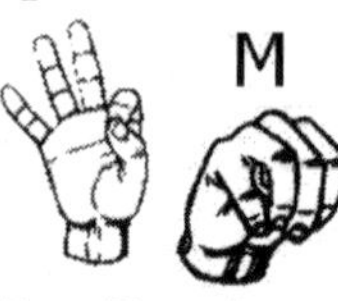

Numeral	Sign	Numeral	Sign
1		6	
2		7	
3		8	
4		9	
5		0	

25. Practice making the ten decimal digits using your fingers.

Which numbers do the following signs represent?

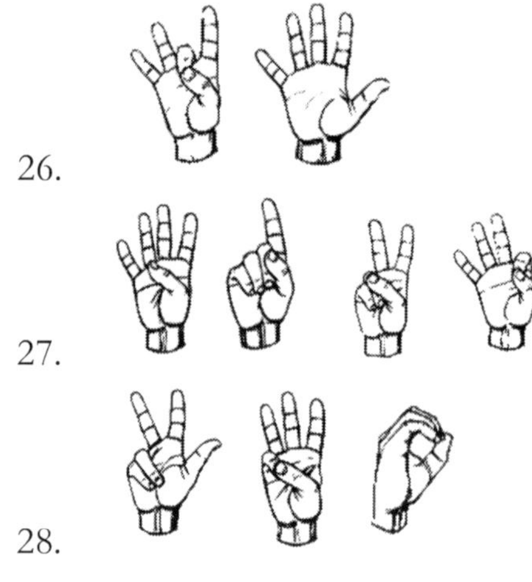

26.

27.

28.

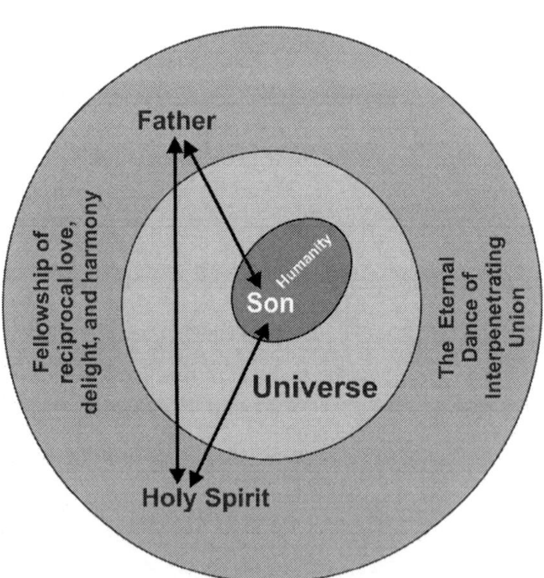

The dance of the
of the Christo-centric universe
is enveloped by the interpenetrating union
that is dance of the onto-relational Trinity

Figure 7. The Christo-centric universe.

1.4 The Utility of Number

Terms, Symbols & Concepts Introduced
1. ≠
2. −
3. ×
4. ÷
5. +
6. <
7. =
8. >
9. Cardinal
10. Classification or unity of concept
11. etc.
12. Matching
13. One-to-one correspondence
14. Ordering
15. Ordinal
16. Presuppose
17. Sequence
18. Tag

The necessity to count in the early history of man is revealed by the need to keep track of domesticated animals (e.g., sheep), to measure or divide land, to keep harvest records, to take census data, and to maintain armies. Farming required counting by keeping track of lunar cycles and the seasons using calendars.

We shall see that all arithmetic operations start with counting. Since God counts, then the ability to count is one of the gifts that God has given man. The action of counting **presupposes**[1] that God has given man three numerical tools to understand His creation:

Three Assumptions

1. **Classification**: Counting presupposes that we can classify a variety of observations under a unity of concept and to endow that unity with a name. For example, we observe three cows, three chickens, three cats, etc. We can classify these various observations under a **unity of concept** or the concept of "threeness."

   The number three is a name for the concept of "threeness." Adam classified the animals. God would not have given Adam this task unless He had given Adam the ability to observe and classify, to articulate a unity of concept. Classification is a way of naming the patterns of God's creation, and a mastery of Arithmetic is the foundation for both observing and naming the rationality of God's creation.

   Ultimately, three (symbol: 3) is a name for a natural number that is one more than two and one less than four. As the theoretical biologist Luke Heaton says, "… the truly essential point is that each number stands in a structured relationship to the other numbers."[2] To see numbers in relationship to other numbers reflects Trinitarian relational dynamics.
2. **Matching**: Counting presupposes that we can match two collections, element for element, and to associate these groups with a name. For example, consider a room full of boys and girls. Tell the group to pair off, i.e., one boy to one girl. Without counting, you will immediately know that the number of boys does not equal the number of girls if one boy or one girl does not have a partner. We match objects regularly. Mathematicians name this matching **one-to-one correspondence**. You know that one of your socks is missing because of this matching ability.
3. **Ordering**: Counting presupposes that we can order collections into a **sequence**[3] and to discover the syntax, i.e., a way of naming, that will permit an infinite extension of these names. For example, after many centuries of using various naming methods, the world now agrees to count in a sequence using the following number symbols: 1, 2, 3, 4, 5, **etc.**[4] We order objects regularly. Each of

---

[1] *Presuppose* means to "suppose beforehand." A presupposition is a basic or preliminary assumption that helps us to order and define the world that God has made.

[2] Luke Heaton, *A Brief History of Mathematical Thought* (London: Constable & Robinson, 2015), p. 301.

[3] Sequence comes from the Latin *sequi,* meaning "to follow." The numbers 1, 2, 3, 4, 5, etc. have an ordered sequence where each number, except 1, is the same amount greater than the one before it as it is the same amount lesser than the one after of it.

[4] *etc.* is the abbreviation of *et cetera*, a Latin phrase meaning "and others," "and so forth," "and so on."

us has a favorite pair of shoes, game, or toy. It is our first choice. *First* is an ordering word. In families, children are known for their birth order: the first-born, the second-born, etc. Ordering enables you to compare sizes of objects. Children learn to order objects by size before they learn to count.

In subsequent lessons, we will discover many laws or the way number relationships work. As we encounter them, remember that they are grounded in the ultimate Law-giver, the Creator God who is onto-relational. Laws reflect the rationality of relationship. We can discover them by thinking carefully about what we are observing. Therefore, we must develop our habits of thinking, about what we observe. Our observations will reveal patterns, and these patterns can be turned into number laws. Learning the elements of Arithmetic will better equip us to learn these patterns and, thereby, fulfill our dominion/vocational callings under God.

THREE EXPRESSIONS OF NUMBER

As man learns to take dominion over the creation, he uses number to answer three practical questions, evidenced by that fact that every culture in history uses number in the following ways:

Table 1

Question	**Expression of Number**	**Operations**	**Modern Symbols**
How much or how many?	**Cardinal:**[5] to express quantity or measurement.[6] Examples: A paper clip weighs one gram. John has $7 more than Susan. Gloria's square garden has sides that measure six feet.	Addition, subtraction, multiplication, and division.	+, -, ×, ÷
How far along in a sequence?	**Ordinal**: to express order. Examples: Mr. Smith's office is on the third floor. My pile of marbles has more marbles than your pile and, therefore, comes before your pile when they are listed according to quantity, largest to smallest.	Greater than, less than.	>, <
Which of many?	**Tag**: to identify different objects. Example: Steve's driver license number is 287643.	Equals, does not equal.	=, ≠

GREATER THAN OR LESS THAN

When we compare two numbers to determine which is lesser (or greater), the symbol < stands for "less than," and the symbol > for "greater than." These symbols were introduced by Thomas Harriot (ca.

[5] A cardinal number is a number on which arithmetic operations turn or depend. *Cardinal* is derived from the Lain *cardo,* meaning "hinge." A cardinal number is therefore an important number. For example, a cardinal in the Roman Catholic Church is a person of importance. In the universe of the Father, Son, and Spirit's making, Christ, the Son, is the supreme hinge, the center of it all, the One who is preeminent in all things (Colossians 1:18).

[6] To measure, we assign a relevant unit quanity as one unit. For example, one centimeter is a unit, or denomination, of length. We then count the number of units that make up the quantity that we want to measure. If it takes eight units to measure a quantity's length, we know its measure is eight centimeters.

1560 1621) in his book entitled *Artis analyticae praxis* (published posthumously in 1631). These symbols did not come into popular use until the 18th century (1700s). It is easy to picture what these symbols mean: the open end always points to the greater number, and the pointed end always points to the lesser number.

Example 1. $1 < 8$ (we say "one is less than eight," not "one is lesser than eight.")
Example 2. $8 > 1$ (we say "eight is greater than one," not "eight is more than one.")

If the digital representation of two numbers is being compared, two cases are possible:

Case 1. If two numbers have different lengths, then the number with more digits is greater than the number with fewer digits.
Example: $345 > 56$, which can also be written $56 < 345$.

Case 2. If any two numbers a and b (Note: We are employing algebraic notation.) have the same length, or the same number of digits, start at the leading digit at the leftmost position and compare digit by digit, place value by place value, from left to right. Check for the first instance of differing digits. If the digit in a is greater than the digit in b, then $a > b$, or $b < a$.
Example: Which is greater, 5251 or 5250?
Comparing digits from left to right:
5 in 5000 = 5 in 5000
2 in 200 = 2 in 200
5 in 50 = 5 in 50
$1 > 0$
Therefore, $5251 > 5250$, or $5250 < 5251$.

ORDINAL REPRESENTATION

Table 2

Cardinal	Ordinal	Shortened Form of Ordinal
1	First	1st
2	Second	2nd
3	Third	3rd
4	Fourth	4th
5	Fifth	5th
6	Sixth	6th
7	Seventh	7th
8	Eighth	8th
9	Ninth	9th
10	Tenth	10th
11	Eleventh	11th
⋮	⋮	⋮
15	Fifteenth	15th
49	Forty-ninth	49th
50	Fiftieth	50th
100	One hundredth	100th
101	One hundred first	101st

20th century AD means "twenty centuries from the birth of Jesus Christ the Word." 20th is an ordinal number. A century, like an hour, is a unit of time. Since one century is 100 years, the 20th century begins 1900 years after the birth of Christ. The first century includes the years 1 to 100. (Note: There is no "zero" year.) The second century covers the years 101 to 200. Therefore, the 20th century includes the years 1901 to 2000 and the 21^{25} century includes the years 2001 to 2100 (Lesson 3.2).

EXERCISES

Define the following phrases and terms:

1. Presuppose
2. Classification or unity of concept
3. Matching
4. Ordering
5. Sequence
6. One-to-one correspondence
7. etc.
8. Cardinal
9. +
10. –
11. ×
12. ÷
13. Ordinal
14. >
15. <
16. Tag
17. =
18. ≠

> Yogi Berra (1925-2015) was a famous New York Yankee baseball player. After his playing days were over, he became a baseball manager. Throughout his life, Yogi has been known for saying many humorous quips. Here are two examples:
> 1. When you come to a fork in the road, take it.
> 2. Pair off in threes.

Give two examples each of ways in which you have performed or seen others perform the following:

19. Classification
20. Matching
21. Ordering

Answer the following question:

22. Explain how, using place value, you can determine whether one number is greater than another number.

Identify the usage of each number or word highlighted in bold as cardinal, ordinal, or tag (The number in one problem will have two valid answers):

23. We will arrive on Flight **8489** tomorrow.
24. "I Believe" placed **last** in the horse race on Saturday.
25. Take Metro Bus **1** to get to downtown.
26. My score on the spelling test was **83**.
27. My father's annual income is **$60,000**.
28. John ran the race in **50** seconds, finishing **second**.
29. At the motor vehicles department, I drew the number **30** and waited a long time before being served.
30. Mrs. Smith said, "We've had **300** customers visit our store today."
31. Jesus, the Word of the Father, said, "I am the **first** and the **last**."

32. The average[1] day's wage in New Testament times was **one** denarius.
33. Baseball slugger Babe Ruth wore **3** on his jersey in the 1920s and 1930s.
34. Mr. Jones lives on **452** Peach Drive.

Using the < and > symbols, compare the following pairs of numbers in two different ways:

35. 11, 3
36. 4, 9
37. 23, 27
38. 423, 418
39. 8718, 8720

Field Project:

40. Procure a small notebook that you can carry with you or use a phone application where you can write and store notes. For the duration of this course, keep track of your encounter with number, cardinals, ordinals, tags, the four operations of arithmetic, and comparisons. Note the date, place, circumstance, and usage.

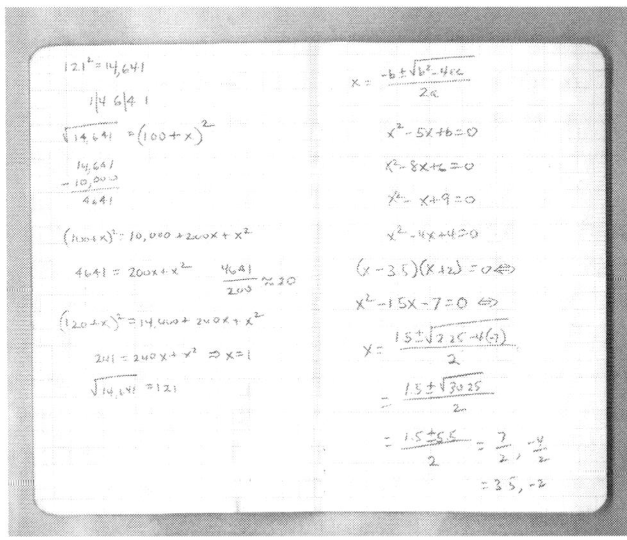

Figure 1. The renowned Moleskine notebook. Source: James D. Nickel.

Table 3: Names of Numbers (Classical Greek and Latin)

	Cardinal		Ordinal	
	Classical Greek	Latin	Classical Greek	Latin
1	*heis, mía, hén*	*unus, una, unum*	*protos*	*primis*
2	*dúo*	*duo, duae, duo*	*deúteros*	*secundus*
3	*treis, tría*	*tres, tres, tria*	*tritos*	*tertius*
4	*téssares, téssara*	*quattuor*	*tétartos*	*quartus*
5	*pénte*	*quinque*	*pémptos*	*quintus*
6	*héx*	*sex*	*héktos*	*sextus*
7	*heptá*	*septem*	*hébdomos*	*septimus*
8	*októ*	*octo*	*ógdoos*	*octavus*
9	*ennéa*	*novem*	*énatos*	*nonus*
10	*déka*	*decem*	*dékatos*	*decimus*
12	*dódeka*	*duodecim*	*dodékatos*	*duodecimus*
100	*hekatón*	*centum*	*hekatostós*	*centesimus*
1000	*chílioi, -ai, -a*	*mille*	*chiliostós*	*millesimus*
10,000	*múrioi, -ai, -a*	*decem milia*	*muriostós*	*decimus millesimus*

> Major conceptual advances in science now require that we recover a view of the universe in which every single thing or event is in fact related to everything else.
>
> Giuseppe Del Re, *The Cosmic Dance* (2000), p. 15.

[1] Average is a word used in statistics, a number approximating the expected value of a group of numbers. We will learn how to calculate this expected value in Step 9.

1.5 Value and Position

> ... the ingenious method of expressing all numbers by means of ten symbols, each symbol receiving a value of position as well as an absolute value ... appears so simple to us now that we ignore its true merits. But its very simplicity ... puts our Arithmetic in the first rank of useful inventions ... remember, it escaped the genius of Archimedes and Apollonius [ancient Greek mathematicians – JN] ....
>
> Pierre-Simon de Laplace (1749-1827).

Terms & Concepts Introduced
1. Algorism
2. Algorithm
3. Expanded notation
4. Rhetoric
5. Standard notation
6. Word names

As we have noted, the system that the whole world now uses is the highly flexible and adaptable decimal positional system.[1] In review, these symbols are named Hindu-Arabic because, during the 5th to 10th centuries AD, mathematicians in India developed the positional system of numbers using ten digits.

The figures for the digits, from one to nine, appear in Indian inscriptions dated to the 3rd century BC, but the principle of place value or zero[2] had not yet been applied. It can be reasonably conjectured that thereafter the Babylonian positional idea filtered through trade channels to India. The earliest known written document making use of this system, written in Sanskrit, is dated 458 AD. We are not sure who the person was who first constructed the decimal positional system, but whoever did it must have experienced a sense of the beauty of structure and order.

Figure 1. Abu Jafar Muhammad ibn Musa al-Khwarizmi. Source: Public Domain.

Figure 2. The lands of Arab conquest. Source: Public Domain.

The Arab military thrust in India resulted in their acceptance and propagation of this system of numeration. In Baghdad in the first decades of the 9th century AD, the Arab mathematician Muhammad ibn Musa al-Khwarizmi (ca. 780-ca. 850), working in a school named the House of Wisdom, wrote the first work in Arabic that taught the Indian method and decimal system. It was entitled *The Book of Addition and Subtraction by Indian Methods*.[3] With the Muslim invasions of Spain and France, scholars there encountered al-Khwarizmi's work. Both scholars and tradesmen, via trade contacts, immediately understood the elegance, simplicity, and efficiency of these methods and, in the 12th century AD, began translating al-Khwarizmi's text into Latin. The method of

---

[1] As we have noted (Lesson 1.3), *decem,* in Latin, means "ten." *Decima* means "a tenth part" or "tithe." In Step 6, we will study fractions, parts of a whole.

[2] The Hindus used the word *sunya,* meaning "empty," for zero.

[3] A Latin translation of this work, prepared by the Englishman Robert of Chester (ca. 1120), was entitled *Liber algorismi de numero indorum* (The Book of al-Khowarizmi on Hindu number).

adding, subtracting, multiplying, and dividing numbers using decimal positional notation became known in Europe as **algorism**, from *algorismus*, the Latin name of al-Khwarizmi. Because of this history, we name the decimal positional system the Hindu-Arabic system. Today, **algorithm**, in the mathematical or computer context, means a method or procedure.

As we have noted (Lesson 1.3), the decimal system consists of ten digits: 0, 1, 2, 3, 4, 5, 6, 7, 8, 9. The English or European symbols for these numbers were derived from the symbols used by the Arabic mathematicians.[4] The beauty of the system is that the placement of each digit has a particular value. For example, in 321, 1 stands for ones (range: 0-9 ones) or one, 2 stands for tens (range: 0-9 tens) or 20, and 3 stands for hundreds (range: 0-9 hundreds) or 300.[5] This place value concept made working with arithmetical operations easier.

Table 1: Place Value in the Decimal System														
Trillions (5th) period			Billions (4th) period			Millions (3rd) period			Thousands (2nd) period			Ones (1st) period		
Hundreds	Tens	Ones	Hundreds	Tens	Ones	Hundreds	Tens	Ones	Hundreds	Tens	Ones	Hundreds	Tens	Ones

We can express very large numbers using the decimal system. We can also write very small numbers, too, but we shall wait to do that until Steps 6 and 7. Earlier, when we discussed the order of units, we saw how we can group successive orders into periods (Lesson 1.3 and Table 1). The largest period we discussed then was the period of the fifth order or the trillions period. Table 2 denotes the names of the higher periods. We rarely use these names today because scientists and mathematicians have developed a shorthand way of expressing very large numbers. In Step 11, we will investigate this convenient way of writing both large and small numbers.

Table 2				
Period	American Name	UK Name	Period	American Name
6th	Quadrillion	Billiard	11th	Nonillion
7th	Quintillion	Trillion	12th	Decillion
8th	Sextillion	Trilliard	13th	Undecillion
9th	Septillion	Quadrillion	14th	Duodecillion
10th	Octillion	Quadrilliard	15th	Tredecillion
			16th	Quattuordecillion
			17th	Quindecillion
			18th	Sexdecillion
			19th	Septendecillion
			20th	Octodecillion

[4] The Arabs of Moorish Spain wrote what were then known as Gobar numerals. Gobar, in Arabic, means "dust." The Arabs did their calculations on sandy dusty slants, i.e., inclined boards. They carried dust or flour in a small bag and traced digits with a finger, stick, or pointed knife on a board dusted appropriately. Later, they used a stylus to write figures on a reusable wax tablet. Then, a slate that could be marked with chalk replaced the wax tablet. Finally, pen and ink were applied to paper. The Gobar numerals are the numerals we use today.

[5] This right to left increase in place value is derived from the ancient design of the abacus. An abacus was an ancient calculator that recorded numbers on a series or strings of beads. See Lesson 2.5.

Table 2				
Period	American Name	UK Name	Period	American Name
			21st	Novendecillion
			22nd	Vigintillion

STANDARD NOTATION

The **standard notation**, using commas, for writing one million, two hundred seven thousand, eight hundred twenty-six is 1,207,826. Reading from left to right, we have:

1 million
2 hundred thousands
0 ten thousands
7 thousands
8 hundreds
2 tens
6 ones

> Modern positional notation, where each digit has a varying value depending upon its position in the representation of a number, is simply the notation of the counting board [the abacus – JN] made permanent.
> Constance Reid, *From Zero to Infinity* ([1955, 1960, 1964, 1992] 2006), p. 3.

EXPANDED NOTATION

Using the symbol + to represent addition,[6] we write this representation in **expanded notation** as follows: 1 million + 2 hundred thousands + 0 ten thousands + 7 thousands + 8 hundreds + 2 tens + 6 ones. More examples:

3412: 3 thousands + 4 hundreds + 1 ten + 2 ones
607: 6 hundreds + 0 tens + 7 ones

WORD NAMES

Word names translate the symbols into words or **rhetoric**.[7] For example:

Standard notation: 5239
Expanded notation: 5 thousands + 2 hundreds + 3 tens + 9 ones
Word name: five thousand two hundred thirty-nine

It is convention not to use the word "and" when you express any number as a word name. For example, the word name for 5239 is not "five thousand two hundred and thirty-nine."

We read the number 30,291,040,027,306,000 as "thirty quadrillion, two hundred ninety-one trillion, forty billion, twenty-seven million, three hundred six thousand." Note that in reading numbers, the name of the ones period is omitted.

[6] The signs that we use in Arithmetic (like + for "plus") have a history of development and refinement. We shall learn more about that history when we explore the basic arithmetic operations.
[7] Rhetoric is more than just writing word names for numbers; it is the effective use of language.

When writing a number in standard notation as a word name, begin at the left with the largest period. Write the hundreds, tens, and ones of each period, in their proper order, followed by the name of that period. Continue this process until you reach the last digit or the ones position of the first period. For example, 8,127,403 is written as eight million, one hundred twenty-seven thousand, four hundred three (Figure 3).

8,127,403

four hundred three

one hundred twenty-seven thousand

eight million

Figure 3.

When writing a word name in standard notation, begin at the left with the largest period. Write the digits for the number named in the period followed by a comma. Continue this process for the next period. If the next period is not named, write three zeroes followed by a comma. Make sure to account for every period. If the largest period is the 5th order, you should write five orders separated by commas. For example, eighty billion, one two hundred twenty-eight thousand, one hundred sixty-six is 80,000,228,166 (Figure 4). Billion is in the 4th period, so you write four periods separated by commas.

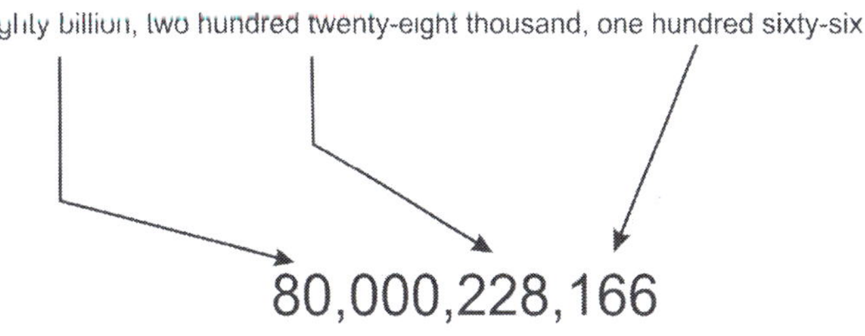

Figure 4.

EXERCISES

Define the following terms:

1. Algorism
2. Algorithm
3. Standard notation
4. Expanded notation
5. Word names
6. Rhetoric

7. Research and determine the size of a typical ancient Roman legion or army.

Read aloud the following numbers:

8. 342
9. 125,004
10. 4,044,034
11. 33,300,330
12. 200,220,022
13. 5,005,500,055

Write the following word names for numbers, using standard notation:

14. One hundred twenty
15. Thirty-seven thousand
16. Sixteen thousand one hundred
17. Fifteen thousand two hundred six
18. Sixty thousand six hundred eight

> And Jesus asked him, "What is your name?" He replied, "My name is Legion, for we are many."
> Mark 5:9

19. Five hundred sixty-three thousand four
20. Ninety-nine billion thirty-seven thousand four

In the following decimal representations (standard notation), identify the number of hundreds, tens, and ones, using expanded notation:

21. 67
22. 280
23. 345
24. 402

Answer the following question:

25. What do you think *decem milia* means? Refer to Table 3 of Lesson 1.4.
26. What place value is hundred ten thousands?
27. What place value is ten thousand thousands?
28. What place value is ten thousand hundred thousands?

Name the cardinal number associated with these prefixes:

29. quad
30. quint
31. sext
32. sept
33. oct
34. non
35. dec
36. undeci
37. duodec
38. tredec
39. quattuordec
40. quindec
41. sexdec
42. septendec
43. octodec
44. novendeci
45. vigint

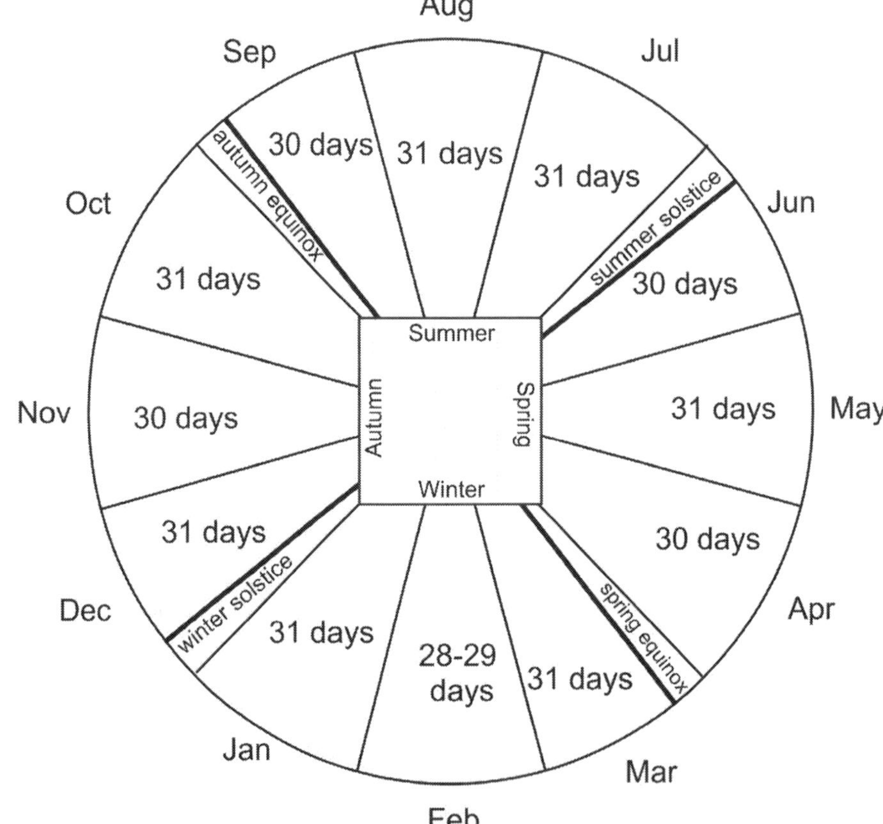

Figure 5. The circle of the year and square of the seasons. The figure depicts the Gregorian Calendar for the Northern Hemisphere. For the Southern Hemisphere, change summer to winder, winter to summer, autumn to spring, and spring to autumn.

STEP 2. THE RHYTHM OF ADDITION

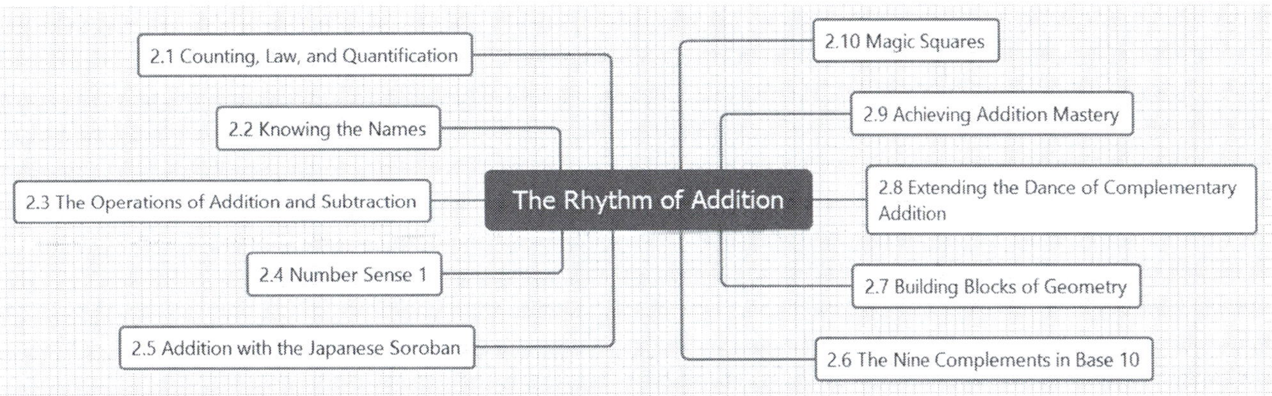

> "Can you do addition?" the White Queen asked. "What's one and one and one and one and one and one and one and one and one and one?" "I don't know," said Alice. "I lost count."
>
> Charles Lutwidge Dodgson (Lewis Carroll), *Through the Looking Glass.*

2.1 Counting, Law, and Quantification

Terms & Concepts Introduced
1. Counting
2. Quantification
3. Scientific Law

Innate in every human being and every culture, ancient and modern, is the sense of number. According to Scripture (Genesis 1:26-28), this is so because of man, male and female, being made in the image of God, in the *imago Dei*. God also made man in His image as a dominion bearer, giving him the responsibility, the authority, to rule over the creation (Psalm 115:16). The prerequisite for wise dominion is an intimate, worshipful, and dependent relationship with the infinite, Triune, Creator God. To correctly understand and use number, we must be in relational submission to our Creator, Sustainer, and Redeemer.

> There is an inherent logic of consistency, you can call it a dance, in the methods of mathematics and this pervasive rationality is revealed at the beginning, in counting and arithmetic.
>
> James D. Nickel

It is only in the light of Christ, the *logos* of the Father, that we can understand anything at all (Psalm 36:9; John 8:12). The fear of the Lord, i.e., the respect and recognition of the joyful presence of the Father, Son, and Holy Spirit, grounded in reception of His unconditional love, is the starting point, the foundation, of all knowledge, including the knowledge of number (Proverbs 1:7; 9:10).

Number represents quantity. Number comes into play when we answer these questions: How much? How many? In the unfolding of history, cultures have developed number systems to help them deal with quantities. The world now uses the decimal system, also known as the Hindu-Arabic decimal system, a numeral system based upon:

1. Ten digits: 0, 1, 2, 3, 4, 5, 6, 7, 8, 9
2. Place value

The Ground of Counting

Counting is foundational to arithmetic. Counting is the numbering of objects. Many of us learned how to count using our fingers. One, two, three, four, five, etc. soon became commonplace. Then we learned how to count objects like coins, toys, marbles, etc. We learned that when we count objects we can count only things that are alike. In counting by object, we cannot mix apples with oranges unless we want to count the total number of apples and oranges.

Figure 1. Stars uncountable, the Sagittarius region of the Milky Way. Source: Wikimedia Commons.

The Triune God counts, too. He counts the number of the stars (Psalm 147:4-5; Isaiah 40:26). He knows this number by the nature of His omniscience, not by sequential numbering, as we must do, if we could count all the stars. He knows the number of the stars instantaneously because He has spoken into existence each star by the Word who is Jesus (Genesis 1; John 1:1-3).

Pick a cloudless night and start counting. Scientists estimate that the number of stars in the visible universe is so large, that numbering them is impossible. By their conservative estimates and assuming we can see all the stars, if we counted one star every second, it would take us ten thousand trillion centuries to finish the enumeration. Not only does God count the stars, but He also names each one (Isaiah 40:26). No wonder the psalmist David replied, "God's understanding is infinite!"[1] The stars of the heavens proclaim the glory of their Maker (Psalm 19:1). The dance of the stars reveals the majestic grandeur of the creative relationship that exists in Father, Son, and Spirit, the glory.[2]

The onto-relational Trinity holds together the magnificent outreaches of space. The ancient creeds pronounced, "We believe in God the Father Almighty." We do not believe in God Almighty for power alone is faceless power, a frightening prospect. It is the Father Almighty who is God the Almighty Creator. Jesus reveals the Almighty Father as One who cares, One who intimately knows the death of the sparrow or the falling of a leaf (Matthew 10:29). This is the God revealed to us in Jesus, One who counts the stars knows every detail about our lives, including the enumeration of hairs on our heads (Matthew 10:30; cf. Psalm 139 and Hebrews 4:13).

> If we catch ourselves saying or thinking "God Almighty," we are absolutizing the transcendence of God. That is why, in the foundational creeds of Christianity, we read, "I believe in God the Father Almighty." The almightiness of God can only be grasped rightly when, in the same breath, we cry, "Abba!"
>
> James D. Nickel

The foundation of counting and number is the nature and character of the Triune God. Think, for a moment, of the situation before the creation. Before time and before the existence of matter, there was only the eternal, perichoretic life of the Triune God, a life of harmonious and satisfying relationship. In the one being of God, we discover a reciprocal self-giving relationship or persons, an interpenetrating perichoresis, a dance of mutual indwelling (John 14-17).[3] The glory of God, therefore, is the reality of the love indwelling the onto-relational Trinity. This Triune glory is the ground and reason of creation. God the Father created all things, visible and invisible, by speaking forth God the Son with the breath of God the Spirit. And why did He create? What plan did the Triune God have in mind? His plan, from before the beginning, was to adopt humanity into the circle of the fellowship He has known from eternity, a holy communion of reciprocal love and harmony, the dance of His delight (Ephesians 1:1-14).

Deuteronomy 6:4 states:

"Hear, O Israel! The Lord is our God, the Lord is one!"

The Hebrew for one is not the mathematical concept of "one" as in "one pencil." In this verse "one" means "unity" or "united" or "altogether."[4] The Lord God is a unity in diversity, unity in relationship. As we have noted, the Hebrew word for God is *Elohim,* a plural word. Note in Genesis 1:26, God says, "Let Us make man in Our image, according to Our likeness ...."[5] God is plural and united; He is many and one.

---

[1] The Hebrew word translated "infinite" in our English Bibles means "cannot be numbered."

[2] Glory, in Hebrew, means "weight or significance."

[3] The three persons in one being fellowship with each other in what human words can best describe as a dance of delight, an interpenetrating love relationship that is so beautiful we call it holy (Psalm 29:2; Psalm 96:9).

[4] The Hebrew word translated "one" speaks of a unity as in one cluster of grapes, a cluster consisting of many individual grapes.

[5] Some Hebrew scholars posit that the "us" in Genesis 1:26 is not referent to the plurality of the Triune God, but to this God addressing the heavenly host with His intentions for mankind.

In John 10:30, Jesus says something that unfolds the meaning of Deuteronomy 6:4, a meaning that revolted the hardened Jewish mind of some His hearers. From the Greek text:

"ἐγὼ καὶ ὁ πατὴρ ἕν ἐσμεν." (I and the Father one we are.)

The Greek word ἕν translated one is neuter, not the masculine εἷς. ἕν means one essence. It is this unity of essence that the Greek text emphasizes as paramount. The New Testament reveals a plurality of persons in the Godhead: Father, Son, and Holy Spirit (Matthew 28:18). The three distinct persons are one in being, and the one in being are three distinct persons, as the early creeds of the Christian church articulate. There is a mystery about the nature of the Godhead, but this mystery is not illogical. The eternal One and the Many has always been, is, and always will be, perfect harmony and fellowship.

The Triune nature of God is the foundation for a true understanding of anything, including the understanding and use of number.

SCIENTIFIC LAW

The Triune God's creation is replete with the one and the many, or unity in diversity. In mathematics, for example, three cows, three horses, three cats reveal a unity (we call it "threeness") in diversity; the unity is shown in different animals or any countable object. As we have seen, counting presupposes that we can classify a variety of observations under a unity of concept and to endow that unity with a name. **Scientific Law**, often written in mathematical form, is a generalization, a unity, generated from many, a diversity of instances, observations or experiments.

In the structure of mathematics, we find many unifying concepts or laws that tie a diversity of mathematical relationships together. We shall discover many of those unifying concepts in the lessons to come. The laws of the created order mathematically expressed and the laws of the framework of mathematics soundlessly and voicelessly echo the commanding speech, the Word, of the Creator.

QUANTIFICATION

Number is a way to describe magnitude or position. Magnitude represents how large or how small, or the quantity or amount, the cardinality principle. We can either count objects or measure objects. The process of counting and/or measuring is **quantification**.[6] Position represents place, location, or order, the ordinality principle.

Numerals are symbols that describe or represent number. Using the Hindu-Arabic numerals, we can classify any and every object of God's created order. For example, the symbol 2 can signify the concrete in terms of two cows, two atoms, two dollars, two meters, two miles, etc.

The whole of creation, visible or invisible, ties together or interconnects. Number reports on this interconnectedness. Read Colossians 1:15-17 carefully, for it reveals that the person of Jesus Christ is the unity, the cohesion, the logic that holds all created reality together. There is unity, purpose, and plan, to all things diverse.

> ... in Jesus Christ we meet the logic that predates the universe, the one rhyme and reason behind it all, the one Word, the Logos, written into the marrow of the cosmos and of human existence within it.
>
> C. Baxter Kruger, *A Discourse on Method: The Communion of the Holy Trinity as the Basis and Logic of Christian Theology* (2002).

We cannot make sense of anything unless we posit the starting point or beginning of plan, purpose, and unity in the Triune God of Scripture as revealed in the Lord Jesus Christ. Christ is, indeed, the ground of the rationality of the universe.

---

[6] *Quantitative* means "of or pertaining to the describing or measuring of quantity."

EXERCISES

Define the following terms:

1. Counting
2. Scientific Law
3. Quantification

What place value does the numeral zero [or zeroes] hold in the following notations?

4. 1024
5. 5,000,008
6. 150,128
7. 390,789,160,118,003
8. 41,002

> The rationality that number brings to light is never independent of its ultimate source, the wisdom of God in Christ. Number discloses the order, the many patterns, the multilevel dance of creation. This dance not only came into being by the Word of God but is held together by Him. Christ is the ground of the rationality of number.
>
> James D. Nickel

Memory work:

9. Memorize Psalm 19:1, Psalm 36:9; Psalm 147:4-5, Isaiah 40:26, John 1:1-5, John 8:12, and Colossians 1:15-17.

Short essay (one to three sentences):

10. Give two examples of God counting.
11. Explain the decimal positional notation system with an example.
12. Explain the difference between God's counting and our counting.

Long essay/answer:

13. As well as you can, explain the perichoretic nature of the Trinity.
14. As well as you can, explain why the fear of the Lord God of Scripture is the foundation, the starting point, for understanding anything truly.
15. Look up and write the Nicene-Constantinopolitan Creed (381 AD) in your notebook. Verbally read the entire confession and note the emphasis on the Trinity.

> We take numbers for granted and use them every day. Numbers are universal: all cultures have them. Yet the idea of number, so obvious to us today, is the result of a protracted process of progressively abstract reasoning. It took humanity thousands of years to advance from literal quantities to the concept of numbers. How do we add things up? By seeing each object simply and exclusively as a unit. By seeing things as entities while rejecting their specific differences.
>
> Denis Guedj, *Numbers: The Universal Language* (1997), p. 13.

> The heavens are telling the glory of God silently. Their speech is the abundance of stars; billions and billions speak of extravagant love.
>
> James D. Nickel

2.2 Knowing the Names

> Who knoweth not the names, knoweth not the subject.
>
> Carl Linnaeus, *Critica Botanica* (1737).

Arithmetic is the science, or knowledge, of number and the art of calculation with numbers. As a science, Arithmetic studies the nature and properties of number, the undergirding laws that show how different numbers relate to each other. As an art, knowledge of arithmetic operations enables a person to apply the structure of numbers to practical or economic purposes.

As we have noted, the ancient Greeks understood arithmetic as the "theory of the structure of numbers." They used another word, *logistica*, to describe the skill of calculation. Our definition incorporates both aspects: theory, the why, and calculation, the how.

Terms & Concepts Introduced
1. Abstract or pure numbers
2. Applied or denominate numbers
3. Concrete number
4. Denomination of a concrete number
5. Integral number
6. Like numbers
7. Number theory
8. Quadrivium
9. Set
10. Trivium
11. Unit
12. Unit of a number
13. Unlike numbers

What is the difference between arithmetic and mathematics? Arithmetic is one part of a much broader discipline of study called mathematics. Mathematics, like a tree, includes many branches of knowledge. Some of the other branches of mathematics that you will usually study are Algebra, Geometry, and Trigonometry. Arithmetic, studied in elementary school, forms the roots of the mathematics tree.

Number, as the centerpiece of arithmetic, is intimately tied to both the rational order and constraints of what we daily observe. Rational order reveals law-like structure and we will encounter many of the laws of arithmetic in future lessons. What are constraints? Constraints mean that 1 + 1 cannot be equal to anything but 2 in the base 10 system.

As we have noted, the cosmos is the rational order of all things, things formed by the Word of the Father, the *logos*, who is Christ. Being all-wise (Romans 16:27), what this God creates reflects His wisdom.

"O Lord, how many are Your works! In wisdom You have made them all ...." (Psalm 104:24).

"...true knowledge of God's mystery, that is, Christ Himself, in whom are hidden all the treasures of wisdom and knowledge." (Colossians 2:2-3).

Number reports on and reflects the wisdom and beauty of this order. Since we are commanded by the Triune God to subject this reality to scientific probing or questioning (Genesis 1:26-27; Proverbs 25:2), an examination which necessitates measuring objects, the rationality number is one of the strategic tools of science.

Basic Definitions

Arithmetic has its own vocabulary, names used that have unique and special meanings. Let's start with some basic definitions.

A **unit** means a single object.[1] Unit comes from unity. We usually attach the word "one" to a unit. Therefore, we often interchange the words "unit" and "one." We speak of one cat, one ball, one day, one month, one year, or one century.

---

[1] The English astronomer and mathematician John Dee (1527-1608) first coined this definition in 1570. The Greek word for single or alone is transliterated *mónos*. From this word, we get *monad* meaning "uniqueness" or "seclusion." Words derived from the unit or monad concept are atone (to make one with), once, union, unanimous, universal, monopoly, monastic, monolithic, monogram, and monotonous.

The unit is the foundation of counting since we always start with one.

If we want to count how many marbles are in a bag, we first assume that there are marbles inside the bag. If the bag is empty, we must embrace another important concept in arithmetic, zero.

Different cultures throughout history have assigned meanings or mystical qualities to unit. Since the unit is the foundation of counting, some cultures have concluded that all things come from one principle or process. Thales of Miletus (ca. 640-546 BC), an ancient Greek philosopher, said that water is the one principle. By Scripture, we know that all things originated in the Triune God. As we have noted, this God is not an absolute unity; He is unity in diversity. The God of Scripture, the Alpha and Omega, the beginning and the end, is One and Three. All things do not come from a solitary One; all things come from and are sustained in the Three in One and One in Three.

What is number? The answer is simple and, as you will discover in your future studies of mathematics, complex. Number, as a noun, is a term we use to describe the rationality order of the cosmos. We can also use of number as an adjective. A number is either a unit or a group, or **set**, of units. We speak of one hat, five boys, ten watches, etc.

We use number as an adjective to answer the questions regarding (1) size, how many, (2) position, in terms of order, or (3) identity or tag (Lesson 1.4).

A number that expresses whole objects is an **integral number**. Integral comes from the Latin word integer meaning "undivided, whole, entire, or untouched." When we speak of two boys, we think of two whole persons, not parts of persons (e.g., their arms, legs, etc.). When we measure things, we often must think in parts. Think about measuring the length of a piece of wood. Let's say it measures two and one-half feet. The two is a whole number, but one-half is part of a whole. The part numbers are fractions and we shall explore these numbers later (Step 6). Soon, we will also talk about a set of numbers named integers (Step 3). When we define the set of integers, we will find that our concept of wholeness will be expanded to include a new idea.[2]

The number twelve is composed of twelve occurrences of one unit.[3] Twenty dollars consists of twenty units of one dollar.[4] Six inches consists of six units of one inch.[5] The **unit of a number** is one of the collections of units that makes up, or constitutes, the number.

CONCRETE AND ABSTRACT

A number can stand alone or it can be connected to a kind of object. When a number is joined to an object, it is a **concrete number**. The object is the number's **denomination**.[6] Denomination means a name for a group of things. The denomination of a number denotes what that number counts or measures. For example, the denomination of twenty dollars is dollars. Examples of concrete numbers are three dollars, five cats, six hammers, six feet, 240 pounds, etc. Counting pencils, we say, "There are four pencils." We do not say, "There are four." When we say to someone, "Give me five," the phrase has no meaning unless we know what five objects are going to be given to us, e.g., five dollars or one hand consisting of five fingers. We must always know the objects that we are counting/measuring and indicate

---

[2] This new idea is revealed in reading a temperature scale. Sometimes the measure of temperature is spoken of in degrees (Lesson 2.7) below zero. The set of integers includes the idea of integral numbers below and above zero (Lesson 3.2).

[3] This conception carries within it the idea of multiplication (Step 4); i.e., $12 \times 1 = 12$.

[4] The English word dollar is derived from *thaler*, a silver coin first minted in 1516 in Joachimsthal, Czechoslovakia. Originally named *joachimsthaler*, in honor of the town, it was later shortened to *thaler*. The symbol for dollar is $. As a unit, we write this sign in front of a number, not after it. For two dollars, we write $2, not 2$.

[5] An inch, coming from the Latin *uncia*, is an ancient unit of measurement meaning "one-twelfth part" (see Step 6 and Step 7). It is approximately equal to the length of the distal phalanx of an adult thumb, the distance from the interphalangeal or top joint to the tip of the thumb. One foot is equal to twelve inches, three feet are equal to one yard, and 5280 feet are equal to one mile.

[6] We shall meet denominate numbers again when we study denominators in fractions (Step 6).

those objects when we say or write the result of our counting/measuring. A concrete number is always used as an adjective. Concrete numbers are what we call **applied numbers** or denominate numbers.

When a number stands alone, without any connection to a denomination, it is an **abstract number**. Abstract numbers are also **pure numbers**.[7] When we speak of the number four, eight, or thirty-five, we are speaking, and thinking, abstractly. An abstract number, therefore, is used as a noun. The study of the structure of abstract numbers is now known as **number theory**. We shall explore some of this structure in future lessons, especially when we discover the various laws or relationships that govern calculation with numbers.

> Once an abstract notion of number is present in the mind, ... the essence of mathematics is seen to be not the numbers themselves but the collection of relationships that exist between them.
>
> John D. Barrow, *Pi in the Sky: Counting, Thinking, and Being* (1992), p. 105.

> Historically, the study of number was a central component of one's education. It originally comprised one of the four subjects of study, the **quadrivium**. The quadrivium plus its foundational counterpart, the **trivium**, formed the basis of liberal arts education.
>
> James D. Nickel

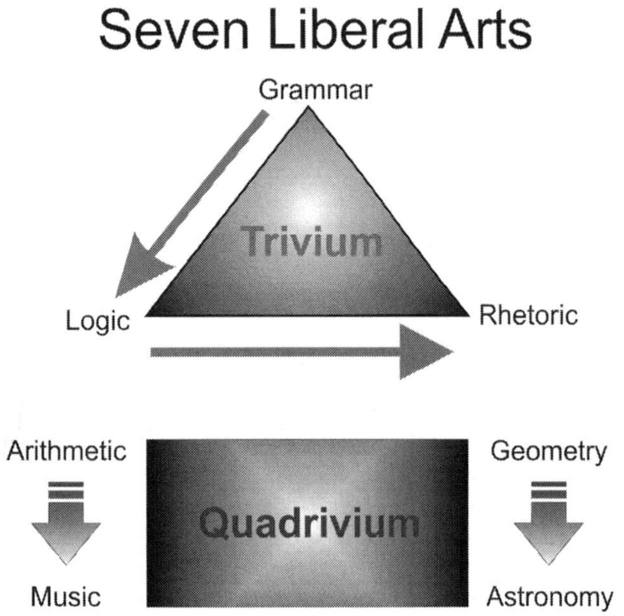

Seven Liberal Arts

Grammar
Trivium
Logic
Rhetoric
Arithmetic
Music
Quadrivium
Geometry
Astronomy

Figure 1.

There is a perichoretic interplay between the concrete and the abstract. A concrete number is number applied. The abstract is in the concrete and the concrete is in the abstract. We can use five dollars, while five, an abstract number, is not much use. However, do not be fooled into thinking that abstract numbers have no use whatsoever. Ponder this quote from the mathematician and philosopher Alfred North Whitehead (1861-1947):

> Nothing is more impressive than the fact that as mathematics withdrew increasingly into the upper regions of abstract thought, it returned back to earth with a corresponding growth of importance for the analysis of concrete fact.... The paradox is now fully established that the utmost abstractions are the true weapons with which to control our thought of concrete fact.[8]

LIKE AND UNLIKE

Like numbers have the same kind of unit, the same type of quantity, or the same denomination. For example, four dollars and eight dollars are like numbers. However, four dollars and eight lire, a unit of

[7] The distinction between abstract and concrete numbers is a modern concept, beginning in the 16th century. Greek mathematicians thought of abstract numbers as related to their structure and connections. Early writers who taught *logistica* were generally not concerned with this distinction.

[8] Alfred North Whitehead, *Science and the Modern World* (London: Free Association Books, [1926] 1985), p. 41.

money in Italy, are not like numbers. Six days and fifteen days are also examples of like numbers. We can also add or subtract like numbers. For example, one dollar added to four dollars gives five dollars. Like numbers may be either concrete or abstract. This is why we can add, in the abstract, two and two to get four.

Unlike numbers have different kinds of units, or denominations, and represent different kinds of quantities. We cannot add or subtract unlike numbers. For example, three apples and four oranges equal three apples and four oranges. As the saying goes, "You cannot add apples and oranges." Why? Apples and oranges are different objects or different denominations. Even though we cannot add apples to oranges, we can change the denomination to pieces of fruit. If we have three apples and four oranges, then how many pieces of fruit do we have? We have seven pieces of fruit

> Grammar doth teach to utter words.
> To speak both apt and plain,
> Logic by art sets for the truth,
> And doth tell us what is vain.
>
> Rhetoric at large paints well the cause,
> And makes that seem right gay,
> Which Logic spake but at a word,
> And taught as by the way.
>
> Music with tunes, delights the ear,
> And makes us think it heaven,
> Arithmetic by number can make
> Reckonings to be even.
>
> Geometry things thick and broad,
> Measures by Line and Square,
> Astronomy by stars doth tell,
> Of foul and else of fair.
>
> Thomas Wilson (1524-1581), *Rule of Reason* (1551).

EXERCISES

Define the following terms:

1. Set
2. Unit
3. Integral number
4. Unit of a number
5. Concrete number
6. Denomination of a concrete number

> Mathematics is *non notationes, sed notiones* (not notations, but notions).
>
> Carl Friedrich Gauss (1777-1855).

7. Applied or denominate numbers
8. Abstract or pure numbers
9. Number theory
10. Like numbers
11. Unlike numbers
12. Trivium
13. Quadrivium

> Our minds ... are capable of forming abstract number concepts because we are capable of thinking about particular, concrete cases.
> Luke Heaton, *A Brief History of Mathematical Thought* (2015), p. 20.

Answer the following questions:

14. (a) How many units are in two? (b) In ten cents? (c) In twenty dollars? (d) In twenty-four acres?
15. (a) What is the unit or denomination in twenty-five cents? (b) In fifteen books?
16. Are three trees and six trees like or unlike numbers?
17. Are three trees and six trees concrete or abstract numbers? Why?
18. Seven and nine are what kind of numbers?
19. What kind of numbers are eleven acres and fifty miles?
20. What kind of numbers are three pens and ten pens?

Name:

21. Two numbers that are like and abstract.
22. Two numbers that are like and concrete.
23. Three numbers that are unlike and concrete.

Glance quickly at each picture group in Figure 2 and then close your eyes.

24. How many objects did you remember in each picture group?
25. Which picture group did you have difficulty remembering the number of objects?
26. What can you conclude from this experiment?

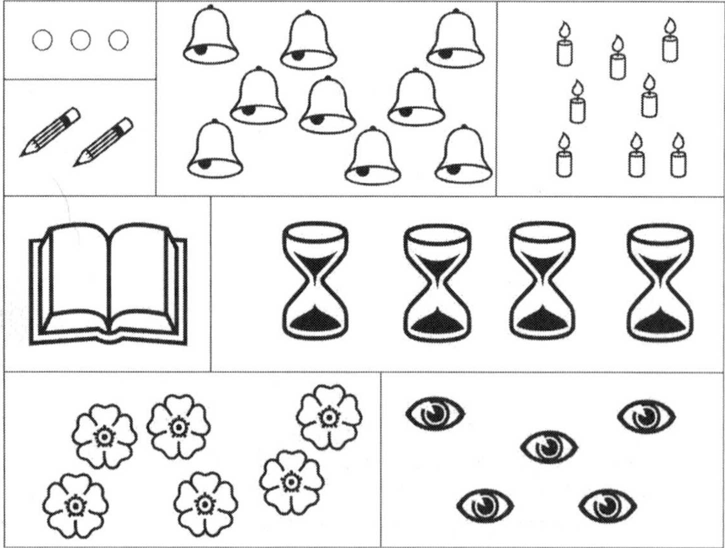

Figure 2.

2.3 The Operations of Addition and Subtraction

Terms, Symbols & Concepts Introduced
1. $\mathbb{N}$
2. ∞
3. Addition
4. Angle
5. Ellipsis
6. Fibonacci Sequence
7. Finite
8. Generalization
9. Hexagon
10. Infinity
11. Inverse operations
12. Natural numbers
13. Regular hexagon
14. Substitution
15. Subtraction

When the Triune God told Adam to name the animals, He told him to observe and classify. By naming, Adam stamped unity upon the individual species of animal he observed. He saw patterns in the animals and named them accordingly.

> ... early forms of arithmetic and geometry [in Ancient Egypt – JN] grew from the demands of canal building: how long, how wide and how deep?
> James Burke, *Connections* (1978), p. 11.

Likewise, we can use the principles of number and space to stamp unity on the many aspects of God's rational order. Stamping unity upon an object is **generalization**, and generalization is a way to economize. Generalization is naming the patterns of God's rational order. For example, although there are differences within the fox species, a fox will always be a fox. When you say, "There is a fox," there is an economy of thought in the word "fox."

Unity is a way of generalizing our experience with objects. When we count three foxes or three coins, we are stamping the abstract unity of "threeness" to these concrete observations. We are stamping a unity to diversity.

> What do we know of the numbers 1, 2, 3 ..., which appear so obvious that we call them *natural*? The limitless sequence of *integers*, seemingly so familiar, is full of mysteries. Natural numbers have all sorts of patterns and rules, some of them unexpectedly regular, and many remarkable properties. Their study constitutes the theory of numbers, called *arithmetic*. The great German mathematician Carl Friedrich Gauss (1777-1855) called arithmetic the "queen of mathematics."
> Denis Guedj, *Numbers: The Universal Language* (1997), p. 61.

Arithmetic is a way of abstracting the rudimentary processes of thinking. Arithmetic is a way, using thought, to invoke the principle of unity in diversity, a principle that the Creator God, being Triune, has embedded in all creation, both in our minds and in the world outside of our minds. When we count objects, we are identifying them; we are sorting them by name. When we count, we are thinking, "All those objects look like oranges, and I'm going to identify them by grouping them, using a numeral."

> The mathematician's instinct is to structure that process of understanding [the patterned nature of creation – JN] by seeking generalities that cut across the obvious subdivisions.
> Ian Stewart, *Nature's Numbers* (1995), p. 13.

Regarding the principle of economy of thought, if we have one hundred coins and we want to name, or record, how many we have, there is no need for us to fill a page with one mark for each coin. With the decimal system, we write just three digits: 100. Indeed, economy of effort!

The perichoretic structure of creation reveals patterns that can be understood in the context of:

(1) Number or Arithmetic
(2) Space or Geometry

(I) NUMBER

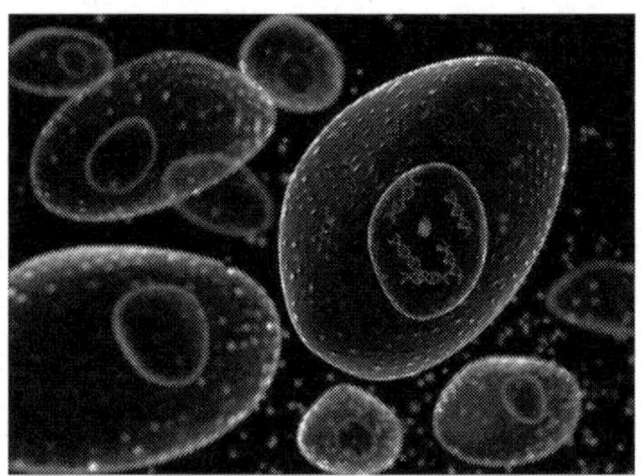

Figure 1. DNA nucleus of organic cells. Source: iStockPhoto.

In English grammar, an **ellipsis**, Greek for omission, is the intentional exclusion from a sentence of one or more words. For example, "He went to … school." The three dots, the ellipsis, mean we have omitted something from this sentence. In mathematics, an ellipsis is used to mean "and so forth," "in the same manner," or "following a definite pattern." For example, when we see $1 + 2 + \ldots + 10$, we should be thinking, "I am adding the numbers one to ten." If the ellipsis occurs at the end, it means that the pattern revealed before the ellipsis continues *ad infinitum*. Therefore, the list of numbers represented by "1, 2, 3, …" means we have an infinite collection of numbers.[1] These are the numbers we use to count and there is no end to them. Because they are natural to counting, we name this list of numbers as the set of **natural numbers**, signified by this symbol: ℕ. The set of natural numbers, the basic building block of arithmetic, also reveals something of the nature of infinity. As you continue your study of mathematics, you will discover how often infinity appears on the scene.

Figure 2. Sunflowers. Source: James D. Nickel.

What do we know about the sequence of natural numbers? We can identify several features:

1. Natural numbers follow each other according to a pattern. Other than one, each number in this sequence is one more than its predecessor.
2. Natural numbers get larger and larger as we continue to list them.
3. Natural numbers have no end. Although there is a first natural number, there is no largest natural number.
4. The set of natural numbers is a prototype example of the nature of numerical order.

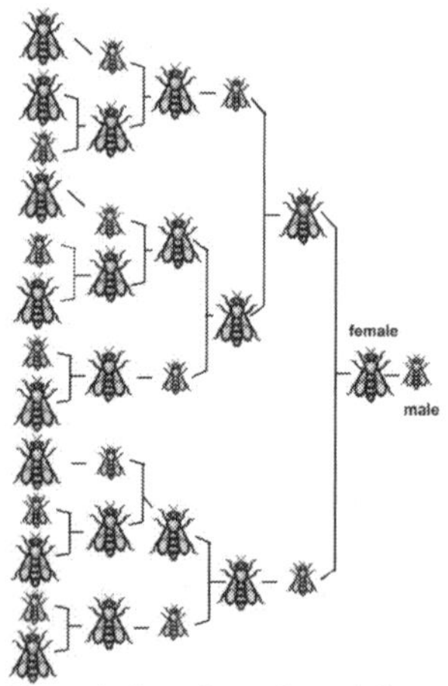

Figure 3. Genealogy of a male bee: Do you see the Fibonacci Numbers?

We use the natural numbers to count and order objects. We can rearrange this set to count other objects. For example, in the biological world, we can understand cell division numerically. We start with one cell. After a certain amount of time, the cell divides into two cells. Then, each of the two cells divides and we now have four. We can illustrate what is happening in a table.

Table 1: Cell Division

Order	0	1st	2nd	3rd	4th	5th	6th	7th	8th
Number of cells	1	2	4	8	16	32	64	128	256

[1] For a mathematician, **infinity**, symbolized by ∞, is not a number; it represents the concept of "increase without bounds or limits." In contrast to the concept of the infinite, **finite** means something that can be completely determined, at least in theory, by counting or measurement.

We can understand many objects of creation (e.g., pine cones, pineapples, sunflowers) using a fascinating sequence of numbers known as the **Fibonacci sequence**. This sequence is named in the honor of Leonard Fibonacci, the man who introduced the Hindu-Arabic decimal notation to Europe in the 13th century.[2] In his book *Liber Abbaci* (1202, second edition: 1228), he added problems known to Hindu mathematicians about rabbit populations and the genealogy of male bees.[3] In Table 2, do you see the pattern?

Table 2: The Fibonacci Sequence

Order	1st	2nd	3rd	4th	5th	6th	7th	8th	9th	10th	11th	12th	13th	14th	15th
Fibonacci Number	1	1	2	3	5	8	13	21	34	55	89	144	233	377	610

(2) SPACE

Figure 4. Honeycomb. Source: iStockPhoto.

Many forms of the created order reveal geometric properties, e.g., circles, parallel lines, polygons, regular polygons, cylinders, angle.[4] Circles can be seen in a rainbow. The banks of a river, when the river runs straight, form parallel lines. Trunks of trees form cylinders. A can of soup that you can buy at your local grocery store is in the shape of a cylinder. Honeybees generate a honeycomb in the form of a **regular hexagon**, six equal angles. An **angle** (Figure 7), in geometry, is formed by two sides starting from the same point. (See Lesson 2.7 for a more precise definition.[5]) In a **hexagon**, there are six such angles. If these angles are equal, then the figure is a regular hexagon.

Figure 5. Rainbow. Source: iStockPhoto.

Figure 6. Circular rainbow. Source: iStockPhoto.

In Lesson 2.7, we will develop the mathematical notion of space.

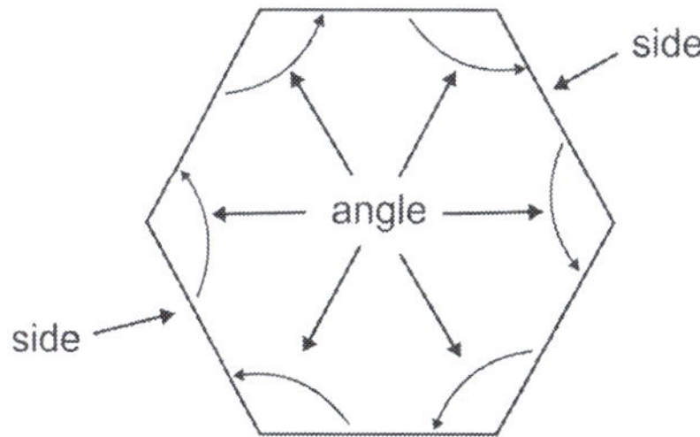

The Triune God has designed the honeybee to construct a honeycomb with a regular hexagon as its foundational shape. In a regular hexagon, all six sides are equal, and all six angles are equal.

Figure 7. Angles and sides.

ARITHMETICAL OPERATIONS

As we have noted, the foundation for arithmetic is the unit. When classifying the patterns of the created order, we separate an object from all other objects by giving it a name and identifying it as a unit that can be repeated, or counted, many times. We can also group units to form a set of units, or a new group. For example, in the decimal system, we group

[2] The French mathematician François Édouard Anatole Lucas (1842–91) gave the name Fibonacci sequence to this list of numbers in honor of Leonardo of Pisa.
[3] The real heritage of Leonardo's book is not this sequence, but the arithmetic of commercial transactions.
[4] These words, except angle, will be defined in later lessons.
[5] The angle concept probably arose from the observation of the angle formed by one's thigh and lower leg or forearm and upper arm. Why? In most languages of the world, the word for the side of an angle means either leg or arm.

ten units to form one set of tens, or 10. We group twenty units to form two sets of tens, or 20. Ten groups of ten form the hundreds unit. We can count, or group, the 25 players playing catch with a round ball to form a new set, i.e, a major league baseball team.

The four elementary arithmetic operations[6] are addition, subtraction, multiplication, and division. In this lesson, we will define only the operations of addition and subtraction.

Before we explain these operations, we must first ask the question, "What do we mean by them?" Their meaning is connected to the created reality. Their meaning answers questions like: Which situations require adding numbers? Which situations require subtracting numbers? There are fine layers of meaning in these responses, and we shall explore them.

ADDITION

It is part of our human constitution, made in the image of God, to understand that objects can be added to a group, collected in containers, or joined together, one group to another group. The meaning of **addition** is "joining two groups together."[7] We can also understand addition as counting forward, one and one and one and one, etc., or, more briefly, as multiple counting. As we have noted, you add objects that are alike.

Figure 8. The Italian mathematician Niccolò Fontana Tartaglia (1499-1557) used the first letter of the Italian più (plus) to signify adding.

For example, to find out how many cows are in a pasture, we count the number of cows, but not the goats unless we want to count the number of four-legged animals in the field. To learn how many apples are on a tree, we count the apples, but not the leaves. To determine how many stars are in the night sky, we count them, but not the Moon or the planets. It is important to remember that you can only count things that are similar or you can only count things that have the same denomination. That denomination must be understood clearly before the counting process commences. If we have \$3 and we earn \$6 more, I have \$9. From \$3, we count forward six units of dollars to get \$9. Using symbols, we write:

$3 + 6 = 9$

The symbol for addition is +. In Latin *et* means "and." At some point in the unfolding of history, around the 13th century in Europe, someone **substituted**[8] the cross sign (+) for the letter "t" in *et*. The sign "+" is read "plus," and means "more." Thus, 3 + 6 is read "three plus six" and means that we add three and six.

SUBTRACTION

The meaning of **subtraction** is the opposite of the meaning of addition, or understood as counting backward. Rather than joining two groups together, in subtraction[9] we remove or we take away one object from a group of objects, or a group of objects from a given group. If I have \$9 and I owe my sister \$3, we take away \$3 from the \$9 to get \$6, the amount of money I have left. Subtraction can also be understood as counting backward. I have \$9 and I count back three units of dollars to get \$6.

Figure 9. This subtraction sign was used by the Greek mathematician Diophantus of Alexandria (ca. 200-284). Our subtraction symbol may have come from a bar medieval traders used to indicate differences in the weights of products.

---

[6] The phrase "arithmetic operation" is a modern term. In the 13th century, the word *species* was used to represent the concept. From the 15th to the 17th centuries, the words *parts* and *acts* were used. In history, a listing of the fundamental arithmetic operations varied from four to eight.

[7] Addition is derived from the Latin *additio* meaning "to add."

[8] To replace a mathematical object (in this case a symbol) by another object is common in mathematics.

[9] *Subtraction* is derived from the Latin *subtrahere* meaning "under" and "to draw." It literally means "to draw away from under" or "to withdraw."

Using symbols, we write.

$9 - 3 = 6$

The symbol for subtraction is the minus sign, i.e., – . It is read "minus" and signifies "less."

Subtraction is the **inverse operation**, the reverse dance step, of addition. One operation undoes the other and vice versa. In the subtraction problem, 9 – 3, we ask the question, "What must be added to three to get nine?" Answer: Six.

Addition and subtraction are ways of counting, counting forward with addition and counting backward with subtraction. That is why arithmetic is sometimes known as "counting arithmetic."

Addition and subtraction are the roots of the mathematics tree. It is a tree whose fruit is the patterned order of God's creation. The root of these roots is the Triune God of Scripture, the ground of counting.

In the subsequent lessons, we will investigate the principles of addition.

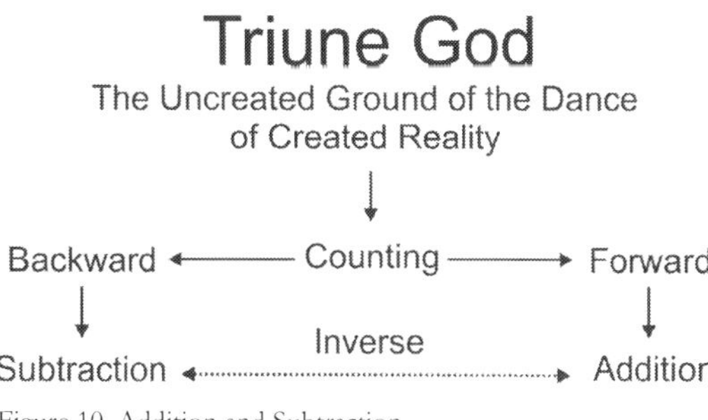

Figure 10. Addition and Subtraction.

EXERCISES

Define the following terms/symbols:

1. Generalization
2. Ellipsis (Define it both grammatically and mathematically.)
3. Infinity
4. ∞
5. Finite
6. Natural numbers
7. $\mathbb{N}$
8. Fibonacci Sequence
9. Hexagon
10. Regular hexagon
11. Angle
12. Addition
13. Substitution
14. Subtraction
15. Inverse operations

Write a short essay:

16. Give two examples of an everyday situation that requires addition.
17. Give two examples of an everyday situation that requires subtraction.
18. Demonstrate how subtraction is the inverse of addition.
19. Demonstrate how addition is the inverse of subtraction.

Counting exercises:

20. Count the number of spirals going one way and the number of spirals going the other way on the pinecone in Figure 11. Make sure that you mark where you start counting.
21. Count the number of spirals going one way and the number of spirals going the other way on the thistle in Figure 12.
22. Count the number of spirals going one way and the number of spirals going the other way on the sunflower in Figure 13. Make doubly sure that you mark where you start counting.
23. What is the name of numbers you counted in the preceding three exercises?

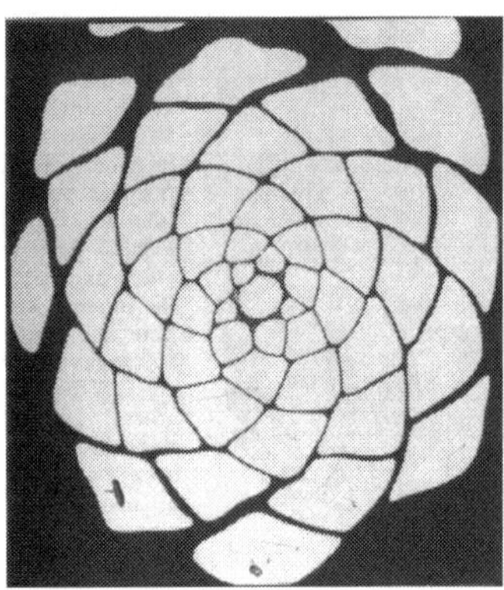

Figure 11. Pinecone. Source: James D. Nickel.

Figure 12. Thistle. Source: James D. Nickel.

Figure 13. Sunflower. Source: iStockPhoto.

2.4 Number Sense 1

Number sense is a way of describing our feel for numbers. This sense is a gift from God that enables us to order, or name, the creation. It gives us the ability to make connections or to see relationships, to visualize numbers revealed in a situation.

Terms, Symbols & Concepts Introduced
1. ‖
2. Addend
3. Binary operations
4. Common denomination
5. Difference
6. Dynamic addition
7. Equation
8. Left side of an equation
9. Minuend
10. Parallel lines
11. Right side of an equation
12. Static addition
13. Subtrahend
14. Sum
15. Terms of an equation

> Our positional number system is perfect and complete, because it is as economical in symbols as can be and can represent any number, however large. ... it allows everyone to do arithmetic.
>
> Modern science and technology may have their roots in antiquity, but they could only flourish as they have in the context of the modern era and in the framework of a number-system as revolutionary and efficient as our positional decimal system, which originated in India. To move mountains, the mind requires the simplest of tools.
>
> Georges Ifrah, *The Universal History of Numbers* (2000), p. 592, 599.

Remember, in real-life situations we never deal with the abstract numbers like five; we encounter concrete or denominate numbers like five coins or five books. Abstraction is the mental process of extracting a number from a given situation. The power of mathematics lies in disconnecting number from things, i.e., to think abstractly. Instead of working with five coins, we consider only the meaning of "five."

Utility of the Decimal System

The power of the decimal positional system lies in its simplicity, order, and elasticity. In review, the decimal position system incorporates three essential features:

1. Zero
2. Base 10
3. A digit will represent a different value based upon its position.

> We are so accustomed to making computations by writing them down that the process seems to us as old as civilization itself. Yet the use of writing for calculation turns out to be a late and unusual practice in human history. Computing by writing could only be accomplished with the invention of positional notation, using zero. The concept of zero was a new and radical invention, one that changed the world, so that ten symbols represent all the numbers in the universe.
>
> Denis Guedj, *Numbers: The Universal Language* (1997), p. 45.

There are many benefits to this structure. First, the decimal positional system is readily comprehensible. With it, we are saved from ambiguity. Because placement counts, the digit 8 in the number 8 means eight ones, while the digit 8 in the number 80 means 8 groups of ten, or eight tens.

Second, the principle of position is a strategy of beautiful simplicity. Every place is available for any digit to occupy. The one exception is a number written like this:

0418

It is not necessary to write the leading zero. In mathematics, a zero written like this is not significant. It does not add any meaning to the number 418.

Every group of digits represents a precise natural number and, conversely, every natural number can be represented by a particular group of digits. The number three hundred twenty-five is written, using digits, as 325. 517 is the unique representation of the number five hundred seventeen.

Third, the placement of digits on a horizontal line makes them easy to read.

Fourth, the number of digits used to represent a number is an indication of the size of the number. In other words, more digits mean a larger number.[1] This link between length and value makes comparing or determining whether a number is greater or less than another number, easy to do. For example, since 10,000 has more digits than 9999, five digits compared to four digits, 10,000 is greater than 9999, or 10,000 > 9999.

Fifth, we have seen that other numeration systems employ the use of letters. In an alphabetical language, you can jumble letters to generate nonsense. For example, the letters "zfdkabm" mean nothing in the English language. Not so with the decimal positional system. Any group of digits will always represent a precise number.

Sixth, the decimal positional system has the capacity to represent any number that we can possibly conceive. Ten digits, the number of fingers on two hands, can represent any number!

> The Triune God created everything, visible and invisible, in, through, and by Christ, the *logos* (John 1:1-3; Colossians 1:15-17). Therefore, all mathematical patterns in the perichoretic structure of creation, numerical or spatial, find their ultimate source of interconnectedness, rationality, order, beauty, and meaning in the second person of the Trinity. The dance of number is grounded in Christ.
>
> James D. Nickel

In future lessons, we shall be investigating the many laws that are embedded in the decimal positional system, laws that beautifully unfold inside this uniform structure. This patterned order reflects the uncreated ground of every facet of rationality, the One who holds all things together, the faithfulness that is the person of the Triune God of Scripture, the faithfulness centered in the God-man, the Lord Jesus Christ. Christ the Word is the ground of the rational structure of creation, a configuration of pattern and order. Since our minds think in terms of pattern and order we can, therefore, grasp, in part, some of the glories of the manifold nature of the faithfulness of the Triune God.

Once we visualize numbers written in the decimal positional system, we can apply the arithmetical operations to generate conclusions about them. Our first exploration will be addition and then we will explore its inverse companion, subtraction (Step 3).

In 13th-century Europe, just after the acceptance of Hindu-Arabic numerals, addition was often named "aggregation." Leonardo Fibonacci used the words "composition" and "collection"[2] to represent addition. Other words for addition were "assembling," "joining," and "summation." In 1496, Gregor Reisch, in his Latin book *Margarita phylosophica*, illustrated addition as follows:

---

[1] Except when we are working with the decimal representation of fractions. See Step 6 and 7.

[2] *Collect* means "to gather together" or "assemble." For example, a teacher collects exams from his students. In church worship, the "collects"* are the prayers made to God by the assembly of saints before the reading Scripture in a communion service.
*(Note that the stress is on the first syllable in the noun. Compare *produce*, verb, and *produce*, noun.)

		Today Known As:
213	*Numerus cui debet fieri additio*	*Addend*
+ 684	*Numerus addendus*	*+ Addend*
897	*Numerus adductus or collectus*	*= Sum*

In 1540, Gemma Frisius Reiner (1505-1555) developed a subtraction problem using Latin.

		Today Known As:
897	*Numerus ex quo subductus which became numerus minuendus (or superior)*	*Minuend*
– 684	*Numerus subducendus which became numerus subtrahendus (or inferior)*	*– Subtrahend*
213	*Numerus residius or differentia*	*= Difference*

Notice the inverse relationship between subtraction and addition. With the operation of addition, two addends are summed while in the operation of subtraction the difference is between the sum, the minuend, and one of the addends, the subtrahend. Because of this "twoness," we name these operations **binary operations**.[3]

ADDITION, SUBTRACTION, AND EQUATIONS

The modern symbols for addition and subtraction, + and –, first appeared in 1456 in an unpublished manuscript by the German mathematician and astronomer Johann Müller (1436-1476), also known as Regiomontanus. In his textbook entitled *Whetsone of Witte* (1557), the English mathematician Robert Recorde (ca. 1510-1558) introduced two **parallel lines** (=) as the symbol of equality.[4]

As we have noted, addition is the method of joining two numbers together; it is a method of finding a number equivalent to two or more numbers. The **sum** is the number obtained by addition. We generally add only two numbers at a time, in binary fashion.[5] These two numbers are **addends**. We have already been using the sign of equality (=). It is read "equals," "equal to," or "equal." Thus, 3 + 6 = 9 is read "three plus six equals nine." It may also be read "three and six are nine."

Subtraction is the process of calculating the difference between two numbers. Subtraction is what you do when someone says to you, "Deduct what I owe and pay me the rest." A subtraction problem consists of three numbers. We start with two numbers and calculate their difference. The **minuend** is the greater, or superior, of the two numbers, or the number upon which the action of subtraction is performed.[6] The **subtrahend** is the lesser, or inferior, of the two numbers, or the number performing the action.[7] The third number is the **difference**, also called the remainder or, better yet, balance. It is the result obtained by subtraction. When placed between two numbers, the minus sign means that the number after it is to be subtracted from the number before it. Thus, 8 – 6 says "six is subtracted from eight."

An **equation** is a statement of equality between two numbers or two quantities. The equal sign separates the quantity on its left from the quantity on its right. We name the quantity on the left of the equal

3 Binary means "consisting of two."

4 *Parallel*, in the Greek, means "beside one another." In geometry, two lines (Lesson 2.7) are parallel if they do not intersect (do not meet), no matter how far they are extended. Think of the two rails of a railroad track as concrete examples of parallel lines. The symbol for "is parallel to" is ||.

5 People with a rare gift of number sense can add three or more numbers at one time.

6 *Numerus minuendus*, in Latin, means "number to be diminished."

7 *Numerus subtrahendus*, in Latin, means "number to be subtracted."

sign the **left side of the equation**. We name the quantity on the right of the equal sign the **right side of the equation**. The numbers on each side are the **terms of the equation**. In the equation $3 + 5 = 8$, there are two terms on the left side and one term on the right. The quantity on the left, $3 + 5$, equals the quantity on the right, 8.

Consider the equation $6 + 2 = 8$. This is an abstract equation. We can assign concrete meaning to this equation using a balance beam scale (Figure 1). On the left-hand scale, we place two weights, one weighing six grams and the other weighing two grams. On the right-hand scale, there is one weight of eight grams. The scale is balanced, or even because the two weights on the left equal the one weight on the right. The equation is read "six plus two equals eight" and means the sum of 6 and 2 is 8. The left side is $6 + 2$ and the right side is 8. 6, 2, and 8 are terms of the equation. Remember the balance-scale picture. It will serve as a useful visual connection.

Figure 1. Scientific balance beam. Source: James D. Nickel.

In the equation $6 + 2 = 8$, 6 and 2 are contained in 8 by the operation of addition and 8 is contained in the sum of 6 and 2. In other words, there is a perichoretic relationship between the addends and the sum.

The symbol for "not equal" is $\neq$. We can use this symbol in an equation to indicate an imbalance. For example, $6 + 3 \neq 8$. Using the balance beam scale, this means that "six grams plus three grams is not equal to eight grams." We have already invoked the symbols of inequality: $<$ meaning "less than" and $>$ meaning "greater than." Using an inequality symbol, we write a true statement: $6 + 3 > 8$. We read it as "six plus three is greater than eight." We can also write $8 < 6 + 3$. We read it as "eight is less than six plus three." To establish equality, we write $6 + 3 = 9$.

DYNAMIC AND STATIC ADDITION

To enhance your feel for numbers and addition, take three pennies and place them in one row on a table.

<table>
<tr><td>O</td><td>O</td><td>O</td></tr>
</table>

Now, add three more pennies to make a new row.

<table>
<tr><td>O</td><td>O</td><td>O</td></tr>
<tr><td>O</td><td>O</td><td>O</td></tr>
</table>

How many pennies are there? Ask yourself, "How did I arrive at the answer?" You could have counted from one to six. This is the "Alice in Wonderland" method of addition, counting one and one and one and one and one and one. Instead of counting from one to six, you could have added three to three. Instead of counting, we have joined two groups of three together to get six.[8]

There is nothing simpler than this, is there? But, we have a fine layer of subtlety because there are really two forms of addition: dynamic and static.

[8] Or, you could have seen that there are two rows and three columns and we could invoke the operation of multiplication, or two times three is equal to six. We shall leave that thought for now, because we will investigate the concept of multiplication in Step 4.

In **dynamic addition**, to join means "to change a situation." We had a group of three pennies and then we joined three more pennies to the original group. We changed the situation. We now have six pennies instead of three pennies. The three pennies have grown dynamically to six pennies. Our denomination is pennies and we add like or **common denominations**.

In **static addition**, joining means "the grouping of types" of different denominations; we have a group of three pennies and we join three nickels to that group. We now have different denominations: three pennies and three nickels. How many coins do we have? Six coins. We have joined types; a penny is a coin and a nickel is a coin. Coin is now the new denomination or the common denomination.

Key words indicating addition:

- Add
- Augment
- Combine
- Enlarge
- Gain
- Grow
- In all
- Increased by
- Plus
- Rise
- Sum
- Total

Here is another example of static addition: If Steve has three pencils and Sherry has four pencils, how many pencils do they have altogether? Although pencil is the common denomination, nothing is growing; the situation has not changed. Therefore, the sum of seven pencils does not represent dynamic addition. It represents static addition.

The distinction between dynamic and static addition will help us better understand the process of subtraction when we get to that lesson.

Now, visualize the six pennies group. You can visualize them as $3 + 3$. You may write the equation, $3 + 3 – 6$. Or, preferably, you just see six. Now add another row of three coins:

O	O	O
O	O	O
O	O	O

How many pennies are there now? Again, you could have used the "Alice in Wonderland" method and counted one to nine; or you could have added three to six. As an equation, we have $6 + 3 = 9$. Visualize these three rows of pennies as a group. Do you see $3 + 3 + 3$? Or, do you see $6 + 3$? Or, better yet, do you just see 9?

Apply this same principle to groupings that total ten pennies, then twelve pennies, and finally twenty pennies. Make as many row/column arrangements as possible, except a single row of pennies or a single column of pennies. Physically coordinate these relationships with real pennies or a good substitute.

Working with these coins gives you a rudimentary understanding of the relationships that exist between numbers. For example, 2, 5, and 10 are related. 3, 4, 2, 6, and 12 also interact quite nicely. 2 and 10, and 4 and 5, are convenient and orderly row/column arrangements for a group of 20 pennies.

In addition to seeing these relationships, let's investigate ways that will speed up your working with addition. The basic technique to learn is to see only the answer. This works in reading words also: when you read a word like "object," you do not say "o" "b" "j" "e" "c" "t". You see the whole word at once and say to yourself, "object." The same procedure must be mastered when working with numbers. For example, when you see $3 + 4$, think "7." Do not say, "Three plus four is seven." When you see $3 + 4$, say "7." This may seem second nature to some of you, but if it is not, start applying this method immediately.

Consider $3 + 5$ and say "8."
Consider $3 + 7$ and say "10."
Consider $2 + 6$ and say "8."

Now consider 2 + 3 + 4. Adding from left to right, say, "5" (2 + 3) and then say "9" (5 + 4). Do not think in your head "5 + 4" when you see 5 + 4. Think "9" when you see 5 + 4.

Let's try another: 3 + 4 + 2. Say "7," then say "9."

Look at the following sets of addends. See the sum in your mind, and say it immediately:

1 + 5	5 + 2
2 + 3	7 + 1
7 + 2	1 + 9
5 + 4	6 + 3
3 + 7	1 + 1
2 + 4	4 + 6

If you just saw the answers, great! If not, try again until you get used to this procedure. Have someone write out a list of these addition problems for you and practice until you can see the answers immediately.

Another way to speed up your working with numbers is to work with them from left to right. For example, when you consider the number 8235, you read it from left to right: eight thousand, two hundred, thirty-five. You do not read it from right to left: five ones, three tens, two hundreds, and eight thousands. For now, just ponder on this approach. In later lessons, we shall provide plenty of practice using it.

In conclusion, here are three ways to develop your number sense.

1. Work from left to right: 56 is "56" not "6, 5." (You probably already do this.)
2. Learn to see relationships: 2, 3, 4, and 6 relate to 12 in a special way.
3. See only the answer: see 6 + 3 as "9."

EXERCISES

Define the following terms, concepts, or symbols:

1. Parallel lines
2. ||
3. Sum
4. Addend
5. Minuend
6. Subtrahend
7. Difference
8. Binary operations
9. Equation
10. Left side of an equation
11. Right side of an equation
12. Terms of an equation
13. Dynamic addition
14. Static addition
15. Common denomination
16. With a short essay, explain the difference between static and dynamic addition.

> The most overrated and at the same time most explosive symbol of mathematics is that equation sign which states that nothing happens unless something is already happening.
> Stanley L. Jaki, *Chance or Reality and Other Essays* (1986), p. 198.

Write these 25 addition problems on a 3" by 5" cards (" is a symbol that represents the denomination of inches). Write one problem per card, and write the sum on the reverse side. Master these sums by looking at each card and immediately seeing the sum in your mind. Then shuffle the cards and continue your practice of seeing these sums.

17. One-digit addends where the sum is less than 10:

1 + 1	2 + 4
1 + 2	2 + 5
1 + 3	2 + 6
1 + 4	2 + 7
1 + 5	3 + 3
1 + 6	3 + 4
1 + 7	3 + 5
1 + 8	3 + 6
2 + 2	4 + 4
2 + 3	4 + 5

18. One-digit addends where the sum equals 10:

1 + 9
2 + 8
3 + 7
4 + 6
5 + 5

Name the sides and the terms of each of the following equations:

19. 13 + 9 = 22
20. 35 + 15 = 50
21. 46 + 9 = 55
22. 44 + 16 = 60
23. 15 + 51 = 55 + 11
24. 13 + 35 = 7 + 41

Figure 2. Hexagonal wall on Interstate 90 in Wallace, Idaho. Source: James D. Nickel.

> In its history, mathematics, as a language, has resolutely pointed to a given order, a specificity of order, in the universe. Naturalistic mathematicians and scientists have done their best to avoid the implications of such order; i.e., the revelation of consistent rationality in the universe points to an ultimate ground of rationality beyond it.
>
> James D. Nickel

2.5 Addition with the Japanese Soroban

Terms & Concepts Introduced
1. Complement
2. Rectangle
3. Soroban

Addition … How can we master this operation? In this lesson, we will study the Japanese Soroban, a tool that hundreds of thousands of students use to perform addition faster than using a calculator!

Figure 1. Chinese version of the abacus. Source: iStockPhoto.

The way this ancient calculation device, the abacus, works will introduce us to a streamlined left-to-right method that will help us add quickly and accurately.

An abacus is a tool possessing a rich and varied cultural history.[1] It has helped people perform the arithmetic calculations of addition and subtraction. The device is constructed in a **rectangular**[2] frame set with rods on which balls or beads are easily moved, up or down. We shall focus on the Japanese version of the abacus, the **Soroban**, a device used in Japan since the 16th century. Watching Japanese students use this device is like watching fingers dance!

Figure 2. Soroban, the Japanese abacus, in cleared position. Source: James D. Nickel.

The preliminary requisite for working with the Soroban is the mastery of two tables. For the addition table, two addends are represented in the first row and first column. To sum 7 and 2, find the row labeled 7 and the column labeled 2. The sum, 9, is in the cell in the intersection of this row and column. You could also find the sum of 9 in the cell in the intersection of the column labeled 7 and the row labeled 2. For now, disregard the blank cells.

Table 1: One-Digit Addition (Sums to 10 or Less)

+	**1**	**2**	**3**	**4**	**5**	**6**	**7**	**8**	**9**
1	2	3	4	5	6	7	8	9	10
2	3	4	5	6	7	8	9	10	
3	4	5	6	7	8	9	10		
4	5	6	7	8	9	10			
5	6	7	8	9	10				
6	7	8	9	10					
7	8	9	10						

[1] The original Greek meaning of abacus was "sand." In Hebrew, the word meant "dust." These meanings suggest that before the introduction of the mechanical abacus, people indicated numbers as dots in the sand. Two of the primary evidences of civilization are literacy and numeracy and almost every early numerate civilization has used some form of counting board or abacus.

[2] A rectangle is a geometric figure consisting of four straight and connected sides with four interior angles of measure 90° each.

Table 1: One-Digit Addition (Sums to 10 or Less)									
8	9	10							
9	10								

In the subtraction table, the first column represents the minuend and first row the subtrahend. Again, disregard the blank cells for now. If the minuend is 9 and the subtrahend is 2, locate the row labeled 9 and the column labeled 2. The difference, 7, is in the cell in the intersection of this row and column.

Table 2: Subtraction with Minuends from 1 to 10

–	1	2	3	4	5	6	7	8	9	10
1	0									
2	1	0								
3	2	1	0							
4	3	2	1	0						
5	4	3	2	1	0					
6	5	4	3	2	1	0				
7	6	5	4	3	2	1	0			
8	7	6	5	4	3	2	1	0		
9	8	7	6	5	4	3	2	1	0	
10	9	8	7	6	5	4	3	2	1	0

Do you see the dance in the structure of both tables? The choreography of these numbers is grounded in the inverse relationship between relating addition to subtraction.

The refined version of the Soroban, in use since 1938, consists of a series of vertical rods and one horizontal separator. For each rod, there is one bead above the separator, in the "heavens," and four beads below, on the "earth." There is space on the rods to move the beads up or down by one position.

Each vertical rod contains beads that act as number counter that incorporates place value. Starting from the rod on the right and going to the left, they represent the first order, ones, the second order, tens, the third order, hundreds, etc.

It would be helpful to purchase a Soroban so that you can experiment with it. Many Internet sites sell the Soroban. You can even install Soroban software applications on your computers.[3]

Before using this device, it needs to be cleared. This is done by tilting it up so all the beads drop down. Then, gently lay it flat and run your right forefinger along the horizontal separator to push up all the single beads.

For the following demonstration, we will picture only part of the Soroban, the four rods starting from the right of the device, or the ones, tens, hundreds, and thousands positions.

[3] There are even Soroban applications that you can use on a smart phone or a tablet computer.

COUNTING FROM 1 TO 10

Look at the rightmost rod, the ones position. Each of the four beads on the earth stands for one. The one bead in the heavens stands for five ones. We are now going to count from 1 to 10 on the Soroban.

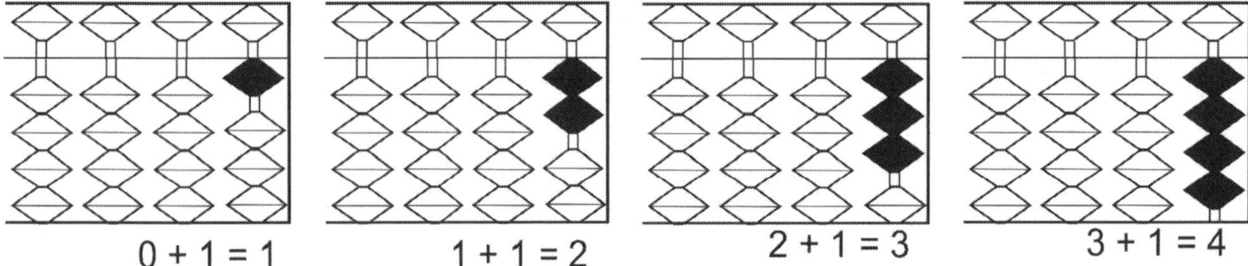

Figure 3.

In Figure 3, to indicate 1, you move the top bead on the earth up one position. In other words, adding one of the earth beads is equivalent to moving it up one position. To indicate 2 (add 1 to 1), move the second earth bead up. To indicate 3 (add 1 to 2), move the third earth bead up. To indicate 4 (add 1 to 3), move the fourth earth bead up.

Next (Figure 4), we want to add 1 to 4 to get 5 as our sum. All four earth beads are up so we cannot add 1. To indicate 5, we move all four beads down (subtract 4) while, at the same time, with another finger, we move the top bead down (add 5). Adding 1 is the same as subtracting 4 and adding 5. *Make sure you see this;* i.e., on the Soroban, addition makes use of subtraction. Note carefully:

$4 + 1 = 4 + (5 - 4) = 4 + 1 = 5$

The operation in parentheses, 5 – 4, represents the double movement of the beads, four earth beads down and one heaven bead down, or subtracting 4 and adding 5.

4 + 1 =
4 + (5 – 4) = 5
Subtract 4, Add 5

Figure 4.

In Figure 5, to indicate 6 (add 1 to 5), move the top bead on the earth up one position. To indicate 7 (add 1 to 6), move the second earth bead up. To indicate 8 (add 1 to 7), move the third earth bead up. To indicate 9, (add 1 to 8), move the fourth earth bead up.

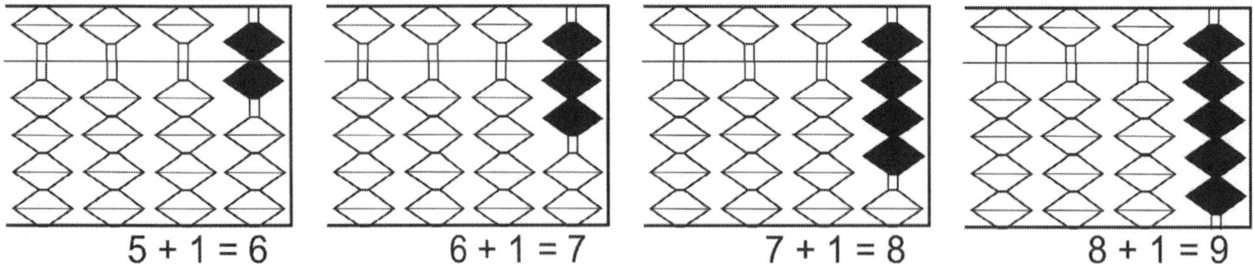

Figure 5.

In Figure 6, to indicate 10 (add 1 to 9), all beads in the ones rod are cleared. Clearing all beads in the ones rod means you move four down on the earth (subtract 4) and one up in the heavens (subtract 5). The result is zero ones. Next, the first earth bead in the tens rod is moved up. This means one ten. To add 1 to 9, we subtract 9 and then add 10. Note carefully:

$9 + 1 = 9 + (10 - 9) = 9 + 1 = 10$

The operation in parentheses, 10 – 9, represents the double movement of the beads, clearing the ones rod and moving one earth bead up on the tens rod, or subtracting 9 and adding 10.

Such are the basic movements, the dance, of the Soroban. Understanding what is happening with the beads is a requirement for understanding the streamlined method of addition that you will learn in subsequent lessons.

$9 + 1 =$

$9 + (10 - 9) = 10$

Subtract 9, Add 10

Figure 6.

To work with the Soroban, you will need to know these facts, facts related to 5 and 10:

$1 + 4 = 5$
$2 + 3 = 5$
$1 + 9 = 10$
$2 + 8 = 10$
$3 + 7 = 10$
$4 + 6 = 10$
$5 + 5 = 10$

COMPLEMENTARY ADDITION

The Soroban makes use of a method called complementary addition as defined in Lesson 2.8. With the Soroban, the **complement** of a given number is the number which, when added to your given number, produces a sum of either 5 or 10 depending on the column, ones, tens, etc., on which you are operating.

There are four complements for 5: 1, 2, 3, and 4. (We omit zero.) The complement of 1 is 4, the complement of 2 is 3, the complement of 3 is 2, and the complement of 4 is 1. Notice the two couplets:

- $1 + 4 = 5$ and $4 + 1 = 5$
- $2 + 3 = 5$ and $3 + 2 = 5$

An important law of mathematics is demonstrated by these two relationships (Lesson 3.7).

There are nine complements for 10: 1, 2, 3, 4, 5, 6, 7, 8, 9. (Again, we omit zero.) The complement of 1 is 9, the complement of 2 is 8, the complement of 3 is 7, and so forth. Notice the four couplets:

- $1 + 9 = 10$ and $9 + 1 = 10$
- $2 + 8 = 10$ and $8 + 2 = 10$
- $3 + 7 = 10$ and $7 + 3 = 10$
- $4 + 6 = 10$ and $6 + 4 = 10$
- $5 + 5 = 10$

Do you see the dance of complements? The Soroban is working with two bases, 5 and 10. We have not discussed base 5 yet. It has 5 digits: 0, 1, 2, 3, and 4. Table 3 shows us how we count in base 5.

Table 3	
Counting in Base 5	**Counting in Base 10**
1	1
2	2
3	3
4	4
10 = one five and zero ones	5
11 = one five and 1 one	6
12 = one five and 2 ones	7
13 = one five and 3 ones	8

Table 3

Counting in Base 5	Counting in Base 10
14 = one five and 4 ones	9
20 = two fives and zero ones	10 = one ten and zero ones
21	11
22	12
23	13
24	14
30 = three fives and zero ones	15
31	16
32	17
33	18
34	19
40 = four fives and zero ones	20 = two tens and zero ones
100 = one 25 (five fives), zero fives, and zero ones	25 = two tens and five ones

The streamlined method we will use for addition in base 10 follows the operation of the Soroban. Before we see how it works, study Figure 7 for a few examples of how we represent 14, 27, and 345 on the Soroban.

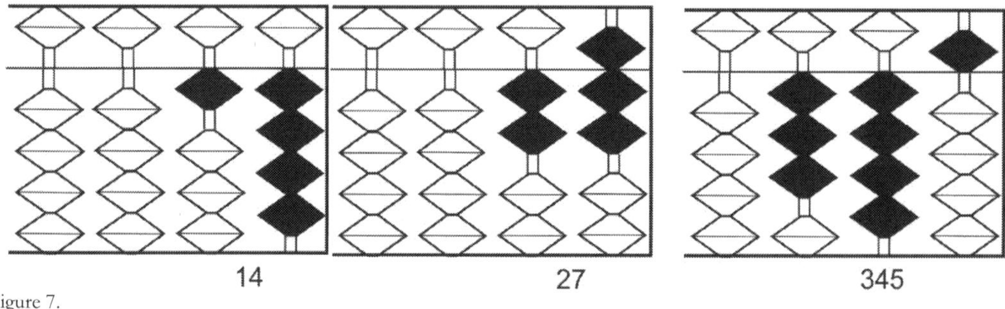

Figure 7.

Let's use the Soroban to demonstrate these thirteen addition problems:

Problem 1. $1 + 2 = 3$
Problem 2. $2 + 5 = 7$
Problem 3. $2 + 6 = 8$
Problem 4. $4 + 1 = 5$
Problem 5. $4 + 3 = 7$
Problem 6. $3 + 7 = 10$
Problem 7. $6 + 4 = 10$
Problem 8. $9 + 4 = 13$
Problem 9. $6 + 6 = 12$
Problem 10. $9 + 7 = 16$

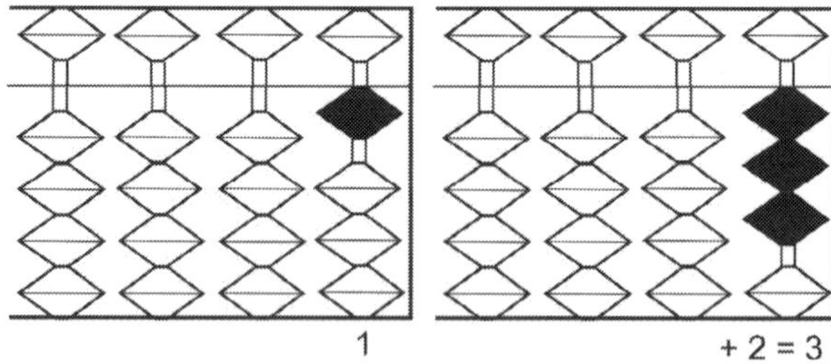

Figure 8.

Problem 11. 25 + 2 = 27
Problem 12. 63 + 24 = 87
Problem 13. 58 + 13 = 71

1 + 2 = 3

In the ones column (Figure 8), we move one earth bead up. To add 2, we move two earth beads up in the same column.

2 + 5 = 7

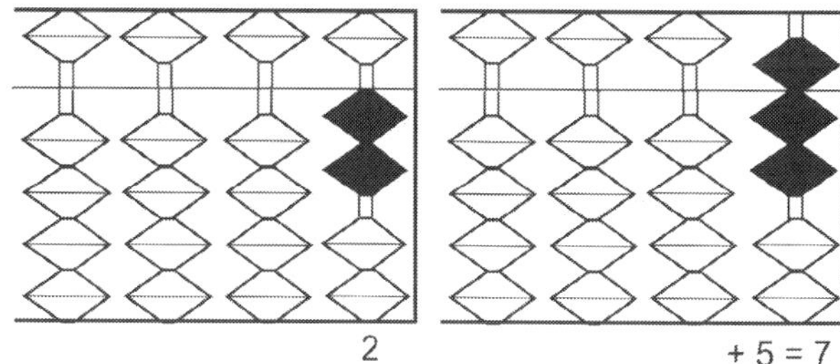

Figure 9.

In the ones column (Figure 9), we move two earth beads up. To add 5, in the ones column, we move the heaven bead down.

2 + 6 = 8

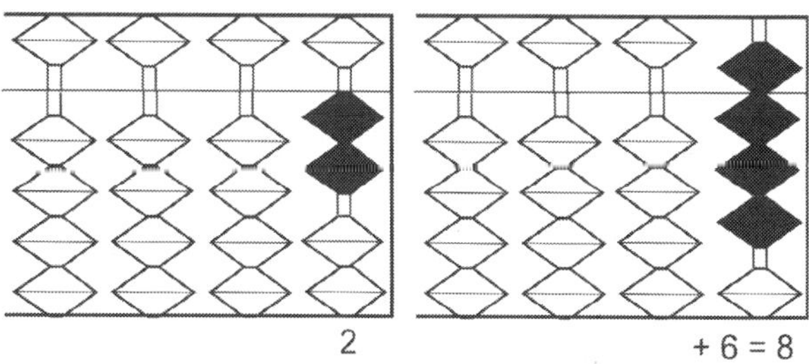

Figure 10.

In the ones column (Figure 10), we move two earth beads up. To add 6, in the ones column we do a double movement: (1) We move the heaven bead down and (2) one earth bead up (5 + 1 = 6).

4 + 1 = 5

We duplicate what we did in our counting exercise (Figure 11).

In base 5, the complement of 1 is 4. So, instead of adding 1, we added 5 and then subtracted 4. We did this:

$4 + 1 = 4 + (5 - 4) = 5$

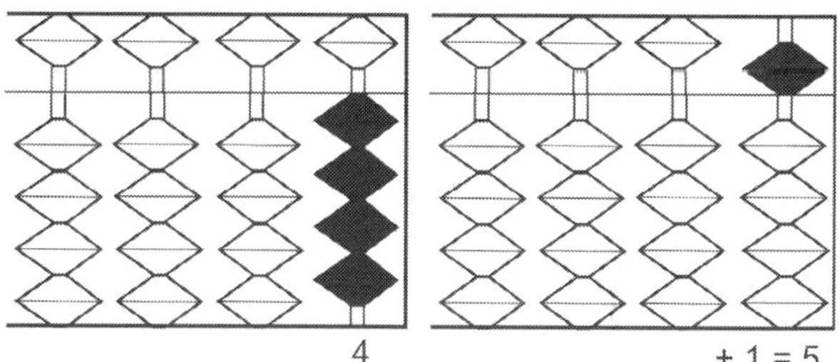

Figure 11.

4 + 3 = 7

In the ones column (Figure 12), we move four earth beads up. To add 3, we note that we have used all the earth beads. We do a double movement since adding 3 is the same as adding 5 and subtracting 2. We move the heaven bead down and two earth beads down.

In base 5, the complement of 3 is 2. So, instead of adding 3, we add 5 and then subtract 2. We did this:

$4 + 3 = 4 + (5 - 2) = 7$

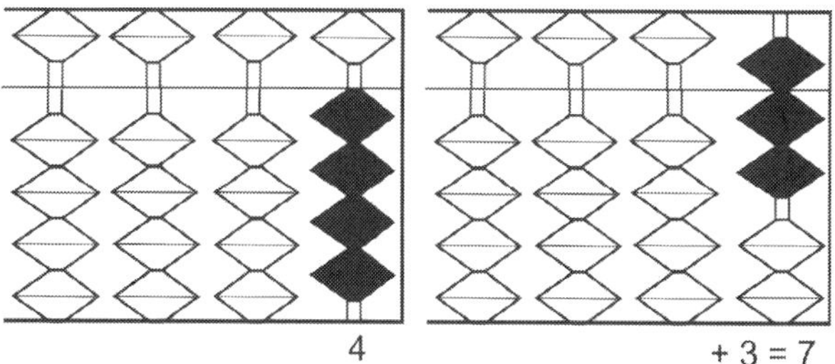

Figure 12.

$3 + 7 = 10$

In the ones column (Figure 13), we move three earth beads up. To add 7, we note that we have only one earth bead and the heaven bead to work with (5 + 1 = 6). We want to add 7, not 6. What do we do? In this case, we work with complements of 10. In base 10, the complement of 7 is 3. So, adding 7 is the same as adding 10 and subtracting 3. We do a double movement: In the tens column, we move one earth bead up (add 10). In the ones column, we move three earth beads down (subtract 3). We did this:

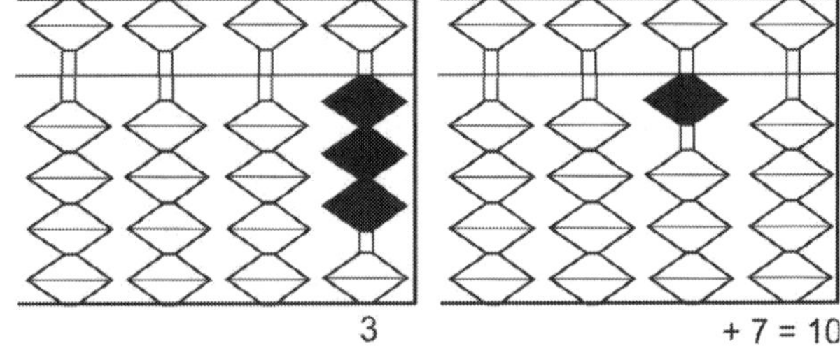

Figure 13.

$3 + 7 = 3 + (10 – 3) = 10$

Let's say that we did not notice that we only could add 6, not 7, in the ones column. We bring the heaven bead down (add 5), and then we notice that you have only one earth bead with which to work. We want to add 2, not 1. So, we move the one earth bead up (add 1). To add the other 1, we add 10 (one earth bead up in the tens column) and then subtract 9 (the heaven bead up and four earth beads down in the ones column). Note that 9 is the complement of 1. Do you see that the first way is easier than the second? We need to learn to see which complement is the appropriate one.

$6 + 4 = 10$

In the ones column (Figure 14), we move one earth bead up and the heaven bead down. To add 4, we note that we have only three earth beads with which to work. We want to add 4, not 3. Again, we work with complements of 10 and the complement of 4 is 6. We note that adding 4 is the same as adding 10 and subtracting 6. We do a triple movement: We move one earth bead up in the tens column (add 10), and one earth bead down and the heaven bead up in the ones column (subtract 6). We did this:

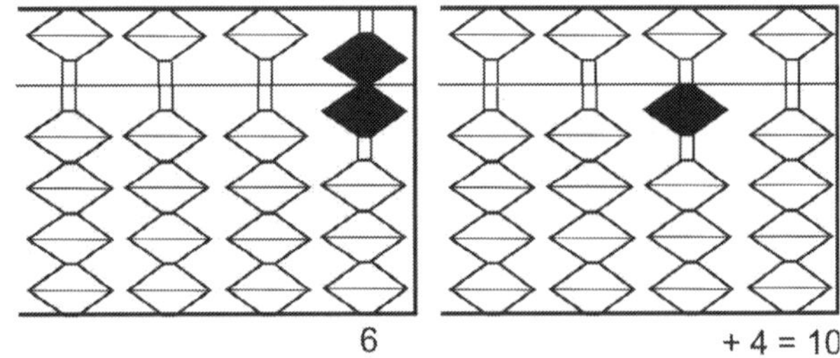

Figure 14.

$6 + 4 = 6 + (10 – 6) = 10$

$9 + 4 = 13$

In the ones column (Figure 15), we move four earth beads up and the heaven bead down. We note that adding 4 is the same as adding 10 and subtracting 6. We do a triple movement: We move one earth bead up in the tens column (add 10), and one earth bead down and the heaven bead up in the ones column (subtract 6). We did this:

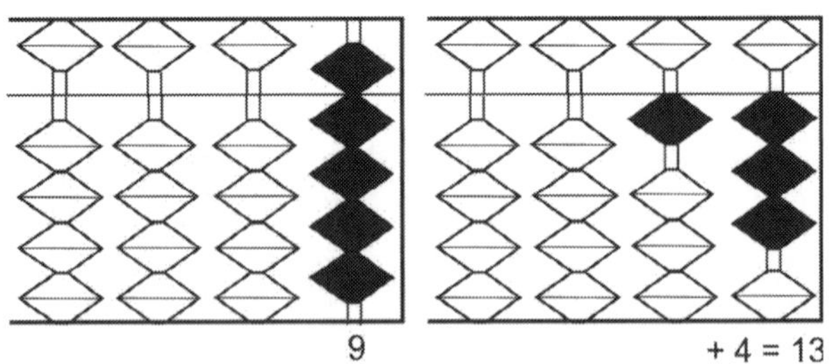

Figure 15.

$9 + 4 = 9 + (10 – 6) = 13$

6 + 6 = 12

In the ones column (Figure 16), we move one earth bead up and the heaven bead down. Adding 6 is the same as adding 10 and subtracting 4. We move one earth bead up in the tens column (add 10) but then we run into an issue with the ones column. We need to subtract 4. We note that subtracting 4 is the same as subtracting 5 and adding 1, or 10 – 4 = 10 – 5 + 1. To subtract 4, in the ones column, we move one earth bead up and the heaven bead up.

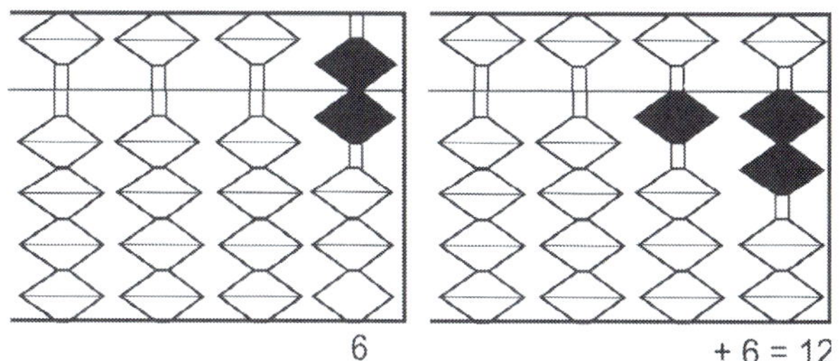

Figure 16.

We have entered an important world with this simple example. Follow along carefully:

6 + 6 = 6 + (10 – 4) = 6 + [10 – (5 – 1)] = 6 + [10 – 5 + 1)] = 12

From this, we can conclude that 10 – (5 – 1) = 10 – 5 + 1. Why? On the left side, we subtract 1 from 5 first since using parentheses tells us to do what is within the parentheses first (Lesson 8.15). We get 10 – 4 = 6. On the right side, we proceed from left to right: 10 – 5 = 5 and 5 + 1 = 6.

9 + 7 = 16

In the ones column (Figure 17), we move four earth beads up and the heaven bead down. Adding 7 is the same as adding 10 and subtracting 3. We move one earth bead up in the tens column (add 10) and then move three earth beads down in the ones column. We did this:

9 + 7 = 9 + (10 – 3) = 16

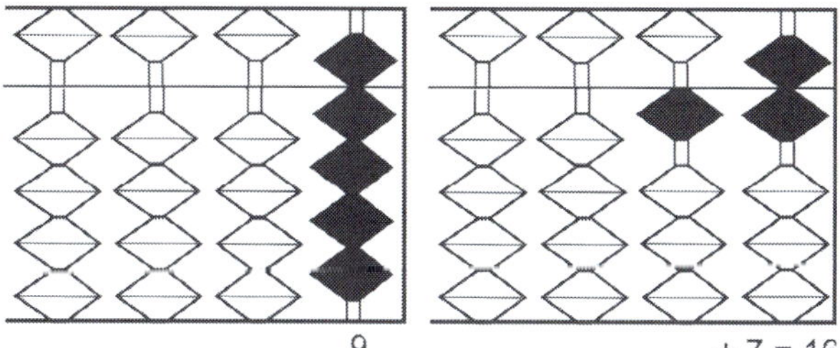

Figure 17.

25 + 2 = 27

First, in the tens column (Figure 18), we move two earth beads up; and in the ones column, we move the heaven bead down. To add 2, we just move two earth beads up in the ones column.

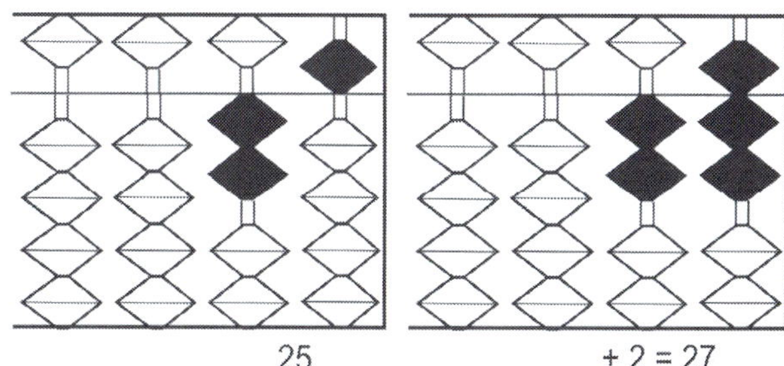

Figure 18.

The next two problems use the left to right method of addition with two-digit addends. This method will form the basis of our streamlined method of addition, a method we will use without the assistance of the Soroban.

$63 + 24 = 87$

In the tens column (Figure 19), we move one earth bead up and the heaven bead down (60) and in the ones column, we move three earth beads up (3): 63 = 60 + 3.

To add 24, we start with the tens column and proceed left to right.

First, we move two earth beads up in the tens column to represent the addition of 20.

Second, we now must add the 4 to the 3 in the ones column. We add 4 by adding 5 and then subtracting 1, the complement of 4 in base 5 or 4 = 5 – 1. We move the heaven bead down (add 5) and one earth bead down (subtract 1).

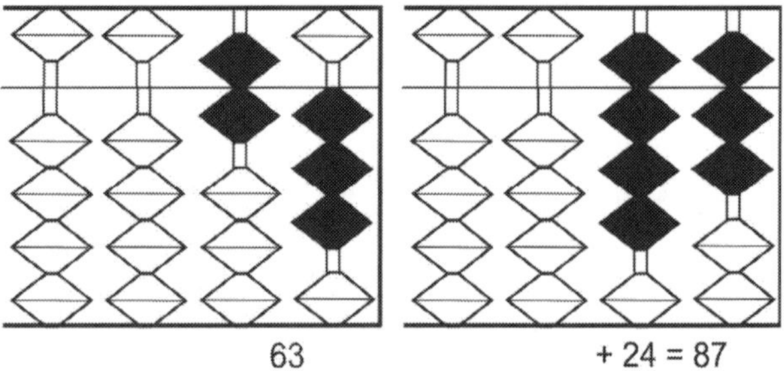

Figure 19.

$58 + 13 = 71$

First (Figure 20), we move the heaven bead down in the tens column and one earth bead down and three earth beads up in the ones column (58).

Second, we add 13, left to right. Since 13 = 10 + 3, we first add 10. We do this by bringing one earth bead up in the tens column.

Third, we must add 3 to the ones column. Instead of adding 3, we will add 10 and then subtract 7 or 3 = 10 – 7. To add ten, we bring one earth bead up in the tens column. To subtract 7, we move one heaven bead up (5) and two earth beads down (2). Nifty, isn't it?

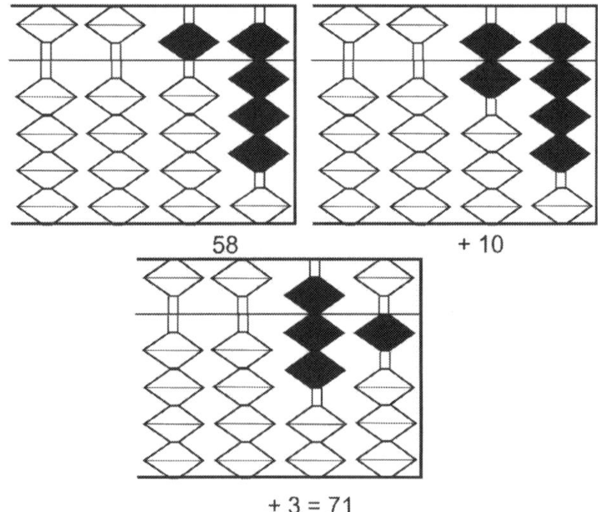

Figure 20.

SUMMARY

Note the Soroban movements: When we add, we move a heaven bead down and earth beads up. When we subtract, we move a heaven bead up and earth beads down.

The reciprocal dance of these movements, down and up for addition and up and down for subtraction, makes sense because subtraction is the inverse of addition.

With practice and some dexterity with your fingers, you can do some lightning-quick calculations with the Soroban. As we noted at the beginning of this lesson, students skilled in its use can find sums quicker than someone else punching buttons on a calculator.

EXERCISES

Define the following terms:

1. Complement
2. Rectangle
3. Explain how the Japanese Soroban is constructed.

Memorize:

4. Table 1 and Table 2 shown at the beginning of this lesson. You should know that 3 + 4 = 7, 5 + 5 = 10, 10 – 8 = 2, 8 – 3 = 5, etc.
5. The complements of 1, 2, 3, and 4 with respect to 5.
6. The complements of 1, 2, 3, 4, 5, 6, 7, 8, and 9 with respect to 10.
7. Write the base 10 numbers from 41-100 in base 5.

> The abacus, in time, became the principal computing machine of Western nations. Its use also spread into the Orient. It was known in China as early as the sixth century A.D. It was in popular use in Japan in the seventeenth century, although doubtless it was known there much earlier. In China the abacus is called the *suan phan*, or arithmetic board; and in Japan, the *soroban*.
>
> Harold T. Davis, "The History of Computation." In *Historical Topics for the Mathematics Classroom* (1969), pp. 91-92.

Soroban exercises:

8. Go over all the examples in your text with your Soroban.
9. Perform all the addition problems in Table 1 on the Soroban.
10. Perform all the subtraction problems in Table 2 on the Soroban. Hint: To subtract 8 from 10 is the same as subtracting 10 from 10 and adding 2, or 10 – 8 = 10 – 10 + 2. To subtract 2 from 10 is the same as subtracting 10 from 10 and adding 8, or 10 – 2 = 10 – 10 + 8.
11. Spend some time experimenting with your Soroban by adding one-digit numbers to one-digit numbers, two-digit numbers to one-digit numbers, and two-digit numbers to two-digit numbers.

For these subtraction problems, practice until you see the difference immediately in your mind:

12. 8 – 2
13. 9 – 1
14. 7 – 3
15. 6 – 3
16. 4 – 2
17. 7 – 1
18. 5 – 1
19. 2 – 1
20. 6 – 1
21. 5 – 4
22. 5 – 3
23. 6 – 2
24. 4 – 1
25. 6 – 4
26. 4 – 3
27. 7 – 2
28. 3 – 2
29. 8 – 1
30. 5 – 2
31. 3 – 1

Figure 21. Hexagonal wall, Wenatchee, Washington. Source: James D. Nickel.

2.6 The Nine Complements in Base 10

Terms & Concepts Introduced
1. Apex
2. Symmetry

Get ready for a revolution in thinking and calculating. We are going to use the principle of the Soroban to teach the method of doing lightning quick sums, either with a pencil and paper or in your head, the streamlined way of addition.[1]

Before we investigate this method, let's first look at our basic addition facts that consist of all possibilities of the sums of one-digit addends. In Lesson 2.4, as homework, we should have mastered the addition facts that sum to less than 10 and the facts that sum to 10. Here they are:

Table 1: Sums Less than 10			
1 + 1 = 2			
1 + 2 = 3			
1 + 3 = 4	2 + 2 = 4		
1 + 4 = 5	2 + 3 = 5		
1 + 5 = 6	2 + 4 = 6	3 + 3 = 6	
1 + 6 = 7	2 + 5 = 7	3 + 4 = 7	
1 + 7 = 8	2 + 6 = 8	3 + 5 = 8	4 + 4 = 8
1 + 8 = 9	2 + 7 = 9	3 + 6 = 9	4 + 5 = 9

There are 20 addition problems to remember for the sums less than 10.

Note, 2 + 5 = 7 and 5 + 2 = 7, so we need to remember only one fact that can be applied two ways. This observation is an important property of addition: it does not matter in which order you add two numbers. You always get the same sum (Lesson 3.7).

Table 2: Sums Equal to 10				
1 + 9 = 10	2 + 8 = 10	3 + 7 = 10	4 + 6 = 10	5 + 5 = 10

There are five more addition problems to remember for sums equal to 10.

In total, there are 25 addition problems to master. These facts are an essential toolkit to success in arithmetic so make sure that you memorize them.

> Arithmetic deals with the rules governing numerical calculations performed with our modern Arabic numerals. Today every schoolboy does sums and other numerical computations that six hundred years ago were performed only by a small handful of advanced scholars.
>
> Denning Miller, *Popular Mathematics* (1942), p. vii.

[1] The source for the principles in this lesson and subsequent lessons on subtraction is Edward Stoddard, *Speed Mathematics Simplified* (Mineola: Dover Publications, [1965] 1994). After World War II, Stoddard noticed how arithmetically proficient Japanese students were. *Speed Mathematics Simplified* is a systematic exposition of the results of his research in the Japanese way of mastering arithmetic operations. See also Scott Witt, *How to be Twice as Smart: Boosting Your Brainpower and Unleashing the Miracles of Your Mind* (West Nyack: Parker Publishing, 1983), pp. 53-73. The ancient Hindus, the developers of the decimal system, used this method and named it "inverse" or "retrograde."

What about sums greater than 10? Let's list them in a way that reveals their structure:

Table 3: Sums Greater than 10							
2 + 9 = 11	3 + 8 = 11	4 + 7 = 11	5 + 6 = 11				
	3 + 9 = 12	4 + 8 = 12	5 + 7 = 12	6 + 6 = 12			
		4 + 9 = 13	5 + 8 = 13	6 + 7 = 13			
			5 + 9 = 14	6 + 8 = 14	7 + 7 = 14		
				6 + 9 = 15	7 + 8 = 15		
					7 + 9 = 16	8 + 8 = 16	
						8 + 9 = 17	
							9 + 9 = 18

There are 20 more addition problems (1 + 2 + 3 + 4 + 4 + 3 + 2 + 1 = 20) to remember for sums greater than 10.

Do you notice a pattern in the way in which they are listed? If so, what is it?[2]

Let's combine these three tables and see if we can notice more patterns embedded in these 45 facts.

Table 4										
									Sum	Addend Pairs
1 + 1									2	1
1 + 2									3	1
1 + 3	2 + 2								4	2
1 + 4	2 + 3								5	2
1 + 5	2 + 4	3 + 3							6	3
1 + 6	2 + 5	3 + 4							7	3
1 + 7	2 + 6	3 + 5	4 + 4						8	4
1 + 8	2 + 7	3 + 6	4 + 5						9	4
1 + 9	2 + 8	3 + 7	4 + 6	5 + 5					10	5
	2 + 9	3 + 8	4 + 7	5 + 6					11	4
		3 + 9	4 + 8	5 + 7	6 + 6				12	4
			4 + 9	5 + 8	6 + 7				13	3
				5 + 9	6 + 8	7 + 7			14	3
					6 + 9	7 + 8			15	2
						7 + 9	8 + 8		16	2
							8 + 9		17	1
								9 + 9	18	1

[2] Pattern recognition is one of the key mathematical skills. Without the ability to recognize patterns, your ability to enjoy mathematics is limited. The good news is that everyone, by practice, can garner the skills of pattern recognition.

What are your observations? First, there is a significant concentration of pairs of addends that sum to 8, 9, 10, 11, and 12. Second, as we move to lower and higher sums, the number of pairs decreases. Third, the easiest sums to remember are those less than 10. Fourth, we have a group of pairs that sum to 10 (1 + 9, 2 + 8, 3 + 7, 4 + 6, 5 + 5). Fifth, note the **symmetry** revealed in the "Addend Pairs" column. Symmetry in mathematics is the beauty of balance. The number of addend pairs has a wonderful balance:

1, 1, 2, 2, 3, 3, 4, 4, 5

Note the pairs: 1 and 1, 2 and 2, 3 and 3, 4 and 4. The number 5 is the **apex**[3] of these pairs. Starting from the top, 5, we reverse the numbers:

5, 4, 4, 3, 3, 2, 2, 1, 1

Note the pairs in reverse after 5: 4 and 4, 3 and 3, 2 and 2, 1 and 1. That's symmetry! Figure 1 shows a few ways to picture this balance of beauty in geometric form. As you continue your mathematical studies, you will encounter many instances that will help you hone your skills at pattern recognition. To detect marvelous and, indeed, surprising symmetries embedded in both numeric and geometric relationships is evidence of a beautiful and pervasive rationality implanted in the form and matter of the universe of the Triune God's making.

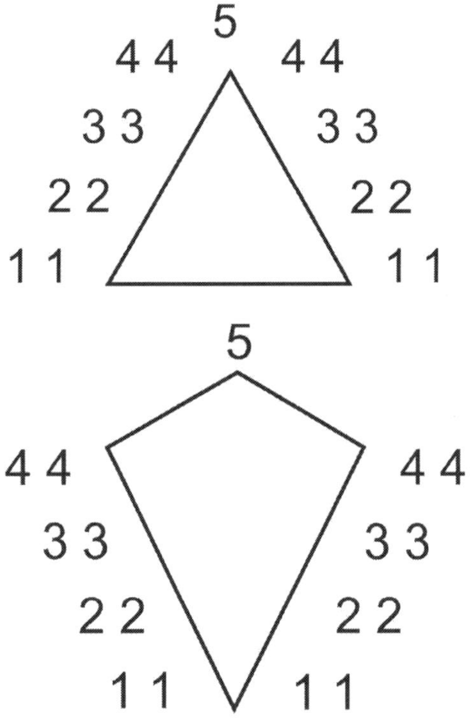

Figure 1.

COMPLEMENTARY ADDITION

For computing sums greater than ten, we invoke what we learned from the Soroban, complementary addition. Complementary addition makes use of the base of our decimal system; i.e., 10. For any natural number less than 10, the complement of that number is the number you must add to it to get 10.

In review, here are the complements of the natural numbers less than 10:

Table 5

Number	Complement	Sum
1	9	10
2	8	10
3	7	10
4	6	10
5	5	10
6	4	10
7	3	10
8	2	10
9	1	10

[3] *Apex*, in Latin, means "tip." It means the highest point of an object.

Note that you only need to remember five pairs since $1 + 9 = 9 + 1$, $2 + 8 = 8 + 2$, $3 + 7 = 7 + 3$, and $4 + 6 = 6 + 4$:

Table 6		
Number	**Complement**	**Sum**
1	9	10
2	8	10
3	7	10
4	6	10
5	5	10

If you know these five pairs, you are ready to do complementary addition. Look at each number in the following list of numbers, and say, in your mind, its complement:

1 5 9 7 8 4 3 2 6

Make sure you recognize the complements immediately. To help you, write the numbers 1 to 9 separately on nine 3″ by 5″ cards. Write the complements on the other side. Sort the cards and have someone show you the cards. Practice until you can say the complements immediately.

ONE-DIGIT ADDENDS WHOSE SUM IS GREATER THAN TEN

Now that you have mastered these five pairs, you are ready to do complementary addition. We are going to find the sums greater than 10 by using complements and simple subtraction.

Let's first consider $6 + 8$.

Step 1. Determine the greater number. In symbols, $8 > 6$ or $6 < 8$. The greater number is 8.
Step 2. Determine the complement of 8. The complement of 8 is 2.
Step 3. Subtract 2 from the lesser number and add 10 to your result.

Here are the steps again, slightly simplified. We want to find the sum of 6 and 8.

Step 1. 8 is the greater number.
Step 2. The complement of 8 is 2.
Step 3. $6 - 2 = 4$; $4 + 10 = 14$

Doing this may seem like a few extra steps, but note how 10 is used. It is easy to add 10 to any number:

Table 7	
1 + 10 = 11	6 + 10 = 16
2 + 10 = 12	7 + 10 = 17
3 + 10 = 13	8 + 10 = 18
4 + 10 = 14	9 + 10 = 19
5 + 10 = 15	10 + 10 = 20

Since it is easy to add 10 to any number, we make use of this simplicity in complementary addition. Remember the steps:

Step 1. Find the complement of the greater number. We choose the greater number because it is easier to subtract the complement of the greater number than the complement of the lesser number.
Step 2. Subtract that complement from the lesser number.
Step 3. Add 10 to your result.

Let's try a few more:

Table 8			
2 + 9	4 + 9	5 + 8	9 + 6
3 + 8	3 + 9	7 + 6	7 + 8
7 + 4	4 + 8	9 + 5	7 + 9
9 + 8	7 + 5	8 + 6	5 + 6

How well did you do? Make sure that you go over these sums again, using the method of complementary addition, if you encountered difficulties. After a while, this process will become second nature to you.

What about 6 + 6, 7 + 7, 8 + 8, and 9 + 9? Since you are summing the same number, take its complement, subtract it from the other number, and then add 10.

For example, consider 7 + 7:

Step 1. The complement of 7 is 3 since 7 + 3 = 10.
Step 2. 7 – 3 = 4
Step 3. 4 + 10 = 14

Consider 6 + 6:

Step 1. The complement of 6 is 4 since 6 + 4 = 10.
Step 2. 6 – 4 = 2
Step 3. 2 + 10 = 12

Let's show 8 + 8 = 16:

Step 1. The complement of 8 is 2 since 8 + 2 = 10.
Step 2. 8 – 2 = 6
Step 3. 6 + 10 = 16

Now, 9 + 9 = 18:

Step 1. The complement of 9 is 1 since 9 + 1 = 10.
Step 2. 9 – 1 = 8
Step 3. 8 + 10 = 18

Here is the three-step thinking involved in adding pairs of one-digit numbers whose sum is greater than 10:

3 + 8 = ? The complement of 8 is 2. 3 – 2 = 1 1 + 10 = 11	7 + 4 = ? The complement of 7 is 3. 4 – 3 = 1 1 + 10 = 11	9 + 8 = ? The complement of 9 is 1. 8 – 1 = 7 7 + 10 = 17
4 + 9 = ? The complement of 9 is 1. 4 – 1 = 3 3 + 10 = 13	5 + 6 = ? The complement of 6 is 4. 5 – 4 = 1 1 + 10 = 11	3 + 9 = ? The complement of 9 is 1. 3 – 1 = 2 2 + 10 = 12
4 + 8 = ? The complement of 8 is 2. 4 – 2 = 2 2 + 10 = 12	7 + 5 = ? The complement of 7 is 3. 5 – 3 = 2 2 + 10 = 12	5 + 8 = ? The complement of 8 is 2. 5 – 2 = 3 3 + 10 = 13

4 + 8 = ? The complement of 8 is 2. 4 – 2 = 2 2 + 10 = 12	7 + 5 = ? The complement of 7 is 3. 5 – 3 = 2 2 + 10 = 12	5 + 8 = ? The complement of 8 is 2. 5 – 2 = 3 3 + 10 = 13
7 + 6 = ? The complement of 7 is 3. 6 – 3 = 3 3 + 10 = 13	9 + 5 = ? The complement of 9 is 1. 5 – 1 = 4 4 + 10 = 14	8 + 6 = ? The complement of 8 is 2. 6 – 2 = 4 4 + 10 = 14
9 + 6 = ? The complement of 9 is 1. 6 – 1 = 5 5 + 10 = 15	7 + 8 = ? The complement of 8 is 2. 7 – 2 = 5 5 + 10 = 15	7 + 9 = ? The complement of 9 is 1. 7 – 1 = 6 6 + 10 = 16

Again, your goal is to commit to all 45 addition facts to memory. Complementary addition can assist you in calculating sums greater than 10, but you want say the sum immediately in your mind; i.e., when you see 7 + 9, 16 comes to mind.

> Once you get used to complementary addition, you will appreciate its power in simplifying the operation of addition. It may take you some time to get used to this process. What is required of you is practice and more practice until you are at home with this technique.
>
> James D. Nickel

EXERCISES

Define the following terms:

1. Apex
2. Symmetry
3. Perform all the addition facts that sum to greater than 10 on the Soroban. (Hint: Start with the lesser number first, and then add the greater number to it using complements.)

Explain the method of complements to add the following sums:

4. 8 + 8
5. 9 + 7
6. 7 + 6
7. 5 + 9
8. 9 + 9

> "How do I get to Carnegie Hall?" asked the passenger.
> "Practice, man, practice," replied the cab driver.

> Mathematics is the most powerful technique for the understanding of patterns.
> Alfred North Whitehead (1861-1947), "Mathematics and the Good" (1942).

9. Practice the following problems until you can answer them immediately (use complements, if you need to, when the sum is greater than 10):

2 + 9	4 + 4	6 + 4	2 + 7	8 + 1	3 + 9	2 + 7	5 + 8	6 + 6	4 + 7
7 + 6	9 + 7	8 + 5	2 + 8	5 + 9	9 + 8	3 + 2	2 + 5	5 + 5	4 + 3

10. Let's try another group. Every time the sum is greater than 10, record 10 with your index finger and remember the digit in the ones position after applying complements. Consider 8 + 5. Record with your index finger by extending only it; this finger represents 1 ten. The complement of 8 is 2, so subtract 2 from 5 and get 3, the digit in the ones position. Add 10 (what you recorded with your index finger) to 3; 10 + 3 = 13. Try this method with these problems. Even if you know these sums already by heart, if the sum is greater than 10, learn to record a ten with your index finger and remember the digit in the ones position and then adding 10 to it. We will make use of the procedure of recording tens in Lesson 2.8.

8 + 7	7 + 3	4 + 9	3 + 3	8 + 1	4 + 3	2 + 9	5 + 2	2 + 2	6 + 9
1 + 1	8 + 7	4 + 5	5 + 6	1 + 6	3 + 8	9 + 5	3 + 5	2 + 8	

11. Here is one final group. Follow the same method for sums greater than 10 as you did in the previous exercise.

1 + 9	9 + 2	3 + 9	5 + 6	7 + 1	9 + 1	2 + 6	5 + 7
2 + 3	1 + 3	2 + 4	8 + 9	6 + 8	9 + 9	2 + 6	4 + 9
7 + 9	9 + 6	8 + 6	3 + 7	3 + 6	5 + 4	4 + 4	8 + 8
8 + 9	9 + 4	7 + 8	3 + 8	1 + 3	4 + 8	7 + 6	5 + 1
2 + 4	7 + 7	1 + 5	6 + 9	1 + 7			

In the previous three activities, you have encountered every possible addition problem involving natural numbers with one-digit addends.

Answer the following problems; they are interrelated:

12. Consider a single-elimination[4] basketball tournament that begins with four teams. How many games need to be played to determine a champion?
13. How many games need to be played to determine a champion with eight teams?
14. How many games need to be played to determine a champion with sixteen teams?
15. State the dance, the rule, that gives you the answer for the number of games played for n teams.[5]
16. How many games need to be played to determine a champion with 4150 teams?

---

[4] Teams that play in a single-elimination tournament are eliminated at the first loss of a match or game.

[5] The Hungarian-born American mathematician Paul Halmos (1916-2006) is the one who found the method to find these answers.

2.7 Building Blocks of Geometry

We have been using 3″ by 5″ cards to help us memorize our addition facts. We will now use these cards, called index cards, to explore some foundational geometric terms.

Terms, Symbols & Concepts Introduced
1. ∠
2. °
3. Δ
4. Acute angle
5. Circle
6. Collinear points
7. Concurrent lines
8. Coplanar points
9. Degrees
10. Intersection
11. Line
12. Line segment
13. Locus
14. m∠
15. Mathematical object
16. Noncollinear points
17. Obtuse angle
18. One dimension
19. Plane
20. Point
21. Protractor
22. Radius
23. Ray
24. Reflex angle
25. Right angle
26. Sides of an angle
27. Space
28. Straight angle
29. Tangent point
30. Tangent ray, line, or line segment
31. Three dimensions
32. Triangle
33. Two dimensions
34. Vertex
35. Zero dimensions

> We owe Geometry in part to ancient Egyptian tax collectors, who used its methods to survey and measure land.

To understand the structure of space, the Greek mathematician Euclid wrote thirteen books about Geometry, *The Elements*. He lived in the Egyptian city of Alexandria approximately 300 years before Christ was born. He started his exposition by defining some foundational geometric notions, all essentially abstract in character.[1]

Point

Euclid began with a point. What is a point? A tiny dot at the end of a sentence. What is a dot? A tiny spot. What is a tiny spot?

We see that we cannot define a point so Euclid defined it as a geometric object[2] that "hath no part" (Book I, Definition 1), i.e., no length, no width, no height. According to this definition, a point, having **zero dimensions**, is nothing! How do we construct geometrical ideas starting from nothing? To Euclid, a **point** indicates location, without any physicality about it.[3] We can visualize, or locate, a corner of a room, naming it as a point. We identify points with capital letters.

Line

Euclid defined a line as having "breadthless length" (Book I, Definition 2). A **line**, therefore, has length, but no width or height. It is **one-dimensional** and unbounded.[4] To Euclid, a line is the path traced out by one moving point.

---

[1] In the modern world of computer graphics, drawing programs such as CorelDraw use the geometric principles articulated by Euclid to help designers create a kaleidoscope of art.

[2] A geometric object is a type of mathematical object. **Mathematical objects** are ideas that are mathematically related and can be defined. For example, a numeral is a mathematical object. An equal sign is a mathematical object. A circle is a mathematical object.

[3] We shall revisit this definition of a point in Lesson 11.4.

[4] A mathematical object is said to be of *one dimension* when a single unit is needed to denote its measure; e.g., the length of a ruler is 12 inches. You are not concerned with height or width. On a ruler, you can locate a single point that reveals the length of the item you are measuring; e.g., a pencil is 4 inches in length.

We can picture a line by imaging the **intersection**[5] of two flat walls, some thread stretched straight, or a ray of light emitted through a small hole. Atlantic Ocean "Liners" are named such because they generally traveled in a straight line from port to port. The word "line" also appears in the name of commercial "air-line" companies.

We name a line in one of two ways:

1. Cursive, lower-case letters: l, m, n, etc.
2. By two points on a given line: $\overleftrightarrow{AB}$ or $\overleftrightarrow{BA}$

We draw a line without endpoints by using double arrows. (See line m or $\overleftrightarrow{AB}$ in the figure.) The arrows indicate the unbounded nature of a line.

A line consists of infinitely many points. To be technically correct, we construct a line from an infinite number of a mathematical objects, points, each of which is physically nothing. Interesting conundrum!

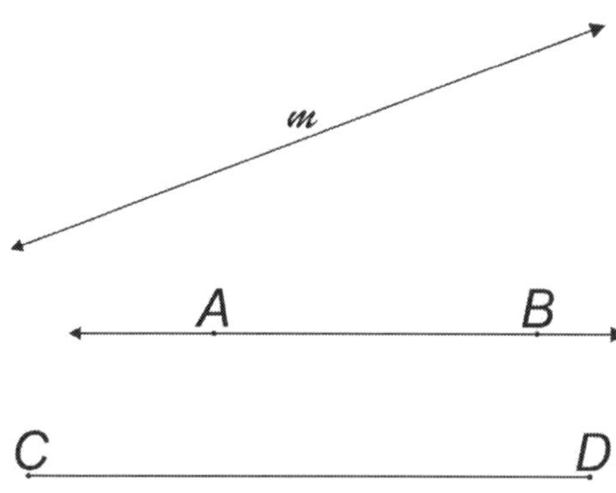

Figure 1. Lines and line segment.

Despite this technical issue, the line is an influential mathematical notion; it is applicable to both geometry and number (Lesson 3.2).

LINE SEGMENT

A **line segment** is a part of a line bounded by two points or endpoints. A line segment is the **locus**[6] of all points on a line between two given points. A line segment is, therefore, bounded. We name a line segment using these two endpoints: $\overline{CD}$ or $\overline{DC}$ in Figure 1. We label the two endpoints using the letters C and D.[7] In 1647, the Italian mathematician Bonaventura Cavalieri (1598-1647) used the bar above CD to indicate line segment CD in his book *Geometria indivisibilibae and Exercitationes geometriae sex*. We shall hear more about Cavalieri near the end of our study (Lesson 14.6).

Ponder this: The number of points on any line segment, there are an infinite number of such points, is the same as the number of points on a line. Note carefully that the line segment can be of any length: 30 miles or 1 inch. Building on the principle of one-to-one correspondence (Lesson 1.4), the German mathematician Georg Cantor (1845-1918) proved this statement in the late 19th century.[8]

PLANE

Any two points in a plane determine one and only one line.

Picture a plane as a map, a flat surface like a wall, a pane of glass, or the surface of water in a still pond. A **plane** has **two dimensions**: length

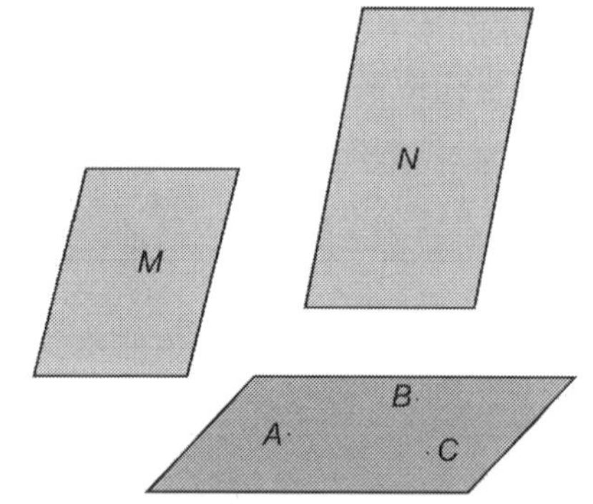

Figure 2. Representation of planes.

[5] Remember the intersection of rows and columns in Table 1 and Table 2 of Lesson 2.5? A road intersection is a junction where one street or road crosses another. The intersection of two walls is a line or, better yet, a line segment. (See next section.) In Geometry, where two lines meet, or where a line and a curve meet, is the point of their intersection. *Intersect*, from the Latin *intersectus*, means "to cut through."

[6] *Locus*, in Latin, means "place." In mathematics, *locus* means the "set of all mathematical objects under consideration or constrained by specific conditions."

[7] The ancient Greeks designated points, lines, and planes by a letter or letters as early as the 5th century BC, as evidenced by the writings of Hippocrates of Chios (ca. 440 BC).

[8] Cantor explored the fascinating notion of an infinite number of infinities (Lesson 13.11).

and width, but no height. Although we picture planes with edges, a geometric plane does not have any edges. Like a line, a plane is unbounded. We name a plane in one of two ways (Figure 2).

1. Capital letter: plane M or plane N.
2. By any three points in the plane: plane ABC.

An index card is a geometric object named a rectangle (Lesson 2.5).[9] A rectangle is a two-dimensional shape that has four interior angles, each measuring 90° where ° is the symbol for **degrees**, a unit of measure or a denomination. Its sides are line segments and the sides opposite each other are parallel and equal in length.

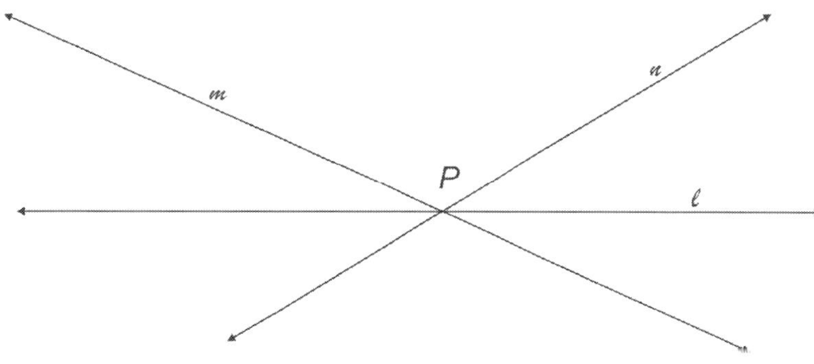

Figure 3. Intersection of three lines.

Points are **collinear** if there is a line that contains all of them; in $\overleftrightarrow{AB}$ (Figure 1), points A and B are collinear points. Points are **noncollinear** if no single line contains them all; in the plane ABC (Figure 2), points A, B, and C are noncollinear points. Points are **coplanar** if there is a plane that contains all of them; in plane ABC, points A, B, and C are coplanar points. Lines are **concurrent** lines if they contain the same point; in Figure 3, point P is the point where lines l, m, and n meet. Line l, m, and n all contain point P, or P is a point on line l, line m, and line n.

A **triangle** (Figure 4) is determined by three line segments connecting three noncollinear points in a plane.[10] We name it using a symbol that looks like a triangle, Δ. In the figure, one name of the triangle is ΔABC. In naming, we can start with any marked point, or vertex (see below), and name the other two verices by either going clockwise or counterclockwise.

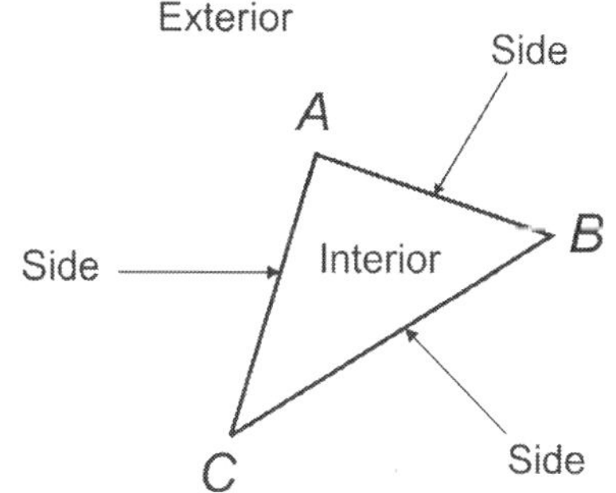

Figure 4. Triangle. Side is sometimes denoted, in older sources, as leg or arm.

Cantor showed that there are an infinite number of points in a plane just as there are an infinite number of points on a line. And, if you could count them, the number of points in the two-dimensional plane and on the one-dimensional line would be equal.

RAYS

Hipparchus (2nd century BC) used basic geometry to estimate the distance from the Earth to the Moon at an amazingly accurate 240,000 miles (384,000 km). Since the Moon's orbital path is oval in shape,[11] its real distance is constantly changing. In 1969, American astronaut Edwin Aldrin (1930-) left

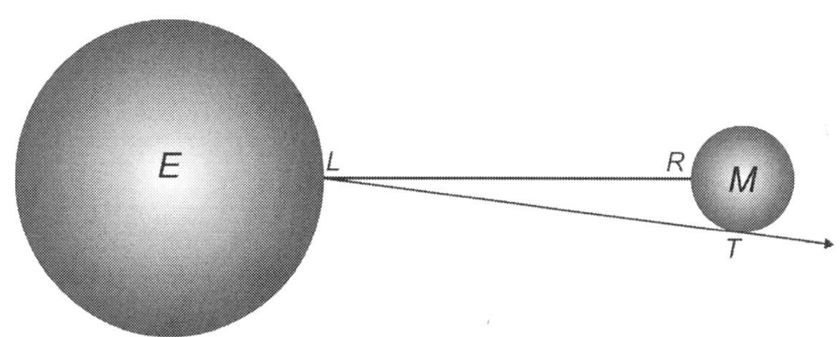

Figure 5. Line segment and ray.

[9] Rectangle comes from the Latin word *rectangulus* meaning "having a right angle."
[10] *Triangle* comes from the Latin *triangulum,* meaning "three angles." Euclid named these figures trilateral, meaning "three sides."
[11] The shape is that of an ellipse (Lesson 8.4).

a lunar laser reflector, point *R* in Figure 5, on the Moon. This device has enabled scientists to calculate the distance between *L*, from the laser on the Earth, to *R*, or *LR*, with a margin of error of less than four inches.

In review (Lesson 2.3), an angle is formed by two sides starting from the same point. To be more precise, an angle is formed by two half-lines, named rays, from the same point. A **ray** is part of a line that extends endlessly in one direction.

The rays that form the angle are its **sides**, and their common endpoint is the **vertex** of the angle.[12] We name rays by their endpoint and by another point on the ray. In Figure 5, we write "ray *LT*", or $\overrightarrow{LT}$, where *L* is the endpoint and *T* is another point on the ray. In the figure, *T* is the **tangent** point,[13] where the laser beam touches the surface of the Moon at one point. $\overrightarrow{LT}$ is the **tangent ray** if it contains one unique tangent point. We can also denote **tangent line segments** and **tangent lines**.

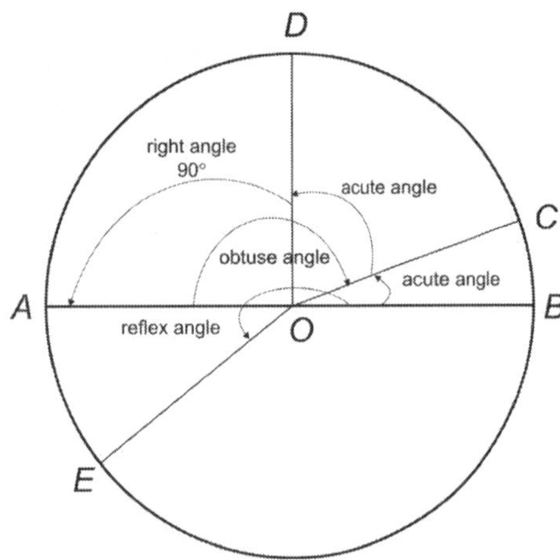

Figure 6. Types of angles.

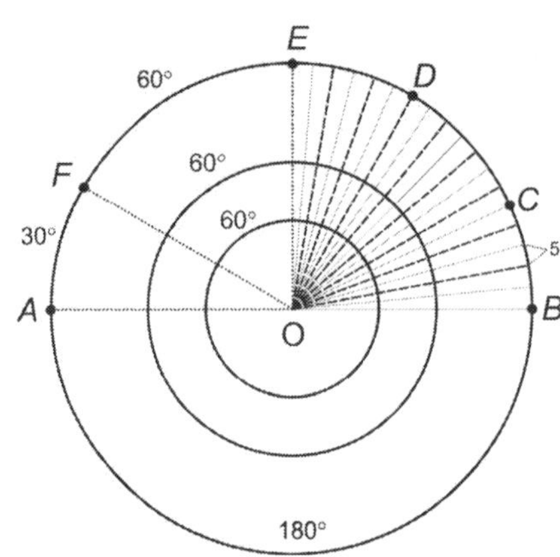

Figure 7. Circle and Degrees.

ANGLES

A **right angle** measures 90°. The measure of an **acute angle** is less than 90°.[14] The measure of an **obtuse angle** is greater than 90° and less than 180°.[15] A **straight angle** measures 180°. A **reflex angle** is an angle that is greater than 180° and less than 360°.[16] By definition, a **circle**[17] contains 360°. This number comes from the ancient Babylonian number system based on 60. (See the homework in Lesson 4.7 for more background.)

A **protractor** is an instrument used for measuring degrees.[18] Usually, a protractor looks like a half circle (Figure 8), and measures 180°, but you can buy a circular protractor which measures 360° (Figure 9). To measure an angle with a half-circle protractor, do this:

Step 1. Find the circle hole on the straight edge of the protractor by locating 90° and follow the line straight down.

---

[12] *Vertere* in Latin means "to turn."

[13] Tangent, from the Latin *tangere*, means "to touch."

[14] *Acutus* in Latin means "needle or sharp."

[15] *Obtusus* in Latin means "dull."

[16] *Reflexus* in Latin means "reflect."

[17] A circle is a curve in a plane generated by one point moving at a constant distance from a fixed point. The constant difference is the **radius** of the circle. *Circulus*, in Latin, is derived from the Greek *kirkos* meaning "ring." As we can see from a circular protractor (Figure 9), a circle is divided into 360 equal units, each unit consisting of one degree

[18] *Protractus* in Latin means "to draw forth or prolong."

Step 2. Place the hole over the vertex of the angle you wish to measure.
Step 3. Line up the zero on the straight edge of the protractor with one of the sides of the angle.
Step 4. Find the point where the second side of the angle intersects the curved edge of the protractor.
Step 5. Read the number that is written on the protractor at the point of intersection. This is the measure of the angle in degrees.

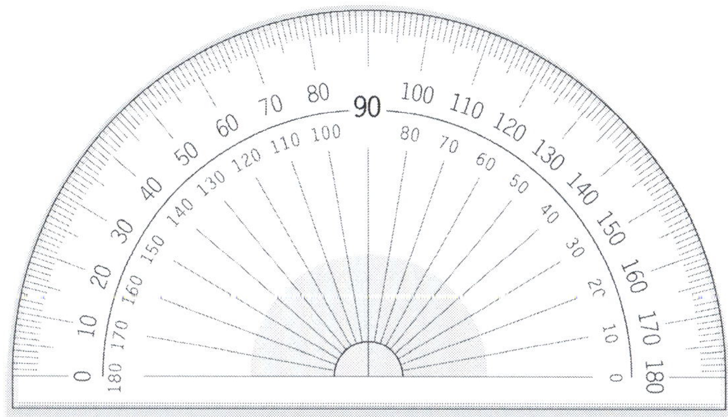

Figure 8. Half-Circle Protractor. Source: iStockPhoto.

Figure 9. Circular Protractor. Source: iStockPhoto.

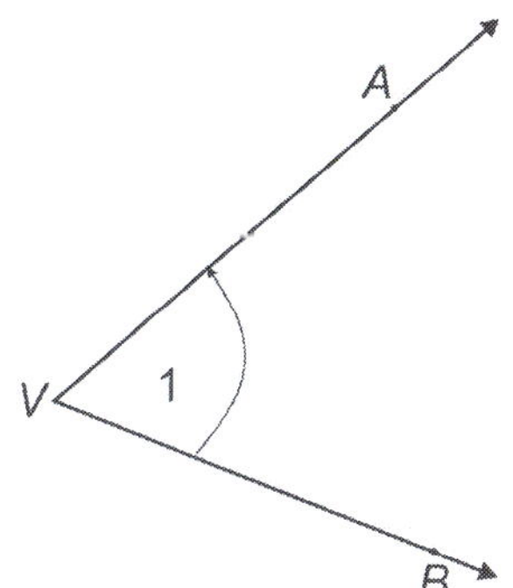

Figure 10. Naming an Angle.

With a straightedge or ruler, draw a few angles, both acute and obtuse, and practice measuring angles with your protractor. Practice measuring angles that open to the right and to the left, so that you get to practice reading both scales of the half-circle protractor.

How do we name specific angles? In Figure 10, the vertex is V and there are two points, A and B, on each side of the angle. We use the symbol $\angle$ to mean "the name of angle."[19] We can name the acute angle four different ways.

1. $\angle 1$
2. $\angle V$
3. $\angle AVB$
4. $\angle BVA$

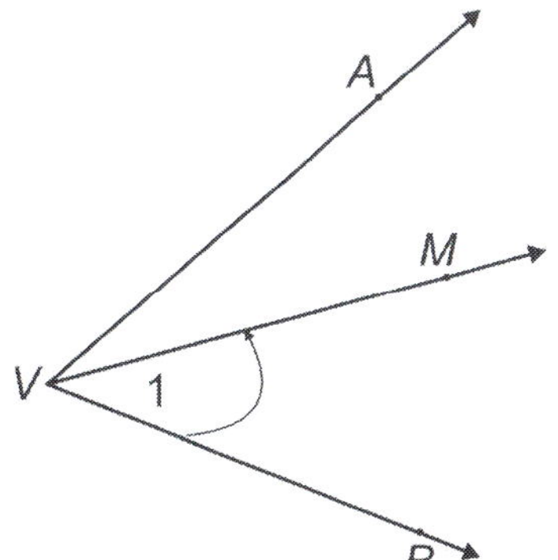

Figure 11. Naming Angles.

If we use three letters to name an angle, the middle letter must always be the vertex of the angle.

In Figure 11, because of the ray VM, or $\overrightarrow{VM}$, we have three possible acute angles. Do you see them? Is $\angle V$ sufficient for naming $\angle 1$? No. $\angle V$ is ambiguous since three angles are possible with this designation. We must use three letters or just the number: $\angle MVB$, $\angle BVM$, or $\angle 1$.

If we measure $\angle MVB$ with a protractor and it is 36°, we signify this measurement as follows

$m\angle MVB = 36°$

[19] The French mathematician and astronomer Pierre Hérigone (1580-1643) used both $\angle$ and < to name angles in *Cursus mathematicus* (1634).

$\angle MVB$ names the angle and $m\angle MVB$ is the measurement of the angle.

SPACE

> It is through deciphering the mathematical patterns carried by light signals that all our knowledge of the space/time universe in its vast or tiny dimensions is derived.
> Thomas F. Torrance, *Theological and Natural Science* (2002), p. 30.

The objects we experience in life are three-dimensional.[20] **Space** is the locus of all points, or the set of all points, in a **three-dimensional** context: length, width, and height.

Again, thanks to Cantor's work, the number of points in any portion of space, there are an infinite number of such points, is the same as the number of points on a line.

Isn't it interesting that we can develop an understanding of the physical world of three-dimensional objects starting with abstract notions – point (zero dimensions), line (one dimension), and plane (two dimensions) – that have no physicality? Such is the way of mathematical knowledge and the relationship of this knowledge to the world of the Triune God's making.

EXERCISES

Define the following terms/symbols/concepts:

1. Point
2. Mathematical object
3. Zero dimensions
4. Line
5. One dimension
6. Intersection
7. Line segment
8. Locus
9. Plane
10. Two dimensions
11. Degrees
12. Collinear points
13. Noncollinear points
14. Coplanar points
15. Concurrent lines
16. Triangle
17. Δ
18. Ray
19. Sides of an angle
20. Vertex
21. Tangent ray, line, or line segment

Figure 12. Source: United States Postal Service.

[20] The physicist Albert Einstein (1879-1955) added a fourth dimension, time, and developed the notion of the space-time continuum (Lesson 13.10). We experience the three dimensions of space, or place, in a framework of time.

22. Tangent point
23. Right angle
24. °
25. Acute angle
26. Obtuse angle
27. Straight angle
28. Reflex angle
29. Circle
30. Radius
31. Protractor
32. $\angle$
33. m$\angle$
34. Space
35. Three dimensions

Figure 13. Space Needle, Seattle, Washington. Source: James D. Nickel.

Answer these questions:
36. How many lines can pass through, or contain, one point?
37. How many lines can contain two separate points in space?
38. In Figure 14, which is longer, AB or CD?

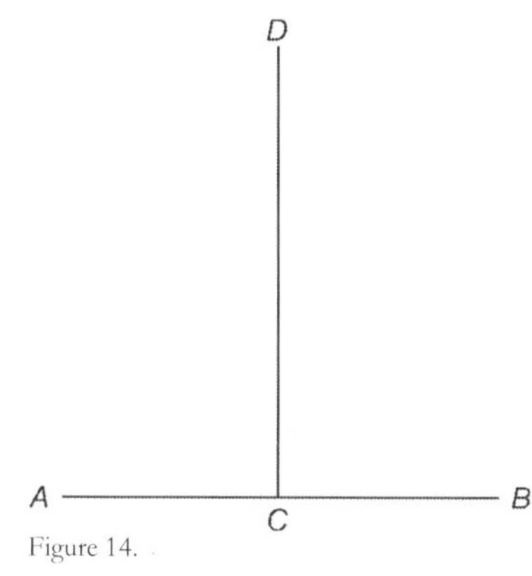

Figure 14.

Illustrate each of the following by labeling two points E and F in your notebook. Then draw this geometric object:
39. $\overleftrightarrow{EF}$
40. $\overrightarrow{FE}$
41. $\overline{EF}$
42. $\overrightarrow{EF}$

Answer these questions:
43. Mark three noncollinear points A, B, and C and draw $\overrightarrow{BA}$ and $\overrightarrow{BC}$.
44. Under what conditions will these two rays together form a line?

What is wrong with this statement?
45. The line segments connecting any three points in a plane make up a triangle.

Field Project:
46. Draw six different acute angles. Measure the angles with a protractor as accurately as you can.

Using Figure 15, find the measure of the following angles without using a protractor. Then denote the type of angle: acute, right, obtuse, or reflex.

47. m$\angle EOB$
48. m$\angle COB$
49. m$\angle DOB$
50. m$\angle EOD$
51. m$\angle EOC$
52. m$\angle FOD$
53. m$\angle AOC$
54. m$\angle AOB$
55. m$\angle BOF$
56. m$\angle FOA$

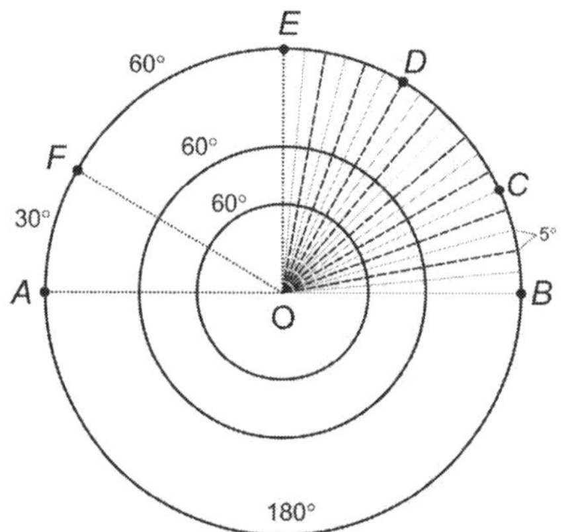

Figure 15. Circle and Degrees.

The ancient Greeks understood arithmetic as the "theory of the structure of numbers." They used another word to describe the skill of calculation: *logistica*. Early Italian writers spoke of logistica as *practica*. The Latin writers of the Renaissance spoke of logistica as "the art of computing" (*ars supputandi*). Dutch writers used the term *ciphering*, particularly between 1500 and 1700, and from this source, through New Amsterdam, came the common use of the word in the early schools of America.

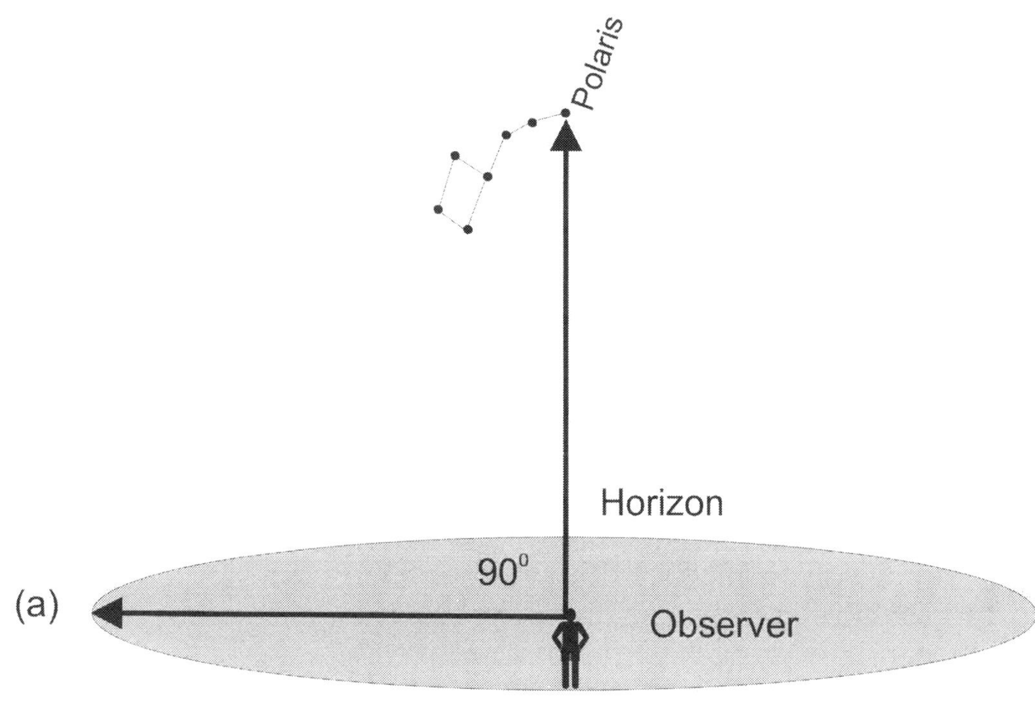
Polaris
Horizon
90^{0}
(a)
Observer

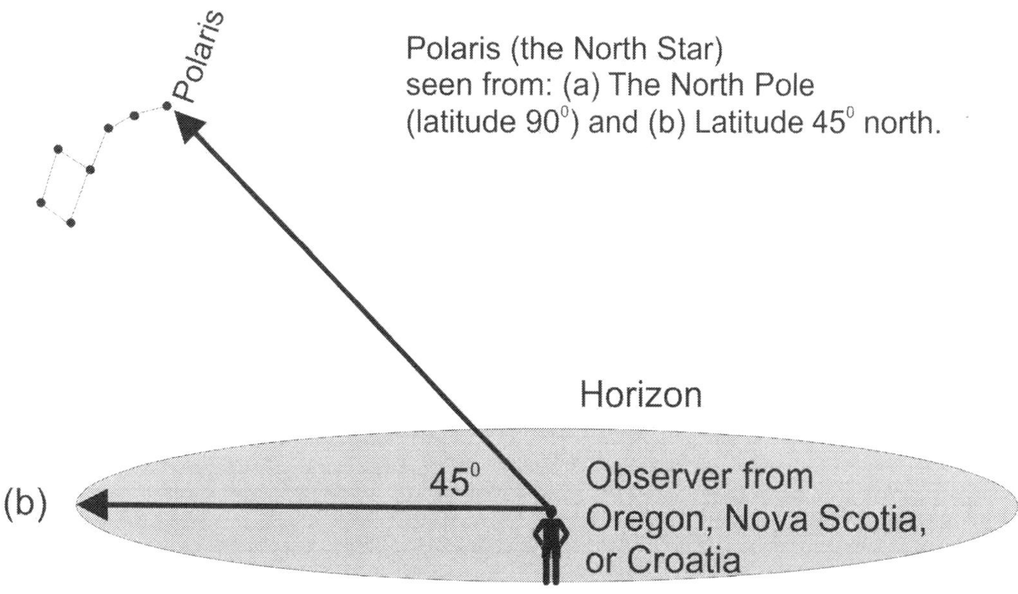
Polaris
Polaris (the North Star)
seen from: (a) The North Pole
(latitude 90^{0}) and (b) Latitude 45^{0} north.
Horizon
45^{0}
(b)
Observer from
Oregon, Nova Scotia,
or Croatia

Figure 16.

2.8 Extending the Dance of Complementary Addition

Terms, Symbols & Concepts Introduced
1. .
2. Casting out nines
3. Centimeter
4. Circumference
5. Complementary addition
6. Diameter
7. Octagon
8. Pentagon
9. Perimeter
10. Quadrilateral
11. Square

In this lesson, we are going to learn to dance with complementary addition, involving (1) more than two addends and (2) addends with two or more digits.

The sum of the lengths of the sides of any figure is its **perimeter.**[1] The measure of the perimeter of a circle, a closed curve, is its **circumference.**[2]

In review, a rectangle is a shape that has four straight sides and four angles each measuring 90°. In geometry, we classify the measurement of the length of a side of a rectangle as one-dimensional. The perimeter of a rectangle is the sum of the lengths of its four sides and it is, therefore, also one-dimensional. If the lengths are in inches, then the perimeter is also in inches.

> Arithmetic is where numbers fly like pigeons in and out of your head. Arithmetic tells you how many you lose or win if you know how many you had before you lose or win. Arithmetic is numbers you squeeze from your head to your hand to your pencil to your paper till you get the answer. Arithmetic is where the answer is right and everything is nice and you look out of the window and see the blue sky – or the answer is wrong and you have to start all over and try again and see how it comes out this time ... If you ask your mother for one fried egg for breakfast and she gives you two fried eggs, and you eat both of them, who is better in arithmetic, you or your mother?
>
> Carl Sandburg (1878-1967), "Arithmetic" (1950).

We cannot separate the perimeter of a rectangle from its dimensionality; they mutually indwell each other and thus reflect the perichoretic nature of number and the dimension of geometric objects. One dimension implies length, and length indicates one dimension.[3] Also, each side of a rectangle mutually indwells, or determines, the perimeter; the perimeter is in each side and each side is in the perimeter. If you change the length of one side, the perimeter changes in like manner because the perimeter of a rectangle is dependent upon the lengths of its sides. Such is the dance of one-dimensional analysis.

> To do math in your head more effectively, it is a good idea to change the way in which we read math problems. Read the problem from left to right, rather than the traditional right to left since it is essentially more important to know the higher numbers in the answer rather than the ending digit.
>
> Arthur T. Benjamin, *The Secrets of Mental Math* (2006), pp. 11-12.

---

[1] *Perimeter* is formed from two Greek words: *peri* meaning "around" and *meter* meaning "measure."

[2] *Circum* is Latin for "circle" and *fer* is Latin for "to carry."

[3] Of course, this mutual indwelling of one dimension is impersonal, unlike the mutual indwelling of the three persons of the Trinity. In number, and in geometry, we are dealing with the rationality of creation in its impersonal, determinate, and inherent form. And, as we have noted, this determinate and proximate revelation of rationality is grounded in the indeterminate and ultimate rationality of the Father, Son, and Spirit.

ONE DIGIT ADDENDS

Let's consider our 3″ by 5″ card. In Figure 1, we have labeled the four vertices of the index card, a rectangle, using the letters *A*, *B*, *C*, and *D*. Note also that "in" is another abbreviation for inches. We name the four sides as follows: $\overline{AB}$, $\overline{BD}$, $\overline{CD}$, $\overline{AC}$. We could also designate the sides like this: $\overline{BA}$, $\overline{DB}$, $\overline{DC}$, $\overline{CA}$.

As we have noted, the distance from point *B* to point *D* is signified *BD*. *DB* would represent the same measure. In other words, *BD* = *DB*. Since this measure is 3″, a concrete number, *BD* = 3″. In similar fashion, *CD* = 5″. Note also that *CD* = *DC*, i.e., the distance between point *C* and point *D* is the same as the distance between point *D* and point *C*. What would be the measure of *AC* and *AB*? Since the opposite sides of a rectangle are equal in length, *AC* = 3″ and *AB* = 5″.

Figure 1.

Let's calculate the perimeter of a 3″ by 5″ card. This calculation equates to adding 3 + 5 + 3 + 5. (Note that we have removed the ″, the inch symbol, so we are now dealing with abstract numbers.) Let's arrange this sum vertically. Note that the four numbers can be arranged in any order.

$$\begin{array}{r} 3 \\ 5 \\ 3 \\ +5 \\ \hline \end{array}$$

Let's walk through this calculation step by step, from top to bottom:

Step 1. 3 + 5 = 8 (first and second addends, a binary operation).

Step 2. Holding 8 in our mind, we add 3 to it (another binary operation): 8 + 3 = 11. Because 11 is greater than 10, we record that we have passed 10 and hold in our mind the number in the ones position, 1 in this case. This mental process is equivalent to noting that 11 = 10 + 1. We record the 10 and keep the 1 in our minds.

Step 3. In Step 2, we held 1 in our mind and we now use 1 to continue the addition: 1 + 5 = 6 (the sum of Step 2, ones position only, and the fourth addend, a binary operation). We have summed the four addends.

We have 6 in our ones position, but we have reached 10, remembered in Step 2. For our final answer, we bring back the recorded or remembered 10 to 6: 6 + 10 = 16. Our sum is 16. 16 is an abstract number. Given our context, the perimeter of the index card is 16″, a concrete number. By our work, we have shown that the abstract is contained in the concrete.

How do we record 10? I suggest two ways: (1) Underline the digit that pushed you over the "10" limit or (2) count with your fingers. Both ways record how many times you have pushed over the "10" limit. On paper, your work would look like this:

$$\begin{array}{r} 3 \\ 5 \\ \underline{3} \\ +5 \\ \hline 16 \end{array}$$

← This is where you passed 10, so you underline the 3 and remember 1, or count one finger.

This is what you should be thinking: "8; 1, record 10; 6; 16." First, you look at the 3 and 5 your mind should say "8." Second, you look at the 3 your mind should say "1, record 10." Third, you look at the 5 your mind should say "6." Fourth, you have 6 in your mind and you have recorded 10 once. Bring back that 10. You immediately have the answer the answer in your mind since 6 + 10 = 16.

Let's arrange these four numbers in a different order:

$$\begin{array}{r} 5 \\ 5 \\ 3 \\ +\,3 \\ \hline \end{array}$$

Adding from top to bottom, this is what you should be thinking: "0, record 10; 3; 6; 16." The first sum is 10. In your mind, you should say "0 and record 10." The second sum is 3, and the third sum is 6. Our answer is 6 ones, and we have passed 10 once. Our answer is 6 + 10 = 16.

On paper, your work should look like this:

$$\begin{array}{r} 5 \\ \underline{5} \\ 3 \\ +\,3 \\ \hline 16 \end{array}$$

Let's calculate the perimeter of Figure 2. It is a **pentagon**.[4] We name this figure pentagon *ABCDE*. Let's calculate using abstract numbers:

$$\begin{array}{r} 6 \\ 2 \\ \underline{6} \\ \underline{7} \\ +\,\underline{9} \\ \hline 30 \end{array}$$

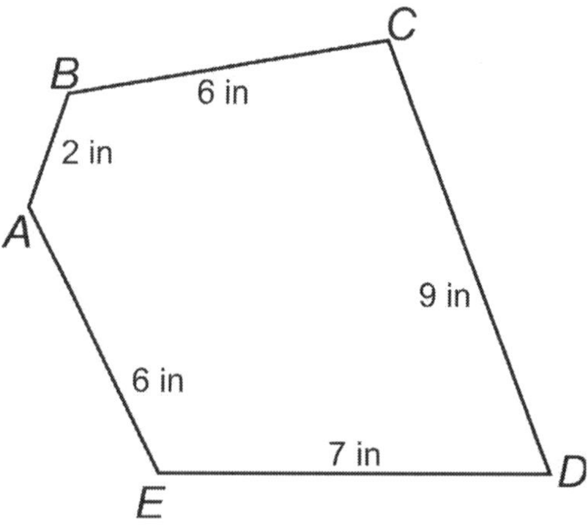

Figure 2.

Adding from top to bottom, this is what you should be thinking: "8; 4, record 10; 1, record 10; 0, record 10; 30." The first sum is 8. The second sum is 4, record 10. The third sum is 1, record 10. The fourth sum is 0, record 10. Our answer is 0 ones and we have passed 10 three times, which equals 30. Our answer is 0 + 30 = 30. The perimeter of this pentagon is 30 inches.

If you use your fingers to record the tens, at the end of this process you should have 0 in your mind and you should also be looking at 3 fingers. The 3 fingers represent 3 tens, or 30. So, your sum is 30. Using fingers to keep track of tens will enable you to do these one-digit sums quickly in your head.

Let's rearrange the five digits in a different order.

[4] A pentagon is a geometric figure having five angles and five line segments as sides. In Greek, *pentagonos* means "five angles."

9
7
6
2
+ 6
30

Adding from top to bottom, this is what you should be thinking: "6, record 10 (or one finger); 2, record 10 (or a second finger); 4; 0, record 10 (or a third finger); 30." The sum is 30.

This is fun! After you get used to it, your speed in adding a list of numbers will increase significantly. Just remember the ones part of your sums and record "10" for every intermediate sum that equals or passes 10.

Let's summarize what we are doing. **Complementary addition** refers first to the three-step process of using complements in adding any pair of single digits. We can extend the dance of complementary addition to include this two-step process of recording and remembering in adding a column of digits. Adding columns of single digits can be summarized using two rules:

Rule 1. Do one step at a time (and remember or record the results of each step).
Rule 2. Never calculate any sum over 10.

Following Rule 1 will lead to good arithmetic. Many problems in arithmetic require multiple steps to get to the answer. Instead of being overwhelmed by the big problem, we learn to break the problem down into steps. Then we do one step at a time.

Rule 2 should please you. In our addition work, there will be no need to calculate any sum over 10. Why 10? 10 is the base of the decimal system, and it has unique and very useful properties. We have capitalized on those properties in such a way as to make any addition problem as easy as adding up to and including 10. Whenever our sum of two digits is greater than 10, we just record a ten, remember the ones digit, and continue the dance.

Add 5, 6, 9, and 7:

5
6
9
+ 7
27

When you add from top to bottom, your thinking should be: "1, record 10 or (one finger); 0, record ten (or a second finger), 7; 27." The sum is 27.

Add 8, 2, 5, 7, 9, 9, 8, 2, 3, and 6:

$$\begin{array}{r} 8 \\ \underline{2} \\ 5 \\ \underline{7} \\ \underline{9} \\ \underline{9} \\ 8 \\ \underline{2} \\ 3 \\ \underline{+\,6} \\ 59 \end{array}$$

Adding from top to bottom, think: "0, record 10 (or one finger); 5; 2, record 10 (or a second finger); 1, record 10 (or a third finger); 0, record 10 (or a fourth finger); 8; 0, record 10 (or a fifth finger); 3; 9; 59." The sum is 59. Neat, isn't it?

Try this one again going from bottom to top. If you want an extra challenge matching the dexterity of your fingers with your mind, attempt these examples using the Japanese Soroban.

TWO-DIGIT ADDENDS AND LEFT TO RIGHT ADDITION

Let's continue this fascinating adventure as we consider two-digit sums. Let's consider the calculation of the perimeter of Figure 4. It has four straight sides, line segments of different lengths, and four points of intersection. This geometric form is a **quadrilateral**.[5] Rectangles (Lesson 2.5), in which opposite sides are equal and each of the four angles measures 90°, and **squares**,[6] for which all four sides are equal and each of the four angles measures 90°, are particular kinds of quadrilaterals (Figure 3).

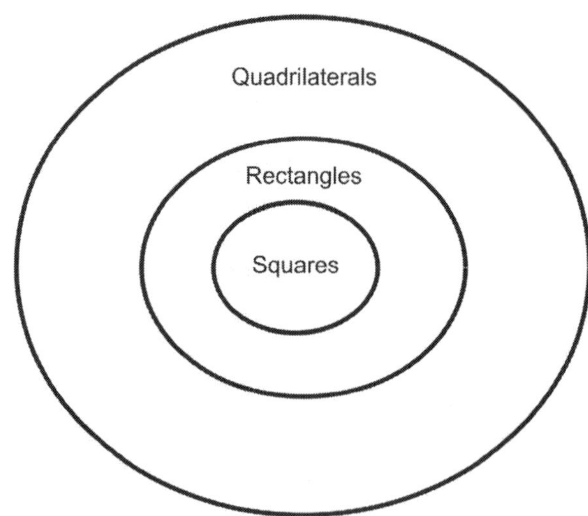

Figure 3. All squares are rectangles and all rectangles are quadrilaterals.

To calculate the perimeter, we add a third rule to our method, adding left to right. Most of you and most of your parents have been taught to do addition problems involving two digits or more from right to left. Following the Japanese Soroban model, we are going to learn to do addition problems from left to right. Why? We read numbers from left to right, why not compute them from left to right? By recognizing this consistency principle, Japanese mathematicians constructed the Soroban in the way that they did. We now add a third rule:

---

[5] The Latin *quadrilaterus* means "four sides." Any four-sided figure in a plane, with four line segments as sides, is a quadrilateral.
[6] *Square* comes from the Latin *quadrare* meaning "to square."

Rule 1. Do one step at a time (and remember or record the results of each step).
Rule 2. Never calculate any sum over 10.
Rule 3. Do multiple digit problems from left to right.

We calculate the sum of 26, 49, 82, and 97, the perimeter of the quadrilateral in Figure 4. We add the digits in the tens column first. We get:

```
 26
 49
 82
+97
---
23
```

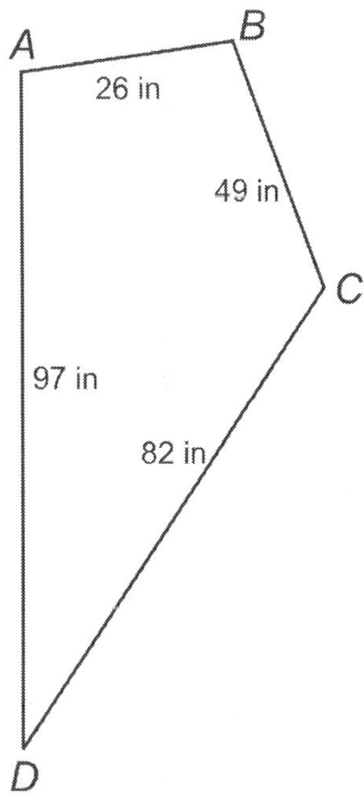

Figure 4.

Step 1. 2 + 4 = 6 (really, 6 tens or 60).
Step 2. 6 + 8 = 14. Record the ten by underlining the 8 and work with 4, which is really 4 tens or 40.
Step 3. 4 + 9 = 13. Record the ten by underlining the 9.
Step 4. Place the 3 (2 tens or 20) under the 9. You've recorded two tens by underlining. Place the 2 one place to the left of 2 since you are adding the tens place value; 2 tens of tens are really 200. What we have really done is this: 20 + 40 + 80 + 90 = 230.

The value of summing the tens column first is that you have an immediate estimate of your answer. You know that the sum is going to be a little more than 230. Also, you have applied what you have learned about one-digit addends. Although we are summing the tens column, you calculate the sum just like you did with one-digit sums: one step at a time; and if a sum is greater than 10, you only record the ten and remember the ones. Since you are working with the tens column, you are recording a hundred and remembering the tens. Even though this process can quickly become a mechanical procedure, always keep in mind what you are doing and what place value you are working with.

Now sum the ones column:

```
  26
  49
  82
 +97
 ---
   2
 234
```

Step 1. 6 + 9 = 15. Record the ten by underlining the 9, and remember the 5 (5 ones).
Step 2. 5 + 2 = 7 (7 ones).
Step 3. 7 + 7 = 14. Record the ten by underlining the 7, and place 4 (4 ones) under the 7. This is why we did not write the last 0 in 230. You've recorded two tens (by underlining). Your sum is 24. Since you are adding the ones place value, 2 tens of ones are really 20; therefore, place 2 one place to the left of the ones column, over the last 3 from the first column. Write it a little smaller.

To review what we have done. We added the tens column first to get 230. Then we proceeded to the ones column. We added it to get 24. 230 + 24 = 254. Our estimate was 230, so our answer checks. Given the context of our problem, the perimeter of the quadrilateral is 254″.

This method is, of course, new to you. With practice, you will find it much easier than the way you have been taught. When you do this automatically, you will think as follows:

For the first column: 6, 4 (line), 4 (line). Write 23 with the 3 aligned under the 9. Do not write the last 0; leave that position blank.

For the second column: 5 (line), 7, 4 (line). Write 4 in the vacant position under 7 and write 2 in the tens position above the 3 in the tens position that you calculated in the tens column. Sum your result 230 + 24 = 254.

ADDENDS OF VARYING LENGTHS

Addends can vary in digit length. When this is the case, make sure that your vertical lineup is in accordance with place value.

Let's say that daily visits to a given website total 915 on Monday, 76 on Tuesday, 525 on Wednesday, 883 on Thursday, and 292 on Friday. What is the total number of visits for the five days?

We first arrange the digits vertically. Which arrangement is correct?

Arrangement 1:

```
915
76
525
883
292
```

Arrangement 2:

```
915
 76
525
883
292
```

Arrangement 1 is incorrect and Arrangement 2 is correct. Why? In Arrangement 1, the 7 in 76 is in the hundreds column, not the tens column. Let's proceed by adding the hundreds column:

```
  915
   76
  525
  883
+ 292
-----
 24
```

We think: 4 (line), 2 (line), 4: 24. First estimate: 2400.

Next, we work the tens column:

915
76
525
883
+ 292

2
247

We think: 8, 0 (line), 8, 7 (line): 27. Second estimate: 2670.
Finally, we engage the ones column:

915
76
525
883
+ 292

2 2
2471

Think: 1 (line), 6, 9, 1 (line): 21. Our sum is now precise: 2691 visits.

CASTING OUT NINES

How do we check our addition work for accuracy? One way is to reorder the addends and calculate the sum again. (See Lesson 3.7 for justification of this reordering.)

The second way is to use a technique that accountants used before the invention of calculators. This technique is **casting out nines**. (In Lesson 5.4 we will investigate why this method works.) In your notebook, sum 146, 928, and 357. Your work should look like this:

146
928
+ 357 Sum: 1431

1 2
1311

To check, add the digits in the sum: 1 + 4 + 3 + 1 = 9. Since the sum is 9, we can "cast it out" and substitute it with zero. Next, add the digits in each of the addends and keep adding until you get a one-digit sum:

1 + 4 + 6 = 11 = 2 (1 + 1)
9 + 2 + 8 = 10 = 1 (We cast out the 9 and add 2 to 8)
3 + 5 + 7 = 15 = 6 (1 + 5)

Now add these sums: 2 + 1 + 6 = 9 or 2 + 1 + 6 = 0 (casting out the nine). We could also cast out nines by finding every instance of 9 in the digits of the three addends combined, or sums equal to 9 in the digits. We do this:

Cast out 9 in 928

Cast out 1 + 8 = 9 in 146 and 928
Cast out 2 + 7 = 9 in 928 and 357
Cast out 4 + 5 = 9 in 146 and 357
Cast out 6 + 3 = 9 in 146 and 357

We have cast out all nine digits in the addends. Our result is zero.

Since zero is the sum in both checks, we have tentative confirmation of the accuracy of our work. This procedure does not guarantee that your answer is always correct. Can you determine why? But, when these two numbers are unequal, you know for sure that you have made a calculation error.

In your notebook, sum 1836, 189, 2618, 356, and 18. Your work:

```
  1836
   189
  2618
   356      Sum: 5017
+   18
  ----
  1 1 3
  3887
```

Adding the digits in the sum, we get 5 + 1 + 7 = 13. Summing 1 and 3, we get 4. To check the sum, add the digits in each of the addends and then sum this total. If the sum is 4, we have a good check of our work:

```
  1836        0
   189        0
  2618        8
   356        5     Sum of digits: 1 + 3 = 4 (Check!)
+   18    +   0
  ----     ----
  1 1 3
  3887       13
```

WORKING WITH DOLLARS AND CENTS

Let's try another example. Let's say that we are buying four books from a catalog, priced respectively at \$5.98, \$3.98, \$4.99, and \$1.99. The postage for this order is \$5.95. What is our cost? Remember, sum the problem step by step, left to right, and always record the tens and remember the ones.

In countries that use dollars and cents, a period (. is the symbol), the decimal point (Lesson 7.1), is placed before the cents or between the dollars and cents. Therefore, \$10.16 is read "10 dollars and 16 cents." Since 100 cents equals \$1.00 (Our currency is named decimal currency.), the cents digits will always occupy two places after the decimal point, and never more than two. If the number of cents is less than 10, a zero must occupy the first place to the right of the decimal point. Thus, "5 dollars and 2 cents" is written \$5.02, and "10 dollars and 9 cents" is written \$10.09. The conjunction "and" is said at the decimal point, and not between place value columns. In arranging for addition, dollars should be written under dollars, and cents under cents, in such order that the decimal points stand in a vertical line. In other words, we line up the decimal points and invoke our left to right method.

We work the dollar-column first:

$$\begin{array}{r} 5.98 \\ 3.98 \\ \underline{4}.99 \\ 1.99 \\ +\,5.95 \\ \hline 18. \end{array}$$

We think 8, 2 (line), 3, 8. The dollar column sum is 18, our first estimate. We move to the column to the right of the decimal point and work the tens column of cents next:

$$\begin{array}{r} 5.98 \\ 3.\underline{9}8 \\ \underline{4}.\underline{9}9 \\ 1.\underline{9}9 \\ +\,5.\underline{9}5 \\ \hline \overset{4}{18.5} \end{array}$$

We think: 8 (line), 7 (line), 6 (line), 5 (line). Our result: $4.50. See it? And, do you see that our second and better estimate is $22.50? Finally, we work with the ones column or cents column:

$$\begin{array}{r} 5.98 \\ 3.\underline{9}\underline{8} \\ \underline{4}.\underline{9}\underline{9} \\ 1.\underline{9}\underline{9} \\ +\,5.\underline{9}5 \\ \hline 18.59 \end{array}$$

(carries: 4 3)

We think: 6 (line), 5 (line), 4 (line), 9. Our result: 39 cents.

What is the total cost? Reading from left to right, we just mentally see that 4 + 18 = 22 or $22. Next, we see that 3 + 5 = 8 so we have 89 cents. The cost of the book order: $22.89.

In your notebook, check your work by casting out nines.

EXERCISES

The problems below make use of the following new terms and symbols:

- An **octagon** is a geometric figure with eight sides, all line segments, and eight angles inside the figure, or interior angles.
- Another symbol for feet is ′ and there are 12 inches in one foot.

Figure 5. Source: James D. Nickel.

- cm is the abbreviation of centimeters, a metric system measurement.[7] One **centimeter** is the approximate **diameter**[8] of a circular shirt button. A centimeter is less than a meter, the approximate height from the bottom of a door to the doorknob, by a factor of 100, i.e., it is 100 times less than a meter. Factor and times are multiplication concepts. (See Step 4.)

Define the following terms or symbols:

1. Perimeter
2. Circumference
3. Pentagon
4. Quadrilateral
5. Square
6. .
7. Octagon
8. Centimeter
9. Diameter

Figure 6. Fossilized sand dollar.
Source: James D. Nickel.

Write a short essay:

10. What are the three rules of complementary addition?
11. Under what circumstances would the casting out nines check be accurate but the sum wrong?

Do the following problems:

12. Create 20 problems containing three one-digit addends and work their sums quickly and accurately in your mind.
13. Examine these sums and verify using complementary addition. Do the work in your head by recording tens with your fingers. Check by casting out nines.

8	8	9	5
3	7	2	6
9	9	6	9
2	2	4	7
+ 5	+ 6	+ 7	+ 8
27	32	28	35

The homework of this lesson and the subsequent lesson contain many types of counting exercises to build number sense, mastery of addition, and speed calculation.

For these counting exercises, make sure that you can verbally do them easily and without too much time for thought. If you must take some time to think about some of your answers, go over the exercises repeatedly until you can confidently do these counts. Incorporate complements in your work. Count by:

14. 2s to 24 starting with 2 (say "2, 4, 6," etc.)
15. 3s to 36 starting with 3
16. 4s to 48 starting with 4
17. 5s to 60 starting with 5
18. 6s to 72 starting with 6
19. 7s to 84 starting with 7
20. 8s to 96 starting with 8
21. 9s to 108 starting with 9
22. 10s to 120 starting with 10
23. 11s to 132 starting with 11

These verbal exercises will require more concentration. Again, go over the exercises repeatedly until you can confidently do these counts. Incorporate complements in your work. Count by:

24. 2s to 19 starting with 1 (say "1, 3, 5," etc.)

[7] The prefix *centi* in Latin means hundred. A century is 100 years.

[8] Diameter comes from the Greek *diametros*, from *dia* meaning "through or across," and *metron* meaning "measure." The diameter is a line segment passing from side to side through the center of a body or figure, especially a circle or sphere.

25. 3s to 28 starting with 1
26. 4s to 37 starting with 1
27. 5s to 46 starting with 1
28. 6s to 55 starting with 1
29. 7s to 64 starting with 1
30. 8s to 73 starting with 1
31. 9s to 82 starting with 1
32. 10s to 91 starting with 1
33. 11s to 100 starting with 1
34. Which exercise, from 24 to 33, was the easiest for you?

Do the following group of exercises verbally. Go over the exercises repeatedly until you can confidently do these counts. Incorporate complements in your work. Count by:
35. 2s to 20 starting with 2
36. 3s to 29 starting with 2
37. 4s to 38 starting with 2
38. 5s to 47 starting with 2
39. 6s to 56 starting with 2
40. 7s to 65 starting with 2
41. 8s to 74 starting with 2
42. 9s to 83 starting with 2
43. 10s to 92 starting with 2
44. 11s to 101 starting with 2
45. Which exercise, from 35 to 44, was the easiest for you?

Verbally count by: (Make sure to incorporate complements in your work.)
46. 2s from 4 to 50
47. 3s from 1 to 43
48. 3s from 6 to 51
49. 4s from 1 to 53
50. 4s from 5 to 45
51. 5s from 1 to 61
52. 5s from 7 to 82
53. 6s from 0 to 72
54. 6s from 2 to 80
55. 6s from 10 to 88

Verbally count by: (Make suyre to incorporate complements in your work.)
56. 2s and 3s alternately from 1 to 31 (Ans. 1, 3, 6, 8, 11, 13, 16, 18, 21, 23, 26, 28, 31)
57. 2s and 4s alternately from 1 to 49
58. 3s and 4s alternately from 0 to 56
59. 2s and 5s alternately from 4 to 60
60. 3s and 5s alternately from 7 to 63

Set up these horizontal addition problems vertically and use complementary addition to determine their sums. Make sure that you align the problems by place value. Check by casting out nines.
61. 7 + 7 + 7 + 77
62. 6 + 8 + 8 + 68
63. 7 + 13 + 13
64. 8 + 9 + 9 + 9

65. 9 + 6 + 6 + 12
66. 7 + 7 + 6 + 5 + 7
67. 10 + 9 + 8 + 7 + 6
68. 3 + 59 + 13
69. 6 + 46 + 105
70. 22 + 11 + 89
71. 10 + 58 + 19
72. 24 + 25 + 35
73. 6 + 6 + 6 + 6 + 6
74. 9 + 9 + 9 + 9 + 9
75. 23 + 105 + 76
76. 17 + 88 + 55
77. 55 + 66 + 77
78. 216 + 15 + 9
79. 36 + 42
80. 53 + 38
81. 65 + 40
82. 53 + 38

For the following exercises, find the sums and check your work by casting our nines.

83. Find the perimeter of an octagon with sides measuring 28″, 36″, 51″, 43″, 57″, 45″, 49″, and 51″.
84. Find the perimeter of a quadrilateral with sides measuring 252′, 198′, 279′, and 243′.
85. Find the total cost of the following book order: one book at $7.99, the second book at $3.99, the third book at $23.99, and postage of $5.95.
86. Find the perimeter of a rectangle whose length is 46 cm and width is 28 cm.
87. Find the sum: 99 + 99 + 99 + 99 + 99
88. Explain a quick way to find the sum in Question 87 in your head without having actually to do the addition work on paper.

Answer this question:

89. Suppose in January you are given a pair of new born rabbits (Figure 7). After two months, they produce another pair of rabbits and they continue to produce a new pair every month after that. Each new pair of rabbits, after two months, generates another new pair and goes on to produce a new pair each month thereafter. If no rabbits die, how many pairs of rabbits will there be in December?

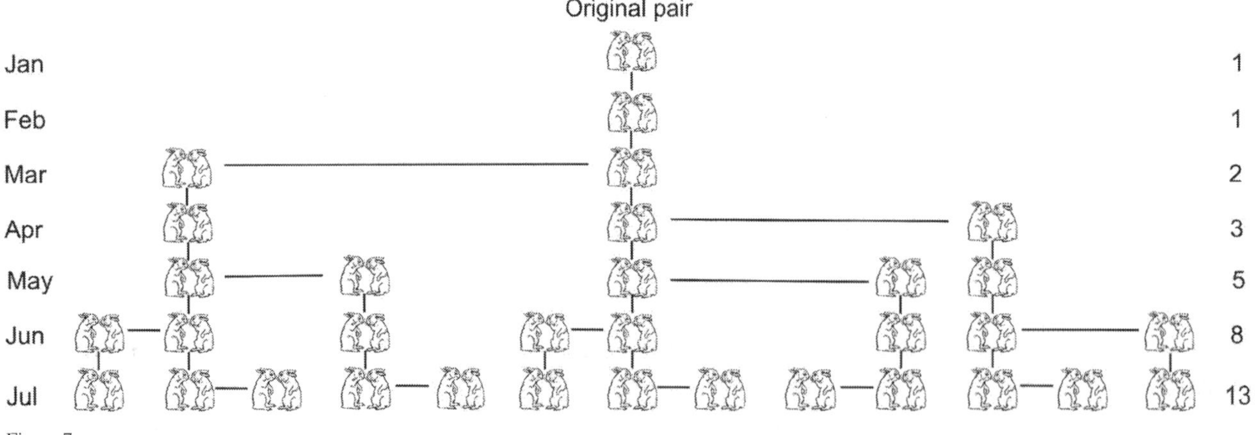

Figure 7.

2.9 Achieving Addition Mastery

In last lesson's exercises, we included several examples of sums involving three-digit addends. With a mastery of our method, sums of digits of any length can be calculated if we continue column by column, left to right, and step by step.

> In 15th-century Italy, logistica was denoted as the "minor art" (*l'arte minore*), and Arithmetic and Algebra as the "major art" (*l'arte maggiore* or, in Latin, *ars magna*).

Let's say that over a period of five months, we have per-month earnings of $327, $456, $622, $718, and $559. How much have we earned in the five-month period? We remove the $ sign so that we are dealing with an abstract sum. We have three columns to add digit by digit. Invoking our left to right method, we start with the hundreds column:

$$\begin{array}{r} 327 \\ 456 \\ \underline{6}22 \\ \underline{7}18 \\ +\,559 \\ \hline 25 \end{array}$$

We think: 7, 3 (line), 0 (line), 5: 25. Our first estimate is $2500.
Next, we engage the tens column:

$$\begin{array}{r} 327 \\ 456 \\ \underline{6}22 \\ \underline{71}8 \\ +\,559 \\ \hline 25\overset{1}{5} \end{array}$$

We think: 7, 9, 0 (line), 5: 15. Our second and better estimate is $2650.
Finally, we add the ones column:

$$\begin{array}{r} 327 \\ 45\underline{6} \\ \underline{6}22 \\ \underline{718} \\ +\,55\underline{9} \\ \hline 25\overset{1}{5}\overset{3}{2} \end{array}$$

We think: 3 (line), 5, 3 (line), 2 (line): 32. Our final and exact sum: $2682. See it? In your notebook, check by casting out nines.

If we remember which place value we are working with, i.e., which column we are adding, any addition problem becomes complementary addition applied to one-digit addends!

A FINE NUANCE

Let's go over a more challenging example of addition. Add 3242, 4332, 1401, 2502 and 8528 in your notebook and check by casting our nines. Your completed work should look like this:

$$\begin{array}{r} 3242 \\ 4332 \\ 1401 \\ 2502 \\ +\,8528 \\ \hline 1\overset{1}{8}9\overset{1}{9}5 \end{array}$$

What is the sum? From left to right, after we add 1 to 8 in the thousands place, we have this:

$$199\overset{1}{9}5$$

What do we do with the addition of 1 to 9 in the tens place? The sum of 1 + 9 in the tens place gives us 0, record 100. We have this:

$$19\overset{1}{9}05$$

Continuing, 1 + 9 in the hundreds place gives us 0, record 1000. We have this:

$$1\overset{1}{9}005$$

We keep going: 1 + 9 in the thousands place give us 0, record 10,000. We have this:

$$\overset{1}{1}0005$$

1 + 1 in the ten thousand place gives us 20,000. Our final sum: 20,005. A quicker way to do this is to look at the initial sum, $199\overset{1}{9}5$, add 1 to 1999 to get 2000 thus generating our final sum: 20,005. Make sure you understand what to do when you see a situation like this. We can indicate this sum as follows:

$$\overset{1}{\overline{1999}}5 = 20,005$$

Let's try one more sum. Let's say that we know the population of four countries:

Country A: 111,848,867
Country B: 97,779,942
Country C: 96,876,249
Country D: 5,696,056

What is the population of the four countries combined?

First, we lay out the digits vertically, carefully aligning the columns:

```
  111,848,867
   97,779,942
   96,876,249
+   5,696,056
```

Second, we proceed to add digit by digit, working from left to right. Work this problem separately in your notebook and then check your work by casting out nines. Your work:

```
  111,848,867
   97,779,942
   96,876,249
+   5,696,056
  1 1 2  2 2 1   1 2
  199,979,994
```

The total population of the four countries is 312,201,114.
To find the sum, we do the following, from left to right:

```
1 1 2  2 2 1   1 2
199,979,994
1 1 1  1 1 1   1
201,190,014
312,201,114
```

Casting out nines:

First addend: Cast out 1 + 8, 1 + 8, 1 + 8; 4 + 6 + 7 = 17 = 8
Second addend: Cast out 9, 9, 9, 7 + 2; 7 + 7 + 4 = 18 = 9 = 0
Third addend: Cast out 9, 9, 7 + 2; 6 + 8 + 6 + 4 = 24 = 6
Foruth addend: Cast out 9; 5 + 6 + 6 + 5 + 6 = 28 = 1
Sum of addends: 8 + 6 + 1 = 15 = 6
Sum: Cast out 3 + 2 + 5; 1 + 2 + 1 + 1 + 1 = 6
Check.

With this method, there is no need to add more than two lines of subtotals if, as you add from left to right, you look ahead to see where you can determine some preliminary sums.

How did you do?

A Shortcut Principle

Let's investigate another speed method that some of you may have already noticed. Here is the principle:

> When adding a group of numbers, always look for two numbers that sum to 10.

These numbers will be one of these five combinations:

1 + 9 or 9 + 1
2 + 8 or 8 + 2
3 + 7 or 7 + 3

4 + 6 or 6 + 4
5 + 5

Whenever you see these pairs, you should mentally tag them as 10 using a line if working on paper, or your finger if adding mentally.

For example, here is an addition problem consisting of one-digit numbers:

3	3
5	5
3	~~3~~
6	~~6~~
4	~~4~~
9	~~9~~
1	~~1~~
8	~~8~~
2	~~2~~
+ 7	+ ~~7~~

We find four groups of ten with 8 left over. Therefore, the sum is 48.

Let's sum 2, 1, 6, 3, 3, 7, 4, 6, 2, and 8. Try regrouping these addends and do the work first in your notebook. Here is the completed work:

2
~~1~~
~~6~~
~~3~~
~~3~~
~~7~~
~~4~~
~~6~~
~~2~~
+ ~~8~~
———
42

$3 + 7 = 10$, $4 + 6 = 10$, and $2 + 8 = 10$. Did you see that $1 + 6 + 3 = 10$?

Now add 7, 1, 8, 2, 3, 2, 6, 3, 6, and 3 in your notebook. Here is the completed work:

~~7~~
~~1~~
~~8~~
~~2~~
~~3~~
2
~~6~~
~~3~~
6
+ 3
———
41

7 + 3 = 10 and 8 + 2 = 10. Did you see that 1 + 3 + 6 = 10?

Remember, there are many ways to regroup digits to get sums equal to 10. All you need to do is flag those with a line or a finger.

The more you do problems like this, the quicker you will be able to see sums adding to 10. Also, I hope that you are seeing how much faster you can do sums using complementary addition. Instead of trying to remember sums greater than 10 in your head, complementary addition is a simple, two-step process of:

1. Recording tens by underlining, and
2. Remembering the ones and adding the next digit.

EXERCISES

The game of bowling consists of ten frames, in which a bowler has two chances per frame to knock down all ten pins. (Note: In the arrangement of rows of pins, 1 + 2 + 3 + 4 = 10. In Lesson 13.13, we will revisit this additive relationship.) If a bowler knocks down all ten pins in those two turns in the 10th frame, an extra bonus roll is awarded. You keep score by adding the number of pins knocked down in each frame. Special bonuses are awarded for strikes and spares. When a strike is bowled, all ten pins are knocked down. We signify a strike by X, the bowler is awarded the score of 10, plus he gets to add the total of his next two rolls to that frame. A spare is when a bowler knocks down all ten pins in two rolls. We signify a spare by /, the bowler is awarded the score of 10, plus the total number of pins knocked down only on the next roll.

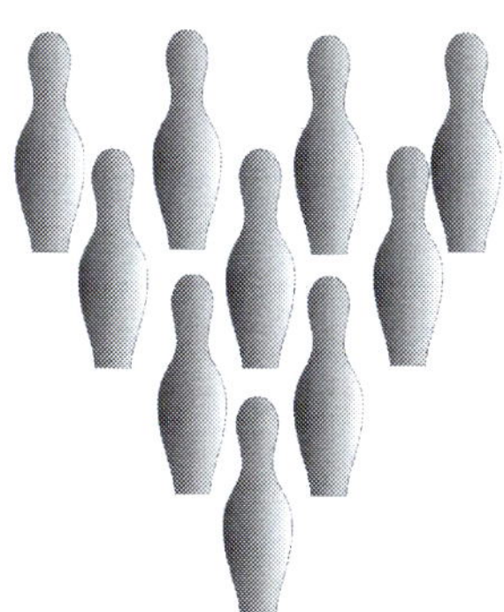

Figure 1.

Name	1	2	3	4	5	6	7	8	9	10	Total

Figure 2. Bowling Scorecard: Ten Frames.

Scoring analysis for Josh, the first six frames, is in Figure 3:

Name	1	2	3	4	5	6	7	8	9	10	Total
Josh	7 2	8 /	2 /	X	9 /	2 3	X	7 /	X	7 3 X	
	9	21	41	61	73	78	98	118	138	158	158

Figure 3. Bowling Scorecard: Ten Frames for Josh.

Frame 1. 7 pins knocked down on the first bowl, 2 on the second. Score: 9.

Frame 2. 8 pins knocked down on the first bowl, 2 on the second (a spare). Score: 10 + 2 (first bowl in 3rd frame) = 12 and the cumulative score is 12 + 9 = 21.

Frame 3. 2 pins knocked down on the first bowl, 8 on the second (a spare): Score: 10 + 10 (first bowl in 4th frame) = 20 and the cumulative score is 21 + 20 = 41.

Figure 4. Source: iStockPhoto.

Frame 4. A strike, all ten pins knocked down on the first bowl. Score: 10 + 9 + 1 (next two bowls in 5th frame) = 20 and the cumulative score is 41 + 20 = 61.

Frame 5. 9 pins knocked down on the first bowl and 1 on the second (a spare). Score: 10 + 2 (first bowl in 6th frame) = 12 and the cumulative score is 61 + 12 = 73.

Frame 6. 2 pins knocked down on the first bowl and 3 on the second: Score: 5 and the cumulative score is 73 + 5 = 78.

Frame 7. A strike, all ten pins knocked down on the first bowl. Score: 10 + 7 + 3 (next two bowls in the 8th frame) = 20 and the cumulative score is 78 + 20 = 98.

Frame 8. 7 pins knocked down on the first bowl and 3 on the second (a spare). Score: 10 + 10 (first bowl in the 9th frame is a strike) = 20 and the cumulative score is 98 + 20 = 118.

Frame 9. A strike, all ten pins knocked down on the first bowl. Score: 10 + 7 + 3 (next two bowls in the 10th frame) = 20 and the cumulative score is 118 + 20 = 138.

Frame 10. 7 pins knocked down on the first bowl and 3 on the second (a spare). Score: 10 + 10 (the bonus bowl in the 10th frame is a strike) = 20 and the cumulative score is 138 + 20 = 158. (Note: if a spare, two bowls, is *not* made in the 10th frame, no bonus bowl is allowed. If a strike is made on the first bowl of the 10th frame, the bowler is awarded two extra rolls.)

1. Write down the score for each of the ten frames for Ben (Figure 5) and calculate his final total.

Name	1	2	3	4	5	6	7	8	9	10	Total
Ben	8 /	9 /	6 /	8 1	X	7 /	X	9 0	X	X X X	

Figure 5. Bowling Scorecard: Ten Frames for Ben.

Invoking complementary addition where needed, verbally add:

2. By 7s from 1 to 71
3. By 7s from 3 to 87
4. By 8s from 0 to 96
5. By 8s from 6 to 102
6. By 9s from 2 to 92
7. By 9s from 10 to 109
8. By 10s from 0 to 120
9. By 10s from 13 to 153

10. By 11s from 1 to 100
11. By 11s from 4 to 92
12. By 12s from 0 to 144
13. By 12s from 3 to 135

Using complementary addition and remembering in your head or with your fingers, rapidly add the following:

14. 4, 6, 5, 3, and 7
15. 6, 4, 8, 2, and 5
16. 10, 9, 5, 3, and 6
17. 7, 3, 10, 9, and 8
18. 14, 5, 3, 6, and 10
19. 13, 5, 6, 10, and 3
20. 12, 10, 2, 0, and 9
21. 27, 3, 10, 8, and 7
22. 36, 12, 7, 4, and 10
23. 11, 12, 10, 9, and 8

Add by repeating the three numbers:

24. Starting from 0: 2, 3, 4, 2, 3, 4, 2, 3, 4, until the sum = 63
25. Starting from 0: 3, 4, 5, 3, 4, 5, 3, 4, 5, until the sum = 84
26. Starting from 0: 2, 4, 6, 2, 4, 6, 2, 4, 6, until the sum = 96

Add alternately, the following pairs of numbers:

27. Starting from 0: 5, 6, 5, 6, 5, 6, 5, 6, until the sum = 88
28. Starting from 0: 6, 4, 6, 4, 6, 4, 6, 4, 6, 4, until the sum = 100
29. Starting from 0: 7, 5, 7, 5, 7, 5, 7, 5, 7, 5, until the sum = 120
30. Starting from 0: 8, 9, 8, 9, 8, 9, 8, 9, 8, 9, until the sum = 119

Count as indicated:

31. By 18s starting from 7 and ending with 187
32. By 13s starting from 8 and ending with 138
33. By 15s starting from 4 and ending with 154
34. By 19s starting from 16 and ending with 206
35. By 17s starting from 120 and ending with 290
36. By 23s starting from 6 and ending with 236

Project:

37. Draw four different angles, two obtuse, and two reflex. Measure the angles with a protractor as accurately as you can.
38. The easiest way to measure the obtuse and reflex angles is with a circle protractor, but you could also do it with a half-circle protractor. Explain how you measure these angles with a half-circle protractor.

Express the following amounts by precise digits and signs (Indicate zero dollars if converting cents only; e.g., 10 cents = $0.10):

39. Eight dollars and twenty cents
40. Forty-four dollars and fifteen cents
41. Ninety-eight cents
42. Nine dollars and sixty cents
43. 29 dollars and 2 cents
44. 99 cents
45. 15 cents
46. 100 dollars and 6 cents

Add the following amounts. Make sure that you line up the decimal points and check your work by casting out nines:

47. $15.04, $8.27, and $0.45

48. \$69.98, \$19.76, and \$98
49. \$9765, \$79.87, and \$1190.99
50. \$899.99, \$9239.76, and \$9.99

Answer the following questions and check your work by casting out nines:

51. Mr. Jones bought a car for \$16,789 and, two weeks later, sold it for \$598 more than its purchase cost. For how much did Mr. Jones sell his car?
52. Mrs. Atkinson paid \$58 for a purse, \$5 and 75 cents for a pen, ten dollars for a pair of gloves, and 89 cents for a soft drink. How much did she pay for all?

Try your newly learned skills with the following set of addition problems. In doing these sums, note that you are combining everything that you have learned so far:

- Some columns involve complements and recorded tens, and some do not.
- Some columns require that you carry tens back to a previous column in the final answer, and some do not.
- Some columns contain digits that you can combine at a glance and some do not.

This is the nature of addition; a variety of patterns and possibilities always exists.

Make sure that you set up the digits vertically and check your work by casting out nines:

53. 207 + 967 + 88 = _______
54. 425 + 816 + 505 + 888 + 999 + 777 = _______
55. Sum 4126, 7819, 2178, 4449, 6278, and 5698
56. Sum 2554, 3162, 5278, and 6117
57. Sum 76,998, 23,686, 56,997, 76,989, 33,898, and 52,899

> I have often been asked by certain youths in whom I have much interest, and who look forward to mercantile pursuits, to put into writing the fundamental principles of arithmetic, commonly called the abacus. Therefore, being impelled by my affection for them, and by the value of the subject, I have to the best of my small ability undertaken to satisfy them in some slight degree, to the end that their laudable desires may bear useful fruit. Therefore in the name of God I take for my subject this work in algorism, ...
>
> *Treviso*, the first printed Arithmetic text, Italy (1478).

Figure 6. Hexagonal snowflake crystals. Source: Dover Publications.

2.10 Magic Squares

Symbol Introduced
1. Subscript notation

Magic squares have appeared throughout history and in a wide variety of cultures.[1] Magic squares are named such because there are so many relationships between the sums of the numbers filling the cells in a square grid. Magic squares are like a square crossword puzzle except that each cell is filled with a different number. What is fascinating about these squares is that the sum of the entries of any row, any column, or any main diagonal is the same.

Let's consider a three by three square grid, with letters representing different numbers:

a	b	c
d	e	f
g	h	i

How do we build a magic square? Following the method developed in ancient China, we will create one form of the Lo Shu magic square. We substitute the letters in this square grid with the numbers 1 to 9 so that the following eight sums are the same; i.e., they all equal k:

Sum of the rows:
$a + b + c = k$
$d + e + f = k$
$g + h + i = k$

Sum of the columns:
$a + d + g = k$
$b + e + h = k$
$c + f + i = k$

Sum of the two diagonals:
$a + e + i = k$
$g + e + c = k$

We start by replacing b with 1:

a	1	c
d	e	f
g	h	i

We shift right one cell and then up one cell. We are outside the square, so we carry the 2 straight down to the lowest square on the bottom row; the 2 is in the i position:

		2
a	1	c
d	e	f
g	h	2

[1] In Medieval Europe, such squares were thought to protect one against the plague.

From 2 on the bottom row, we shift one cell to the right and then up one cell. We are outside the square, so we carry 3 to the left to the first column; the 3 is in the *d* position:

		2	
a	1	c	
3	e	f	3
g	h	2	

From 3, shift one cell to the right and then up one cell. Since this cell is already occupied, we write the 4 directly underneath the last number written; we substitute *g* with 4:

		2	
a	1	c	
3	e	f	3
4	h	2	

From 4, shift one cell to the right and then up one cell. We substitute *e* with 5:

		2	
a	1	c	
3	5	f	3
4	h	2	

From 5, shift one cell to the right and then up one cell. We substitute *c* with 6:

		2	
a	1	6	
3	5	f	3
4	h	2	

From 6, shift one cell to the right. We are outside the square, so we move left to the first column, the cell filled with *a*. We move up one cell. We are outside the square, so we move down to the last row. The cell is occupied by 4, and when this happens, we write the 7 directly underneath the last number written; we substitute *f* with 7:

		2	
a	1	6	
3	5	7	3
4	h	2	

From 7, shift one cell to the right. We are outside the square so we move left to the first column, the cell filled with 3. We move up one cell and substitute *a* with 8:

		2	
8	1	6	
3	5	7	3
4	h	2	

From 8, shift one cell to the right and then up one cell. We are outside the square so we move down to the last row and substitute *h* with 9:

8	1	6
3	5	7
4	9	2

What is *k*, the magic sum? Note that one diagonal is the sum of 4 + 5 + 6. We will revisit how to quickly calculate sums like this in Lesson 13.13.

In your notebook, apply the same procedure with a five by five cell grid. Each cell is labeled with **subscripted** letters, a_1 to a_{25}. You want to fill the cells with the numbers 1 to 25, starting with 1 in the middle of the first row, in the cell a_3:

a_1	a_2	1	a_4	a_5
a_6	a_7	a_8	a_9	a_{10}
a_{11}	a_{12}	a_{13}	a_{14}	a_{15}
a_{16}	a_{17}	a_{18}	a_{19}	a_{20}
a_{21}	a_{22}	a_{23}	a_{24}	a_{25}

Here is the completed work:

17	24	1	8	15
23	5	7	14	16
4	6	13	20	22
10	12	19	21	3
11	18	25	2	9

What is the magic sum? Test it with all rows, columns, and the two diagonals. Note that one diagonal is the sum of 11 + 12 + 13 + 14 + 15. (Refer, again, to Lesson 13.13.)

EXERCISES

Answer these questions:

1. Write *x* subscript 3.
2. Using the method taught in this lesson, create a seven by seven magic square with the numbers 1 to 49.

Using complementary addition, sum each row, each column, and the two diagonals of each of the following magic squares showing that the sum is the same in every case:

3. Another version of the three by three Lo-Shu magic square using only the digits from 1 to 9. (See Figure 1.)

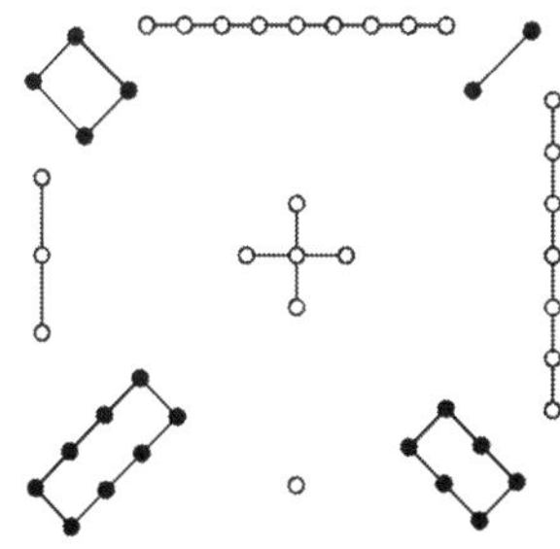

Figure 1. Ancient Chinese representation of Lo-Shu. Source: Public Domain.

4	9	2
3	5	7
8	1	6

4. A four by four magic square developed by the German painter and sculptor Albrecht Dürer (1471-1528). The two main diagonals are 16, 10, 7, 1 and 4, 6, 11, and 13.

16	3	2	13
5	10	11	8
9	6	7	12
4	15	14	1

5. An eight by eight magic square developed by American statesman Benjamin Franklin (1706-1790).

14	3	62	51	46	35	30	19
52	61	4	13	20	29	36	45
11	6	59	54	43	38	27	22
53	60	5	12	21	28	37	44
55	58	7	10	23	26	39	42
9	8	57	56	41	40	25	24
50	63	2	15	18	31	34	47
16	1	64	49	48	33	32	17

6. Double-six magic square (six rows, six columns).

28	4	3	31	35	10
36	18	21	24	11	1
7	23	12	17	22	30
8	13	26	19	16	29
5	20	15	14	25	32
27	33	34	6	2	9

Answer these questions:

7. Answer the magic triangle question in Figure 2.
8. Explain the magic about the magic hexagon arrangement in Figure 3.

Gather 49 coins of the same denomination; i.e., 49 pennies, 49 nickels, 49 dimes, etc. Place one coin in the center of a table and tile the other 48 coins around it as tightly as possible. You should create several layers of coins on the table.

9. How many coins immediately surround your initial coin?
10. Describe the packing arrangement of these tiled coins.

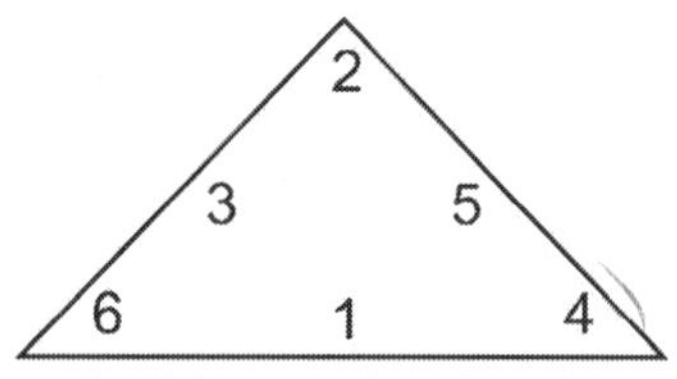

Figure 2.

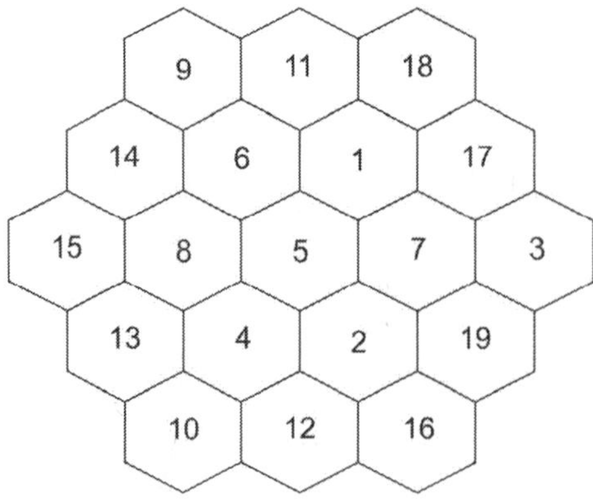

Figure 3.

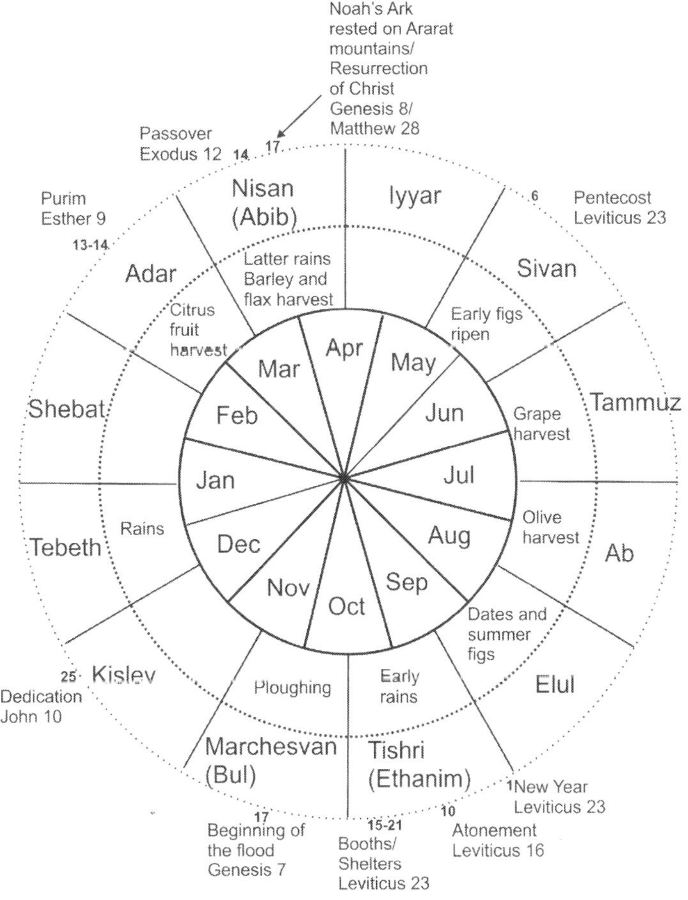

The Hebrew culture knew of the "dance before God" (aka. rejoicing) ... just look at their feast day celebrations and their weddings. The Hebrew word for rejoice means to "twirl around." The whole of creation is wired to this dance and number reports on it. Both creation and redemption speak to something incredibly wonderful and perichoresis touches an infinitesimal part of the molecular tip of the diamond of this glory.

James D. Nickel

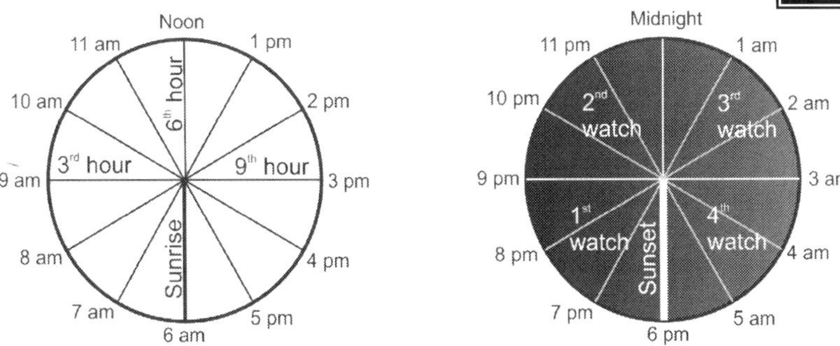

Figure 4. The Hebrew calendar and the structure of the Hebrew day and night.

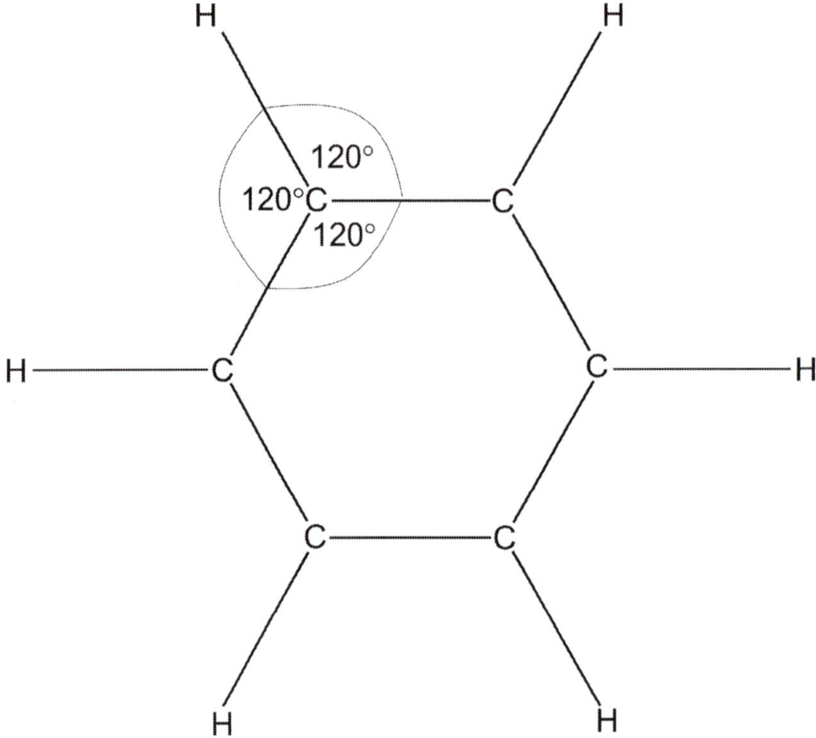

Figure 5. Skeletal formula detail of benzene, C_6H_6 (six carbon atoms united to 6 hydrogen atoms).

Figure 6. Source: United States Postal Service.

STEP 3. REVERSING THE STEPS

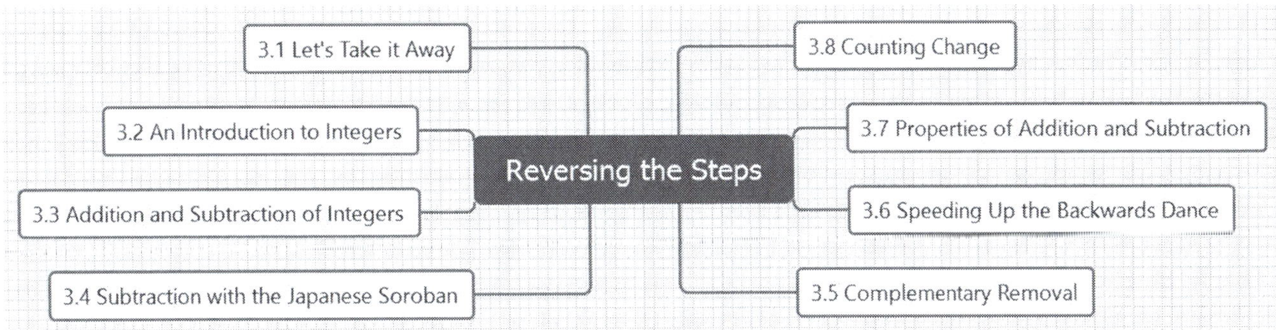

> It was the voice of God in creation that called forth the laws of nature. That is to say, all laws of nature, all its intelligible order, are to be regarded as dependent on the Word of God as their ultimate source and ground. Even physical law points beyond itself to a transcendent ground of intelligibility in the Mind or Word of God the Creator, and it is upon that ground that its constancy as well as its order reposes.
>
> Thomas F. Torrance, *Theological and Natural Science* (2002), pp. 36-37.

3.1 Let's Take it Away

Terms & Concepts Introduced
1. Equivalent expression
2. Expression
3. Numeric expression

We have already encountered some elements of subtraction when we learned how to add using the Japanese Soroban and with the addition facts for single digit sums greater than 10.

> "And if you take one from three hundred and sixty-five, what remains?" "Three hundred and sixty-four, of course." Humpty Dumpty looked doubtful. "I'd rather see that done on paper," he said.
>
> Charles Lutwidge Dodgson (Lewis Carroll), *Through the Looking Glass.*

Let's review what we did with these facts by considering 9 + 8. First, we calculated the complement of 9 which is 1 (10 – 9 = 1). Second, we subtracted 1 from 8 (8 – 1 = 7). Finally, we added 10 to calculate the sum: 10 + 7 = 17.

Using complements, we changed 9 + 8 to 10 + 7, or:

$$9 + 8 = 10 + 7$$

The left side of this equation, 9 + 8, has two terms. We added 1 to the first term and subtracted 1 from the second term to generate 10 + 7. The left side, 9 + 8, is an **expression** that is **equivalent** to the expression on the right side, 10 + 7.[1] Why? Subtracting 1 and adding 1 nets zero. The equation remains balanced, and 9 + 8 = 10 + 7 is a true statement.

Figure 1. 7 – 4 = 3

Three Subtraction Concepts

There are three concepts that explain the meaning of the operation of subtraction:

1. Removal.

   To remove means to "to take away, get rid of, eliminate, diminish, rebate, extract, do away with." "Remove" is the antonym of "add." In dynamic subtraction, if we remove part of a group, then the operation of subtraction tells you how much is left. Just like dynamic addition, dynamic subtraction involves a change in a situation. If I have seven blocks and I take away four blocks, then three blocks remain (Figure 1).
2. Whole-part.

   Let's say that in a classroom of 87 students, 47 are girls. How many students are boys? The answer is 87 – 47 = 40. In whole-part problems, we know the amount of two objects: the total of all objects and the number of objects of one of two types. Based upon the whole and the part, we are asked to find the number of objects of the other type.
3. Comparison.

   If Claire has $250 in her checking account and her brother Peter has $200, how much more money does Claire have than Peter? (Note: the phrase "how much more" is one of comparison or contrast.) The answer is $250 – $200 = $50. If I have a stack of seven blocks and a stack of three blocks, then the stack of seven blocks, by comparison, is four more than the stack of three blocks.

As with addition, we can only subtract units of the same order. With place value, you can subtract only ones from ones, tens from tens, hundreds from hundreds, etc. Also, every subtraction problem can be

[1] An expression in arithmetic is a symbol or a combination of symbols that, when taken together by invoking arithmetic operations, represents a value. For example, 2 + 5 is an expression that represents the value of 7. When the symbols used are numerals, the expression is **numeric**. An equivalent expression is an arithmetic expression that is equal to another arithmetic expression.

checked by addition. To check this subtraction problem 7 – 4 = 3 (minuend – subtrahend = difference), we add 3 + 4 = 7 (difference plus subtrahend = minuend).

We, therefore, understand subtraction as finding the missing addend. In other words, what number must be added to 4 to get 7? The answer: 3.

Because subtraction is finding the missing addend, this operation, as we have already noted, is the inverse of addition. Since we can understand addition as counting forward, then subtraction is counting backward. Because of the nature of this inverse operation (subtraction is reversing the steps of addition), every addition problem can be rewritten as two distinct subtraction problems. Given the addition problem 3 + 4 = 7, we can write:

7 – 4 = 3
7 – 3 = 4

Table 1: Subtraction

Minuend	Subtrahend								
–	**1**	**2**	**3**	**4**	**5**	**6**	**7**	**8**	**9**
1	0								
2	1	0							
3	2	1	0						
4	3	2	1	0					
5	4	3	2	1	0				
6	5	4	3	2	1	0			
7	6	5	4	3	2	1	0		
8	7	6	5	4	3	2	1	0	
9	8	7	6	5	4	3	2	1	0
10	9	8	7	6	5	4	3	2	1
11		9	8	7	6	5	4	3	2
12			9	8	7	6	5	4	3
13				9	8	7	6	5	4
14					9	8	7	6	5
15						9	8	7	6
16							9	8	7
17								9	8
18									9

BUILDING BLOCKS OF SUBTRACTION

We can construct a subtraction table that is analogous to the addition tables that we have already seen (Lesson 2.6). If you remember the 45 addition facts, mastery of subtraction facts is as easy as "adding 1 to 2." The vertical numbers in bold represent the minuends, and the horizontal numbers in bold the subtrahends. The numbers within the table are the differences (remainders or balances). Note the patterned structure, the dance, of the table.

EXERCISES

Define the following terms:

1. Expression
2. Numeric expression
3. Equivalent expression
4. Write a short essay: Explain the three meanings of subtraction and provide a practical example of each.
5. Memorize the 81 subtraction facts shown in the Subtraction Table 1 (not including those with zero balance–which should be obvious). Use 3″ by 5″index cards to help if necessary.

3.2 An Introduction to Integers

Table 1: Subtraction and Integers

		Subtrahend																	
	–	**1**	**2**	**3**	**4**	**5**	**6**	**7**	**8**	**9**	**10**	**11**	**12**	**13**	**14**	**15**	**16**	**17**	**18**
Minuend	**1**	0																	
	2	1	0																
	3	2	1	0															
	4	3	2	1	0														
	5	4	3	2	1	0													
	6	5	4	3	2	1	0												
	7	6	5	4	3	2	1	0											
	8	7	6	5	4	3	2	1	0										
	9	8	7	6	5	4	3	2	1	0									
	10	9	8	7	6	5	4	3	2	1	0								
	11	10	9	8	7	6	5	4	3	2	1	0							
	12	11	10	9	8	7	6	5	4	3	2	1	0						
	13	12	11	10	9	8	7	6	5	4	3	2	1	0					
	14	13	12	11	10	9	8	7	6	5	4	3	2	1	0				
	15	14	13	12	11	10	9	8	7	6	5	4	3	2	1	0			
	16	15	14	13	12	11	10	9	8	7	6	5	4	3	2	1	0		
	17	16	15	14	13	12	11	10	9	8	7	6	5	4	3	2	1	0	
	18	17	16	15	14	13	12	11	10	9	8	7	6	5	4	3	2	1	0

We concluded the last lesson with a subtraction table. In the above table, note the addition of columns and the inclusion of the correct differences in the lower portion.

What about the answers in the blanks? What are the differences in these cases?

Note first, for the differences written, the minuend is greater than or equal to the subtrahend. For example, in the problem $15 - 8 = 7$, $15 > 8$.

To find the differences in the empty cells in Table 1 the minuend is less than the subtrahend. For example, in the problem $8 - 17$, $8 < 17$. Since $17 - 8 = 9$, we know that 9 must play a role in the difference $8 - 17$.

Terms, Symbols & Concepts Introduced
1. $\mathbb{Z}$
2. $-\mathbb{Z}$
3. $+\mathbb{Z}$
4. Absolute value
5. BC
6. BCE
7. CE
8. Celsius
9. Equidistant
10. Fahrenheit
11. Great circle
12. Integers
13. Latitude
14. Longitude
15. Magnitude
16. Number line
17. Opposite
18. Origin on the number line
19. Prime meridian
20. Scalar
21. Set builder notation
22. Vector
23. W
24. Whole numbers

God "arranged everything according to measure, number, and weight" (Wisdom 11:20) and ... He made man in His image, endowing him thereby with a mental ability to see precisely this arrangement.

Stanley L. Jaki, *Numbers Decide and Other Essays* (2003), p. 39.

DIRECTION NUMBERS

> When the minuend is less than the subtrahend, the operation of subtraction requires us to contemplate negative numbers.

We must employ the concept of direction to find these differences. The direction of all the differences listed in the table are positive. The direction of all the differences not listed in the table are negative. Positive and negative indicate direction.

In 200 BC, the Chinese used negative numbers in accounting procedures. They used red sticks on a counting board to represent a debit and black sticks to represent a credit. The Hindus of India used negative numbers in calculations in documents dated from 200 BC to 200 AD. In the 3rd century AD, the Greek mathematician Diophantus described these negative numbers as "absurd." In the 7th century AD, the Indian astronomer Brahmagupta (598-ca. 665) used negative numbers to represent financial debt. The positive (+) and negative signs (-) were first employed in 15th century German warehouses to show when a container of goods was over or under a standard weight. Mathematicians accepted the validity of negative numbers with the publication of the Italian Girolamo Cardano's (1501-1576) *Ars Magna* in 1545.

Figure 1. Thermometer. Source: iStockPhoto.

Thermometers, devices that measure of heat or temperature, use direction numbers. If you live in the United States, this device will be graduated in degrees **Fahrenheit**[1] and some thermometers may also show degrees **Celsius**.[2] Most countries do not use degrees Fahrenheit to measure temperature. Their scales are graduated in degrees Celsius only.

On a Fahrenheit scale, a temperature of 32° represents the molecular condition at which water freezes.[3] If it rains at this temperature, the rainwater will turn into snowflakes. On the Celsius scale, the temperature of 0° represents the freezing point of water and a temperature of 100° represents the boiling point of water. Both scales are direction scales. We refer to temperatures above 0° as "25 degrees" or "54 degrees." When the temperature goes below 0°, we may say "negative 15 degrees" or "negative 22 degrees." Note the use of convention: We do not say that temperatures above 0° are "positive 25 degrees" or "positive 54 degrees." The positive is assumed if we do not use or say it. If the temperature is below 0°, we mention the negative; e.g., we say, "Negative 15 degrees" or we write "-15°."

We measure history with direction numbers. Our dating system is centered in the Incarnation, specifically, the birth, of the Lord Jesus Christ. **BC** means "before Christ" and AD is the abbreviation of the Latin phrase *Anno Domini*, meaning "in the year of our Lord" (Lesson 1.1). In citing positive numbers, we usually do not use the plus sign; likewise, we usually do not use AD when saying or writing dates after the birth of Christ. When we say someone was born in 1917, it is assumed that we mean 1917 AD. Dates before the birth of Christ have BC as a suffix. For example, we know that the Greek mathematician and geometer Euclid lived around the year 300 BC.

---

[1] The German-born physicist Gabriel Daniel Fahrenheit (1686-1736) devised the Fahrenheit degree scale.

[2] This temperature scale is named after the Swedish astronomer Anders Celsius (1701-1744). He set the standard for the measurement of water freezing and boiling but his original scale showed water freezing at 100° and boiling at 0°. Why? To avoid negative temperatures. This scale is also known as centigrade because the freezing and boiling point of water is 0°C and 100°C respectively.

[3] Water boils at 212° on the Fahrenheit scale.

There is no zero in the accounting of time based on BC and AD (Lesson 1.4). Zero marks a point, not a number (Table 2).

Table 2												
Temperature	-5°	-4°	-3°	-2°	-1°	0°	1°	2°	3°	...	99°	100°
Dating	5 BC	4 BC	3 BC	2 BC	1 BC	1 AD	2 AD	3 AD	4 AD	...	100 AD	101 AD

As we can see, if the temperature is -5° in the morning and it rises 8° by the afternoon, then, counting from left to right, it will be 3°. But, if a girl is born on 1 January 5 BC, on 1 January 4 AD, counting from left to right, she will be 8 years old. In our dating system, we start a new millennium AD on the first day of January; e.g., the first millennium started on 1 January 1, the second millennium on 1 January 1001, and the third millennium on 1 January 2001.[4]

In recent years, some people began to use **BCE** and **CE** as date directions. BCE means "before the common era" and CE means "common era." The word common era has replaced the "Christ" (in BC) and "year of our Lord" (in AD). This subtle change in the way we date events reflects a telling worldview change. Some no longer see history in the context of the Incarnation of Christ, the Word made flesh. We would expect this change from people who govern their lives in unbelief concerning the significance and relevance of the Incarnation of Christ, but their unbelief cannot negate the reality of the Incarnation!

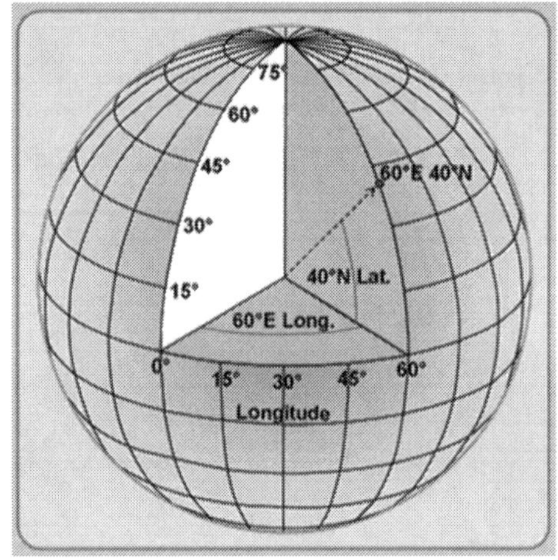

Figure 2. Latitude and longitude. Source: Wikimedia Commons.

In mapping the sphere of the Earth, we use the lines of **longitude**[5] and **latitude** (Figure 2).[6] Longitude lines are curved lines on the sphere that pass through both the North Pole and the South Pole. These curves, actually circles whose

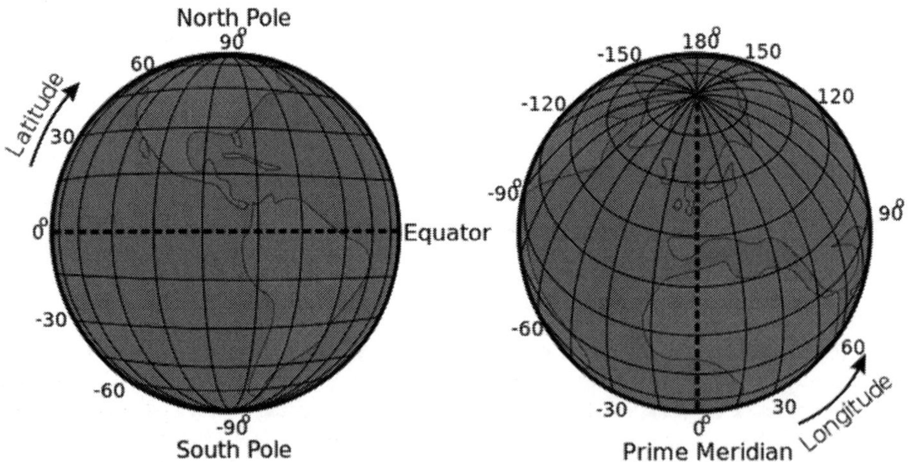

Figure 3. Latitude and longitude (-30° latitude means 30°S and -30° longitude means 30°E). Source: Public Domain.

[4] Century naming is the same. The 2nd century began on 1 January 101, the 19th century began on 1 January 1801, the 20th on 1 January 1901, and the 21st on 1 January 2001.

[5] *Longitude* is Latin meaning "length."

[6] *Latitude* is Latin meaning "broad or width."

center is the center of the Earth, are **great circles**.[7] The 0° longitude line is centered on Greenwich, England, the **prime meridian**.[8] If we move east of Greenwich on our global map, we will see the longitude lines marked; for example, "30°E". Moving west from the prime meridian, we will see longitude lines marked; for example, "30°W" (or -30°). In this case, the letters E and W are direction indicators for East and West.

0° latitude is the Earth's equator.[9] Latitude is the angular distance of a place north or south of the Earth's equator. You can see latitude lines marked north or south of this line; for example, "30°N" or "30°S" (or -30°). In this case, N and S are direction indicators for North and South.

In the 7th century, bookkeepers in India used positive numbers to keep track of assets and negative numbers to account for liabilities. Today, accountants use two columns to represent what they call credits and debits, representing account directions. If I deposit money in a bank, I am crediting my account or adding money to it. If I write a check, I withdraw money from my account. I am debiting my account or subtracting money from it. The colors black and red are used to indicate credit and debit, respectively. If my bank account has a positive or credit balance, I am said to be in the black. If my bank account has a negative or debit balance, I am said to be in the red.

Direction numbers are also used to refer to yards gained or lost in football. They are used to represent measurements up or down in Physics, and prices up or down in the stock market.

As you can see, the same mathematical concept, direction numbers, may represent many aspects of creation and our life. Directions numbers are all around us!

In the cases we have mentioned, we could use the positive (+) and negative (-) signs. For example: credits +, debits -; north +, south -; east +, west -; BC -, AD +; above freezing +, below freezing -.

We can think of negative numbers as the result of subtraction where the minuend is lesser than the subtrahend.

It is commonly agreed to leave the positive sign off numbers. That is, 7 means +7 and we say, "positive 7." We usually omit "positive" and just say, "7." When the negative sign is in front of a number, e.g., -9, we say "negative 9" or "opposite 9." Direction numbers are common in mathematics and science. Scientists and mathematicians call direction numbers **vectors**.[10]

Earlier, we defined the set of natural numbers, 1, 2, 3, 4, 5, …, and the symbol ℕ that represents this set. Mathematicians use **set builder notation** to either describe or list members of a set, enclosed within braces {} and separated with commas. Using this notation:

$$\mathbb{N} = \{1, 2, 3, 4, 5, \ldots\}$$

If we add zero to ℕ, we generate the set of **whole numbers**, symbolized by W. Using set builder notation:

---

[7] A great circle is the shortest path than can be taken between two points on a sphere. It is equivalent to a line on a plane surface. The geometry of a sphere is a different kind of geometry when compared to the geometry Euclid developed.

[8] Meridian is derived from a Latin word meaning "noon" or "midday." A meridian is defined as a great circle of the Earth that passes through the North and South poles and any given point on the Earth's surface. *Prime* is Latin for chief or primary. Meridian is synonymous with longitude.

[9] The equator is a great circle since its center is the center of the Earth.

[10] A vector is a quantity that has both magnitude and direction. In Latin, vector means "carrier" and it is appropriate because it means a quantity that is being carried somewhere, in some direction. A ***scalar***, taken from a Middle English word meaning "the scale of a balance," is a quantity that has only magnitude. "60 miles per hour" is a scalar quantity. If we state that a car is traveling 60 miles per hour northeast, we are using vectors. Vector imaging is used in graphics programs like CorelDraw to create and modify diagrams and pictures.

W = {0, 1, 2, 3, 4, 5, …}[11]

THE NUMBER LINE

The set of whole numbers can be represented geometrically on a line, the **number line**, a mathematical object similar to a ruler or measuring stick.[12]

Zero is the **origin** on the number line. We mark the other natural numbers at equal intervals (Figure 4):

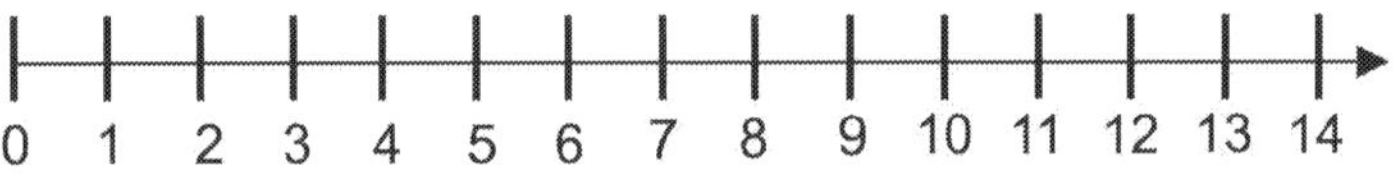

Figure 4. Whole numbers.

The arrow pointing to the right indicates the *ad infinitum* nature of the set of natural numbers; i.e., there is no largest natural number. The only difference between this representation and the temperature scale is the missing negative numbers. To include these negative numbers, we can turn the temperature scale on its side (Figure 5):

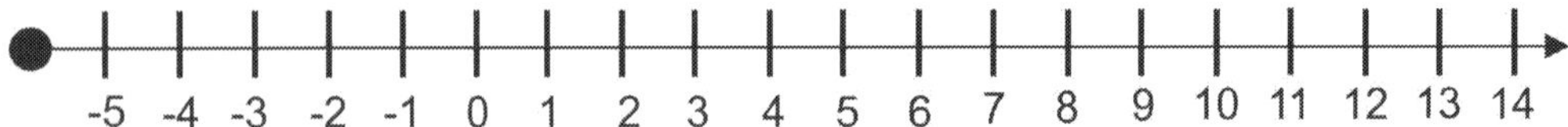

Figure 5. Temperature scale.

If we remove the bulb on the left, a repository for mercury, we have a complete number line (Figure 6):

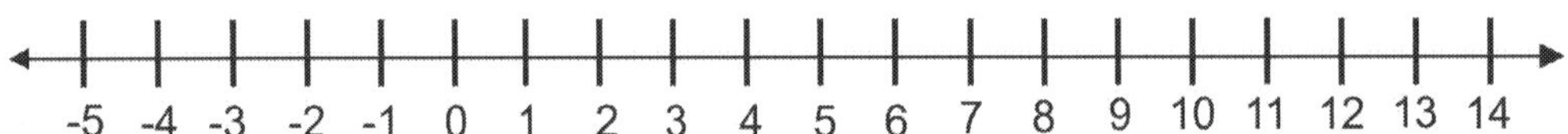

Figure 6. Integers.

On the number line, we have an arrow pointing both to the left and to the right. This double arrow indicates the *ad infinitum* nature of a new set of numbers, the **integers**. As we discovered in Lesson 2.2, another word for whole is integer.[13] In mathematics, the set of integers is defined, using set builder notation, as follows:

$$\mathbb{Z} = \{\ldots, -5, -4, -3, -2, -1, 0, 1, 2, 3, 4, 5, \ldots\}$$

[11] The symbol $\mathbb{N}_0$ is sometimes used to represent whole numbers. The W symbol is not universally recognized by mathematicians.

[12] The French mathematician René Descartes (1596-1650) was the first to introduce the idea that numbers could be represented by points on a line.

[13] *Integral*, in Latin, means "whole" or "undivided."

THE INTEGERS

ℤ stands for the German word *Zahl,* meaning "number." As the number line shows, we represent the negative integers (-1, -2, -3, etc.) to the left side of zero.

The arrow pointing to the left means that the negative integers keep getting larger and larger in the negative direction as we move left just as the right arrow indicates that numbers keep getting larger and larger in the positive direction as we move right.

Figure 7. Integers.

Therefore, the set of integers consists of three parts:[14]

1. Positive integers, the set of counting or natural numbers, or, in symbols, +ℤ or ℕ
2. Zero
3. Negative integers or, in symbols, -ℤ

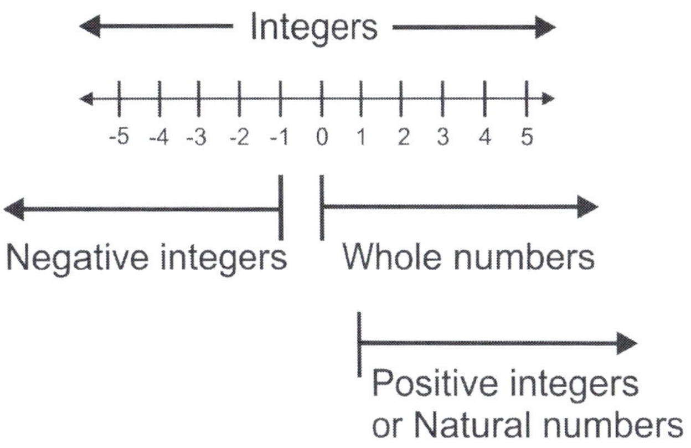

Note also that every natural number can be written as in integer if you include the positive sign. For example, 3 is the symbolic way to write a natural number and +3 is the symbolic way to write the same number as an integer. Note the two symbols, + and 3. The symbol 3 indicates **magnitude**[15] and the symbol + indicates direction. Therefore, +3 represents a vector. To write the symbol 3 should tell you that 3 is both a natural number and an integer since the natural numbers are contained in the integers.[16]

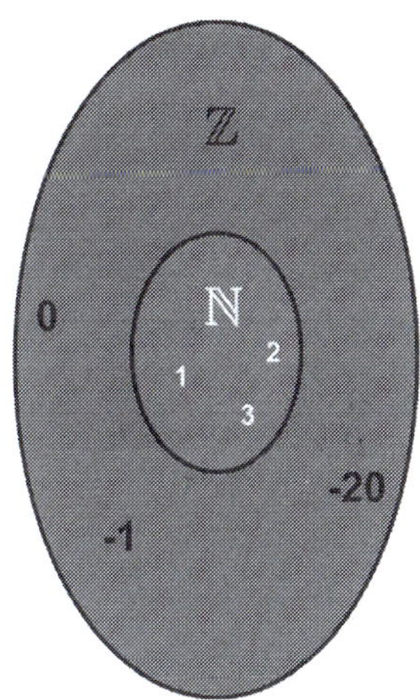

Figure 8. ℕ is contained in ℤ.

The number line is a tool to help you understand number systems. As we have noted, the origin of the number line is 0. The **opposite** of a number is that number on the other side of 0 that is equally far, or equidistant, from 0 as the original number. The opposite, or negative, of 6 is -6. The opposite, or negative, of -3 is 3. We write:

$-(-3) = 3$

We say -3 as "negative 3" or "opposite 3," not "minus 3," since "minus" is the word used for the operation of subtraction.

COMPARISONS

With the aid of a number line, we can easily compare integers. If an integer a lies to the right of another integer b on the number line, then $a > b$. If an integer c lies to the left of another integer d on the number line, then $c < d$. For example, $-4 > -8$ since -4 lies to the right of -8 on the number line, and $4 < 8$ since 4 lies to the left of 8 on the number line. Think temperatures in degrees and these comparisons make sense:

---

[14] Note: You might see the symbol $\mathbb{Z}^+$ used instead of +ℤ , or $\mathbb{Z}^-$ instead of -ℤ.

[15] Magnitude, in Latin, means "great." It represents size or quantity.

[16] The mathematical meaning of containment will be discussed later (Lesson 9.5). For now, note that these two sets of numbers, the natural numbers and the integers, because ℕ is contained in ℤ, or ℕ dwells in ℤ, reflects the perichoretic nature of these two number systems.

-4° > -8°
4° < 8°

ABSOLUTE VALUE

With the set of integers comes a new concept, **absolute value**. The absolute value of a number is the magnitude, the size, of the number, in effect, its distance from 0 on the number line. Just as the distance between two objects can never be negative, the absolute value of a number can only be positive, or zero. For example, -5 is five units from 0 on the number line. The absolute value of -5 is 5. Next, note that 18 is 18 units from 0 on the number line. The absolute value of 18 is 18. Finally, 0 is 0 units from 0 on the number line. The absolute value of 0 is 0.

How far apart is 14°F from -5°F (Figure 9)? Or, what is the difference, or distance, between 14°F and -5°F? The distance between the two is 19°F. Why? Starting from 14°F, we count down until we get to -5°F. To get to the point marked 0°F, the distance traveled is 14°F. To get from 0°F to -5°F, the distance traveled is 5°F. Therefore, the distance between 14°F and -5°F is 14°F + 5°F = 19°F.

Absolute value, always positive or zero, indicates distance or displacement, a scalar quantity, not direction and distance, a vector quantity. The symbol for absolute value is $|x|$ where the letter x represents a number that can be the result of an arithmetical operation or just a number (positive, zero, or negative).[17] The absolute value of any arithmetical operation or any integer is always positive or zero. Whatever number we consider, we ask ourselves, "How far is that number from the origin, 0, on the number line?" Our answer to this question is the absolute value of that number.

Think of absolute value as a rule, a procedure that acts upon an input.[18] We have three cases: (1) If the input is positive, the method returns the positive number as the output. (2) If the input is zero, the method returns zero as the output. (3) If the input is negative, the method returns the opposite of the negative number as the output. In symbols, we write:

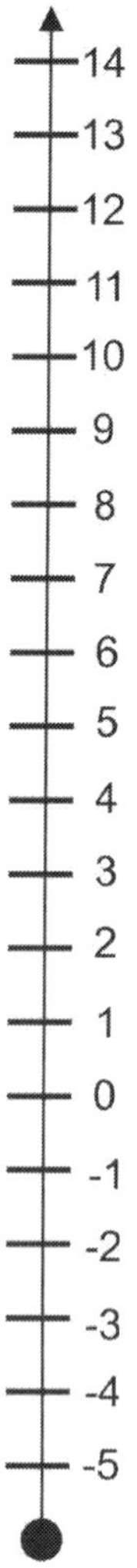

Case 1. $|x| = x$ if $x > 0$
Case 2. $|x| = 0$ if $x = 0$
Case 3. $|x| = -x$ if $x < 0$

Figure 9.

Here are some examples:

Example 1. $|-5| = 5$ (Note: We say "the absolute value of negative five equals 5.")
We have Case 3. First, we substitute x with -5. Next, we apply the absolute value method to -5 to get the output, 5. In symbols, we write:
$-(-5) = 5$

Example 2. $|5| = 5$
We have Case 1. First, we substitute x with 5. Next, we apply the absolute value method to 5 to get the output, 5.

Example 3. $|-17| = 17$ We have Case 3.
Example 4. $|8 - 4| = |4| = 4$ (Note: We say "the absolute value of eight minus four is four.")

---

[17] Using the letter x is the same as using $\mathbb{Z}$, $\mathbb{N}$, or W in the sense that these letters represent an idea, a mathematical object. The key to Algebra is the use of letters as representations. When we see the symbols "$|x|$" we say, "the absolute value of x."
[18] In mathematics, a rule is a function. See Lesson 10.2.

We first subtract 4 from 8 to get 4. We have Case 1. First, we substitute x with 5. Next, we apply the absolute value method to 5 to get the output, 5.

Example 5. $|12 - 12| = |0| = 0$
After we subtract 12 from 12, we have Case 2. First, we substitute x with 0. Next, we apply the absolute value method to 0 to get the output, 0.

We can state two properties from the definition of absolute value:

Property 1. $|x| = 0$ or $|x| > 0$ (Either one or the other case is always true.)
Property 2. $|-x| = |x|$
Examples:
When x is positive: $|-3| = |3|$ since $|-3| = 3$ and $|3| = 3$
When x is negative: $|-(-3)| = |-3|$ since $|-(-3)| = |3| = 3$ and $|-3| = 3$

Table 3: Subtraction and Integers

Minuend	Subrahend																	
–	1	2	3	4	5	6	7	8	9	10	11	12	13	14	15	16	17	18
1	0																	
2	1	0																
3	2	1	0															
4	3	2	1	0														
5	4	3	2	1	0													
6	5	4	3	2	1	0												
7	6	5	4	3	2	1	0											
8	7	6	5	4	3	2	1	0										
9	8	7	6	5	4	3	2	1	0									
10	9	8	7	6	5	4	3	2	1	0								
11	10	9	8	7	6	5	4	3	2	1	0							
12	11	10	9	8	7	6	5	4	3	2	1	0						
13	12	11	10	9	8	7	6	5	4	3	2	1	0					
14	13	12	11	10	9	8	7	6	5	4	3	2	1	0				
15	14	13	12	11	10	9	8	7	6	5	4	3	2	1	0			
16	15	14	13	12	11	10	9	8	7	6	5	4	3	2	1	0		
17	16	15	14	13	12	11	10	9	8	7	6	5	4	3	2	1	0	
18	17	16	15	14	13	12	11	10	9	8	7	6	5	4	3	2	1	0

EXERCISES

Define the following terms/phrases:

1. Fahrenheit
2. Celsius
3. BC
4. CE
5. BCE
6. Latitude
7. Longitude
8. Great circle
9. Prime Meridian
10. Magnitude
11. Vector
12. Scalar
13. Number line
14. Origin on the number line
15. Set builder notation
16. W
17. Whole numbers
18. $\mathbb{Z}$
19. Integers
20. $+\mathbb{Z}$
21. $-\mathbb{Z}$
22. Opposite
23. Equidistant
24. Absolute value

25. Using pattern recognition and what you know of the number line, fill in the blanks of Table 3.

Which value of x makes the following equations true?

26. $|-928| = x$
27. $|76| = x$
28. $|-32| = x$
29. $|125 + 89| = x$
30. $|45 - 10| = x$
31. $|0 - 45| = x$
32. $|1 - 45| = x$
33. $|4 - 8| = x$

> ... the Greeks considered geometry the only acceptable form of mathematics, and since distance cannot be negative, they had no use for negative numbers.
> Jan Gullberg, *Mathematics: From the Birth of Numbers* (1997), p. 72.

True or False:

34. Every whole number is a natural number.
35. Every natural number is an integer.
36. 0 is a natural number.
37. Every integer is whole number.
38. 0 is an integer.

Which number is greater than the other?

39. -8 or -5?
40. -13 or -21?
41. 21 or -13?
42. 6 or -6?

Which number is less than the other?

43. -18 or -19?
44. -5 or 5?
45. 22 or -26?
46. -22 or -21?

47. Answer the two questions in the text box.

> God made the integers; all the rest is the work of man.
> Leopold Kronecker (1823-1891).
>
> Answer the following questions:
>
> 1. Did the Triune God make the integers, the set $\mathbb{Z}$, or did the onto-relational God establish a rational order of creation, a language fabric of the universe, that integers, the creation of man, recognize and report?
> 2. Did God create 1, 2, 3, etc., or did man invent symbols to represent the concept of oneness, twoness, threeness, etc., concepts inherent in the created order?

3.3 Addition and Subtraction of Integers

The earliest written evidence of the use of negative numbers in European mathematics is *Ars Magna*, The Great Art. It was written by the Italian mathematician Girolamo Cardano (1501-1576) and published in 1545. Even into the 17th century, French mathematicians such as René Descartes (1596-1650) and Blaise Pascal (1623-1662) were not convinced that negative numbers were real. The German philosopher and mathematician Gottfried Wilhelm Leibniz (1646-1716) thought negative numbers would lead to absurd conclusions, but he recognized their pragmatic use in calculations. It was only in the 18th century that mathematicians finally accepted negative numbers as valid members of what we now know as the set of integers.

Terms, Symbols & Concepts Introduced
1. $\Rightarrow$
2. $\Leftrightarrow$
3. Additive inverse
4. Algebraic Definition of Subtraction
5. Converse
6. Logical equivalence
7. Parallelogram
8. Rhombus

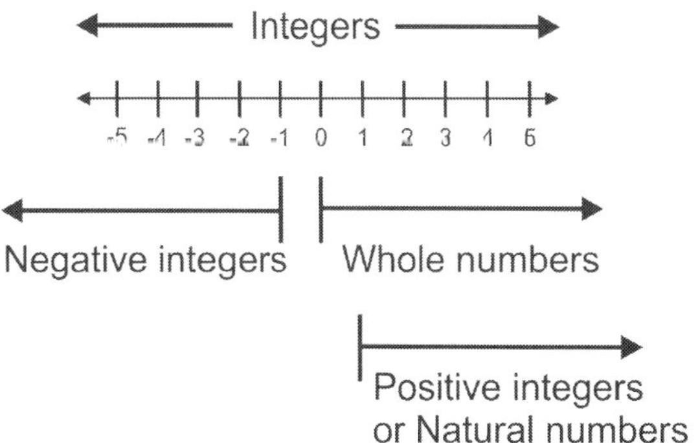

Figure 1. Integers.

How do we add and subtract integers? To start our investigation, we first use the number line to answer these questions:

- Which integer, if any, is halfway between 8 and 12?
- Which integer, if any, is halfway between -4 and 2?
- Which integer, if any, is halfway between -2 and 0?
- Which integer, if any, is halfway beween 1 and 2?
- Which integer, if any, is halfway between -1 and 1?
- Which integer, if any, is halfway between -2 and 2?
- Which integer, if any, is halfway beween -5 and 5?

> Arithmetic studies what particular numbers do, algebra what any numbers do.
>
> Walter Warwick Sawyer (1911-2008), *Vision in Elementary Mathematics* (1964), p. 31.

As we have noted, when zero is halfway between two integers on a number line, those two integers are opposites. -1 and 1 are opposites since both are 1 unit in distance from the origin. -2 and 2 are opposites since both are two units distance from the origin. In general, using a letter to replace any specific number (i.e., algebraic notation), $-a$ and a are opposites since both are a units distance from the origin.

With the arithmetic operation of addition, the **additive inverse** of a given number is that number that, when added to the given number, gives the sum of 0. The sum of opposites is always 0. For example, 1 + (-1) = 0. (Note: We say, "one plus negative one equals zero.") Note the use of parentheses. We could write 1 + (-1) = 0 as 1 + -1 = 0. We surround -1 with parentheses to separate the addition sign from the negative sign. Remember, when we write -1, the symbol in front of 1 does not mean the operation of subtraction; it means negative or opposite.[1] The additive inverse of 1 is -1 and the additive inverse of -1 is 1. In general, the additive inverse of a is $-a$ and the additive inverse of $-a$ is a.

[1] In typesetting, we note this difference by the length of the dash. For subtraction, the dash is longer; e.g., 8 – 7 = 1. To indicate negative direction, the dash is shorter; e.g., -1.

Note:

Table 1		
Sum	**Observation**	**Conclusion**
$1 + (-1) = 0$	$1 + (-1) = 0 \Leftrightarrow 1 - 1 = 0$	$1 + (-1) = 1 - 1$

The symbol in column two, $\Leftrightarrow$, stands for **logical equivalence** and we will use it many times in this text.[2] When we write $1 + (-1) = 0 \Leftrightarrow 1 - 1 = 0$, the symbol $\Leftrightarrow$ represents a double movement where $\Rightarrow$ means "implies." Here is the dance of this double movement:

First movement: $1 + (-1) = 0 \Rightarrow 1 - 1 = 0$

Second movement (**converse**[3] of the first movement): $1 - 1 = 0 \Rightarrow 1 + (-1) = 0$

We use the symbol $\Rightarrow$ as a shorthand way to represent "if-then" statements. "$1 + (-1) = 0 \Rightarrow 1 - 1 = 0$" is a shorthand way of stating "if $1 + (-1) = 0$, then $1 - 1 = 0$." "$1 - 1 = 0 \Rightarrow 1 + (-1) = 0$" is a shorthand way of stating "if $1 - 1 = 0$, then $1 + (-1) = 0$." We will use equivalence statements throughout this text. Let's continue the table we started:

Table 2		
Sum	**Observation**	**Conclusion**
$2 + (-2) = 0$	$2 + (-2) = 0 \Leftrightarrow 2 - 2 = 0$	$2 + (-2) = 2 - 2$
$3 + (-3) = 0$	$3 + (-3) = 0 \Leftrightarrow 3 - 3 = 0$	$3 + (-3) = 3 - 3$
$4 + (-4) = 0$	$4 + (-4) = 0 \Leftrightarrow 4 - 4 = 0$	$4 + (-4) = 4 - 4$
$5 + (-5) = 0$	$5 + (-5) = 0 \Leftrightarrow 5 - 5 = 0$	$5 + (-5) = 5 - 5$
$a + (-a) = 0$	$a + (-a) = 0 \Leftrightarrow a - a = 0$	$a + (-a) = a - a$

The last row is a general statement where the letter a represents any integer. What does "$a + (-a) = 0 \Leftrightarrow a - a = 0$" mean in words? The sum of any integer and its opposite is zero. Or, when the minuend is equal to the subtrahend, the difference is zero. Notice that from this equivalence, we can change a subtraction problem into an addition problem, or, as noted, $a - a = a + (-a)$. Hold this observation for later amplification.

If we wish, we can prefix 1, 2, 3, 4, 5, etc. with the + sign to indicate direction, but it is not necessary. By our definition of additive inverse, -6 is the additive inverse of +6, or more simply, 6; and 6 is the additive inverse of -6.

We will make use of additive inverses when we develop rules for addition and subtraction of integers. Since integers are vectors, i.e., they have magnitude and direction, we must consider signs. With addition, there are four possibilities:

Possibility 1. We can add a positive integer to a positive integer. (We already know how to do that.)

The next three possibilities are new ideas:

Possibility 2. We can add a negative integer to a negative integer.

Possibility 3. We can add a negative integer to a positive integer where the magnitude, or absolute value, of the negative integer is greater than the magnitude of the positive integer.

Possibility 4. We can add a negative integer to a positive integer where the magnitude, or absolute value, of the negative integer is less than the magnitude of the positive integer.

---

[2] In Step 13, we shall investigate logical equivalence as it relates to different types of mathematical statements.

[3] In general, the converse is a statement that is the reverse of another statement. Converse, from the Latin *convertere*, means "to turn around." We will explore a more precise definition of converse in Step 13. In short, logical equivalence means that a statement and its converse are both true.

We are going to use the number line to demonstrate the solution methods. We start with the first addend, positive or negative, and locate it on the number line. If the second addend is positive, we move right that many units. If the second addend is negative, we move left that many units.

EXAMPLE OF FIRST POSSIBILITY: 2 + 3 = ?

Find 2 on the number line and move right 3 units.

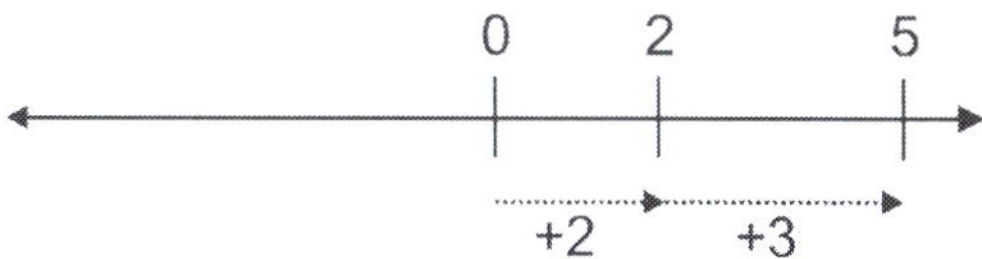

Figure 2.

Therefore: 2 + 3 = 5

EXAMPLE OF SECOND POSSIBILITY: (-2) + (-3) = ?

(Note: We say "negative two plus negative three.") Find -2 on the number line and move left 3 units.

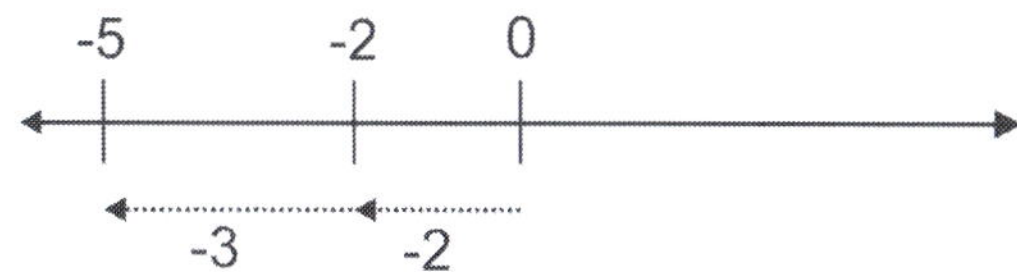

Figure 3.

Therefore: (-2) + (-3) = -5

EXAMPLE OF THIRD POSSIBILITY: (-8) + 3 = ?

(Note: We say "negative eight plus three.") Find -8 on the number line and move right 3 units.

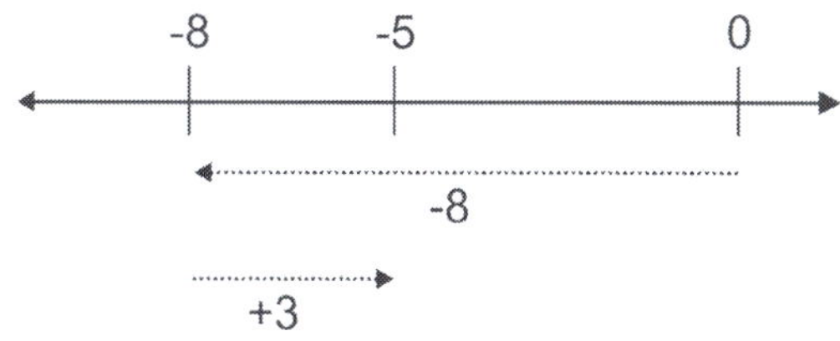

Figure 4.

Therefore: (-8) + 3 = -5

EXAMPLE OF FOURTH POSSIBILITY: (-2) + 5 = ?

(Note: We say "negative two plus five.") Find -2 on the number line and move right 5 units.

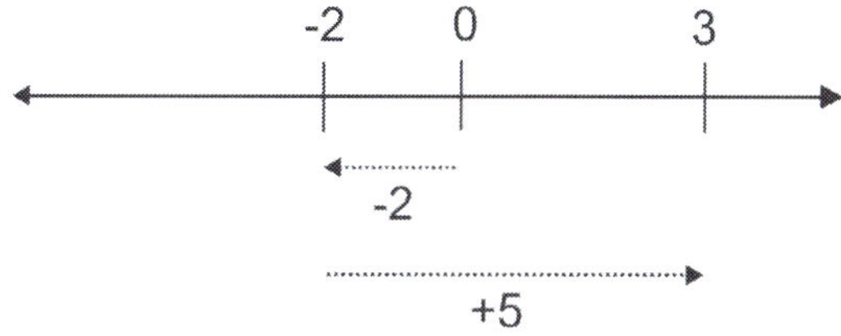

Figure 5.

Therefore: (-2) + 5 = 3

The first two possibilities tell us what happens when we add two integers having the same sign, either positive or negative. Note that we just add their magnitudes and keep the common sign. We can state this observation as a rule, a rule for determining the magnitude and a rule for determining the sign.

ADDING TWO INTEGERS THAT HAVE THE SAME SIGN

1. To determine the magnitude, add their magnitudes.
2. To determine the sign, keep the common sign.

We have two possibilities remaining and they involve the sum of two integers whose signs are opposite. To find the magnitude of the sum, we determine the absolute values of the addends and then subtract the lesser from the greater. What about the sign of the sum? The sign of the sum is the sign of the integer whose absolute value is greater. Therefore, our law for these cases involves subtraction.

ADDING TWO INTEGERS THAT HAVE THE OPPOSITE SIGN

1. To determine the magnitude, calculate the difference between the absolute values of the two numbers.
2. To determine the sign, use the sign of the integer with the greater magnitude or absolute value.

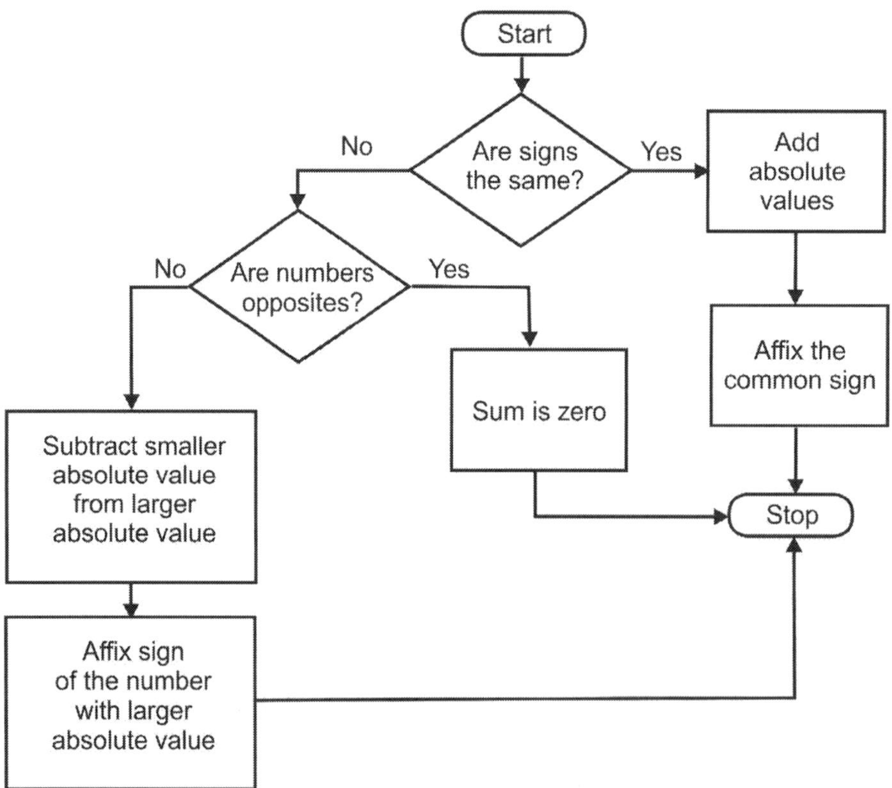

Figure 6. Flowchart for the addition of integers. The rectangles define an operation to perform. The **rhombi**, quadrilaterals with four equal sides and opposite sides parallel (i.e., a **parallelogram**), define a decision to make.

Finally, let's consider how the operation of subtraction works with integers. Let's say that we have a stack of 13 boxes and we want to remove eight boxes. How many boxes remain? Five boxes. We have found the difference between these two numbers, as shown in Figure 7.

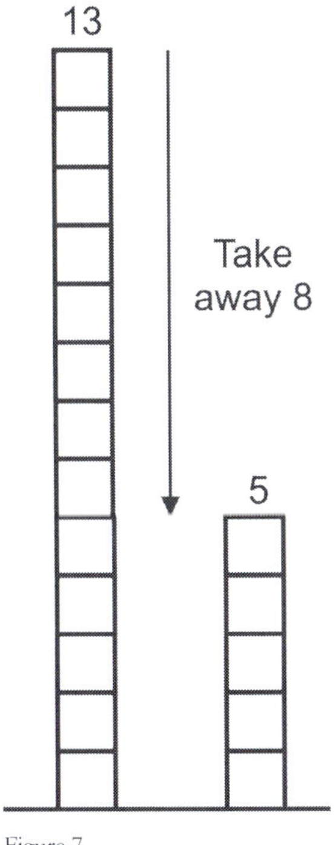

Figure 7.

Let Figure 8 represent temperature in degrees Celsius. We ask, "What is the difference between 13°C and -5°C?" In other words, what is the distance, on a number line, between 13 and -5? We investigated this procedure in our last lesson, and our answer should be 18.

The two problems we illustrated, excluding the concrete context, are:

Problem 1. 13   8 – 5 (Note: We say, "thirteen minus eight equals five.")
Problem 2. 13 – (-5) = 18 (Note: We say, "thirteen minus negative five equals eighteen.")

Figure 8.

Put on your Sherlock Holmes investigative hat and inspect. First, recall our previous definition of opposites. We know that the opposite of 8 is -8 and the opposite of -5 is 5. In Problem 1, let's change the subtrahend, 8, to its opposite, -8. Since we were subtracting 8, to get the same answer, we must add -8. Let's try it:

13 – 8 = 13 + (-8)

We invoke our law of addition, and our answer is:

13 + (-8) = 5

For Problem 2, the opposite of -5 is 5 so we add 13 to 5:

13 – (-5) = 13 + 5 = 18

Confirmation!

We have taken the difference of two positive integers and the difference of a positive integer and a negative integer. Let's try our procedure to see how it works when we find the difference between two negative integers.

Problem 3. (-5) – (-13) = ?

The opposite of the subtrahend, -13, is 13. We add using the laws of addition of integers. We get:

(-5) – (-13) = (-5) + 13 = 8

Let's now exchange the minuend with the subtrahend in Problem 3 and see what we get.

Problem 4. (-13) – (-5) = ?

The opposite of the subtrahend, -5, is 5. We add using the laws of addition of integers. We get:

(-13) – (-5) = (-13) + 5 = -8

Note that when we change the subtrahend with the minuend and subtract, the difference is not the same, even if both the minuend and subtrahend are positive.

We note:

$8 - 3 = 5$, but $3 - 8 = 3 + (-8) = -5$

Let's try one more:

Problem 5. $3 - (-5) = ?$

The opposite of the subtrahend, -5, is 5. We add using the laws of addition of integers. We get:

$3 - (-5) = 3 + 5 = 8$

ALGEBRAIC DEFINITION OF SUBTRACTION

Since we can convert any subtraction problem into an addition problem, we can state a general rule for this, the **Algebraic Definition of Subtraction**. If we substitute any two numbers by the letters *a* and *b*, to find $a - b$, we find the opposite of the subtrahend and then add it to the minuend invoking the appropriate addition rule, or:

$$a - b = a + (-b)$$

(Note: We say, "*a* minus *b* equals *a* plus negative *b*" or "*a* minus *b* equals *a* plus the opposite of *b*.")

If we substitute *b* with *a*, or if $a = b$, then:

$a - a = a + (-a) = 0$
(This is what we concluded earlier in our discussion of opposites.)

In the next lesson, we will use these rules in our exploration of subtraction using the Japanese Soroban.

ABSOLUTE VALUE AND DISTANCE

If we want to find the distance between any two points on the number line, the result will be positive. We subtract one number from the other and take the absolute value of the result. In symbols, the distance between points *a* and *b* on the number line is:

$|a - b| = |b - a|$ (Note: the two expressions generate the same number.)

For example, we can compute the distance between 18 and -9 in two ways:

$|18 - (-9)| = |18 + 9| = |27| = 27$
$|-9 - 18| = |-27| = 27$

EXERCISES

Define the following terms/symbols:

1. Additive inverse
2. $\Rightarrow$
3. $\Leftrightarrow$
4. Converse
5. Logical equivalence
6. Rhombus
7. Parallelogram

8. State the algebraic definition of subtraction in words and symbols.

True or False:

9. The additive inverse of 15 is -15.
10. The additive inverse of -2 is 2.
11. Adding a number to its opposite sometimes gives the sum of zero.
12. Adding a number to its opposite never gives the sum of zero.
13. Adding a number to its opposite always gives the sum of zero.
14. All quadrilaterals are rhombi.
15. All rectangles are parallelograms.
16. All rhombi are squares.
17. All rectangles are squares.
18. All squares are quadrilaterals.
19. If a geometric figure is a parallelogram, then it is a quadrilateral.
20. If a geometric figure is a quadrilateral, then it is a parallelogram.
21. $a, b \in +\mathbb{Z}$ and $b > a \Rightarrow a - b = -(b - a)$.

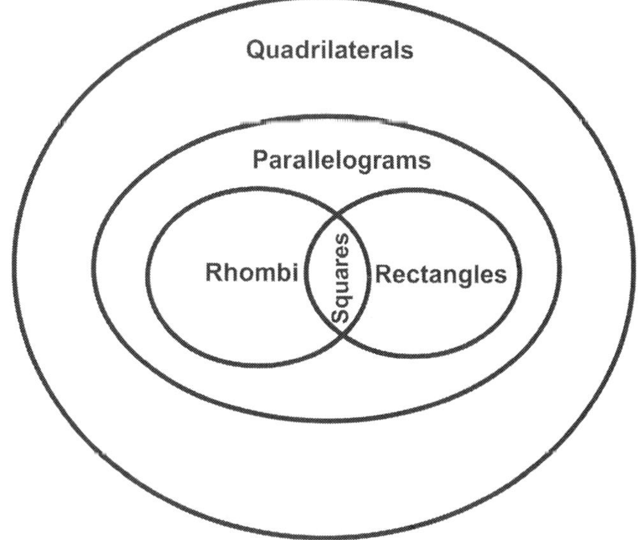

Figure 9. All squares are both rectangles and rhombi. Rectangles, squares and rhombi are parallelograms. A parallelogram is a quadrilateral.

Write the relevant rule for the following procedures:

22. We want to add two integers that have the same sign.
23. We want to add two integers that have opposite signs.
24. We want to subtract any two integers.
25. Complete the following table. The minuend is in the first row and the subtrahend is in the first column. Use the rules explained in this lesson to confirm every entry. Notice the beautiful symmetry, the dance.

–	-10	-9	-8	-7	-6	-5	-4	-3	-2	-1	0	1	2	3	4	5	6	7	8	9	10
-10																					
-9																					
-8																					
-7																					
-6																					
-5																					
-4																					
-3																					
-2																					
-1																					
0																					
1																					
2																					
3																					
4																					
5																					

–	-10	-9	-8	-7	-6	-5	-4	-3	-2	-1	0	1	2	3	4	5	6	7	8	9	10
6																					
7																					
8																					
9																					
10																					

Answer the following question:
26. Why is there no integer halfway between 1 and 2?

We write the opposite of a number by affixing the short dash prefix to the number, i.e., -. Therefore, we read -2 as "the opposite of 2." Answer this question:
27. Affix the dash prefix to -2 to symbolize the opposite of -2. [Hint: Write -2 as (-2) and affix the short dash prefix before (-2).]

Answer the following questions:
28. What is -(-2)? Read as "what is the opposite of the opposite of 2?"
29. What is -(-(-2))? (Hint: Work from the inner parentheses.)
30. What is -(-(-(-2)))?
31. What is $-(-a)$?
32. What is $-(-(-a))$?
33. What is $-(-(-(-a)))$?
34. If $a = 3$, what is $-(-(-a))$?
35. If $a = -3$, what is $-(-(-a))$?
36. If $a = 3$, what is $-(-(-(-a)))$?
37. If $a = -3$, what is $-(-(-(-a)))$

Find the following differences using the Algebraic Definition of Subtraction:
38. 7 – 5
39. 5 – 7
40. 9 – 3
41. 3 – 9
42. -7 – 5
43. -5 – 7
44. -9 – 3
45. -3 – 9
46. 7 – (-5)
47. 5 – (-7)
48. 9 – (-3)
49. 3 – (-9)
50. -7 – (-5)
51. -5 – (-7)
52. -9 – (-3)
53. -3 – (-9)
54. 6 – 6
55. 6 – (-6)
56. -6 – (-6)
57. -6 – 6

Find x:
58. $|3 + (-8)| = x$
59. $|-8 - 4| = x$
60. $|5 - 2| = x$
61. $|-5 - (-7)| = x$
62. $|6 - 9| = x$
63. $|13 - 13| = x$
64. $|5 - (-2)| = x$

3.4 Subtraction with the Japanese Soroban

Subtraction is the inverse of addition. For addition, given a number, we increase it by another number. In subtraction, given a number, we decrease it by another number. We can add or subtract negative numbers which reverses the concepts. Instead of an increase, the operation of addition produces a decrease. Instead of a decrease, the operation of subtraction generates an increase.

With the Japanese Soroban, we can demonstrate this dance of inverses. Remember, there are always two inverses for every addition problem.

Addition problem: $4 + 3 = 7$
Two inverses:
$7 - 3 = 4$
$7 - 4 = 3$

In the following examples, we will demonstrate only one inverse with the Soroban, and show how we use complements in subtraction.

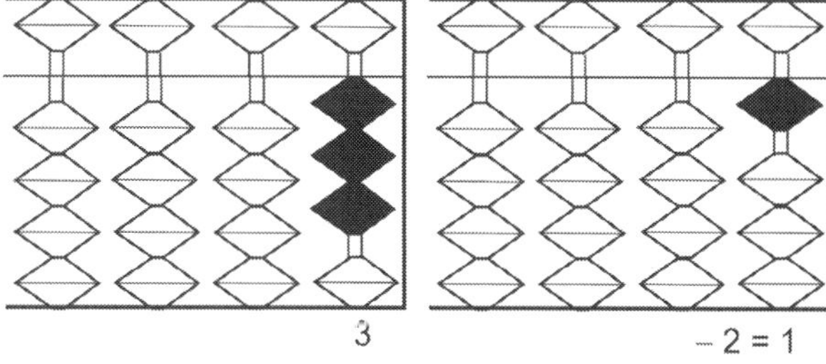

Figure 1.

$3 - 2 = 1$

We move three earth beads up in the ones column (Figure 1). To subtract 2, we move two earth beads down in the ones column. The difference: 1.

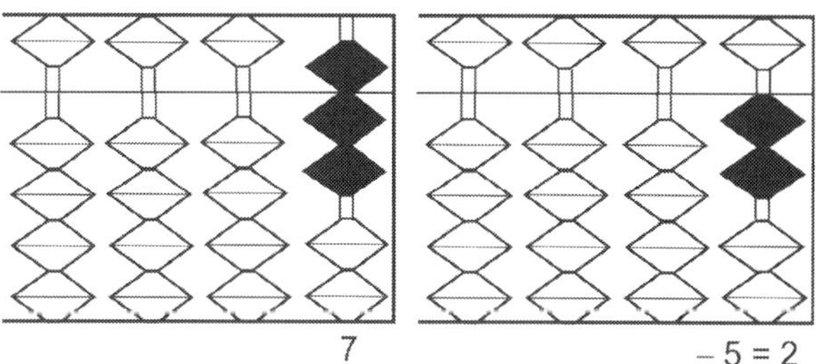

Figure 2.

$7 - 5 = 2$

We move the heaven bead down, and two earth beads up in the ones column (Figure 2). To subtract 5, we move one heaven bead up in the ones column. The difference: 2.

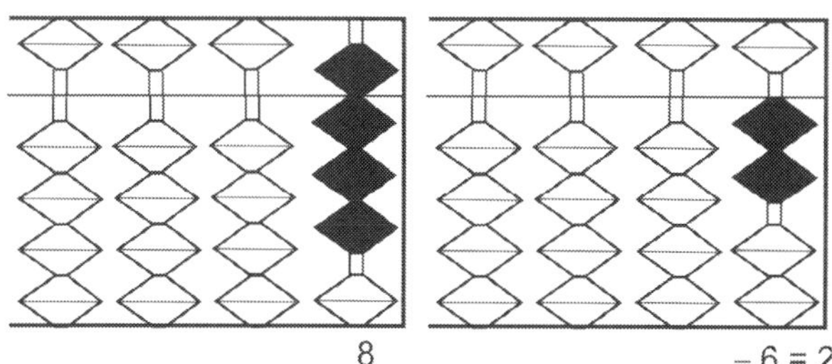

Figure 3.

$8 - 6 = 2$

We move the heaven bead down, and three earth beads up in the ones column (Figure 3). To subtract 6, we move one heaven bead up, and one earth bead down on the ones column. The difference: 2.

$5 - 1 = 4$

We move the heaven bead down in the ones column (Figure 4). Subtracting 1 is the same as subtracting 5 and adding 4. We do a double movement: In the ones column, we move the heaven bead up (subtract 5) and four earth beads up (add 4). The difference: 4.

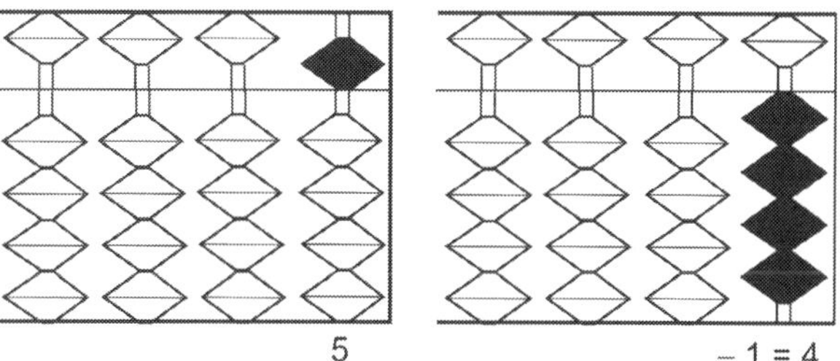

Figure 4.

We did this:

$$5 - 1 = 5 - (5 - 4) = (5 - 5) + 4 = 0 + 4 = 4$$

We could also rewrite subtraction as addition, or:

$$5 - 1 = 5 + (-1)$$

Then, we note:

$-1 = 4 - 5$, or:
$-1 = 4 + (-5)$

Do you see the movement of the heaven bead up (subtract 5, or add negative 5) and four earth beads up (add 4) in the last equation?

$7 - 3 = 4$

We move the heaven bead down, and two earth beads up in the ones column (Figure 5). Subtracting 3 is the same as subtracting 5 and adding 2. We do a double movement: In the ones column, we move the heaven bead up (subtract 5) and two earth beads up (add 2). The difference: 4.

We did this:

$7 - 3 = 7 - (5 - 2) = (7 - 5) + 2 = 2 + 2 = 4$

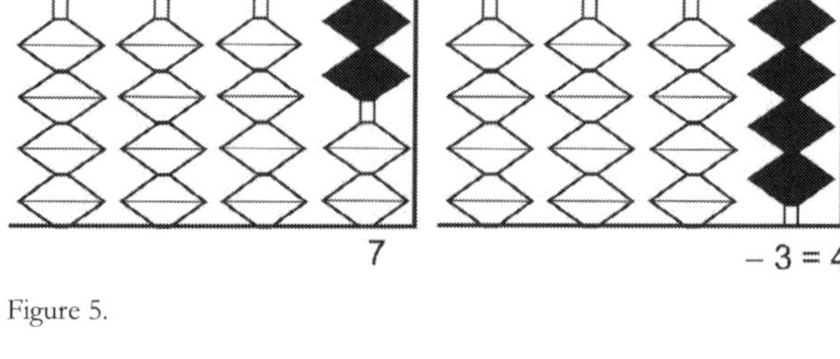

Figure 5.

We could also rewrite subtraction as addition, or:

$7 - 3 = 7 + (-3)$

Then, we note:

$-3 = 2 - 5$, or:
$-3 = 2 + (-5)$

Do you see the movement of the heaven bead up (subtract 5, or add negative 5) and two earth beads up (add 2) in the last equation?

$10 - 7 = 3$

First, we move one earth bead up in the tens column (Figure 6). Subtracting 7 is the same as subtracting 10 and adding 3. We do a double movement: In the tens column, we move one earth bead down (subtract 10) and in the ones column, we move three earth beads up (add 3). The difference: 3.

We did this:

$10 - 7 = 10 - (10 - 3) = (10 - 10) + 3 = 3$

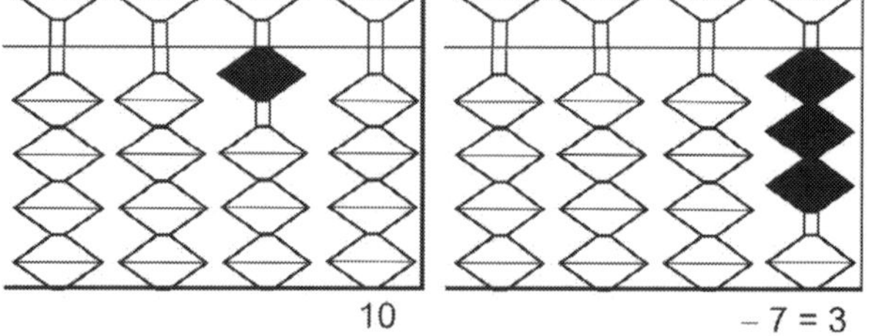

Figure 6.

We could also rewrite subtraction as addition, or:

$10 - 7 = 10 + (-7)$

Then, we note:

$-7 = 3 - 10$, or:
$-7 = 3 + (-10)$

Do you see the movement of one earth bead down in the tens column (subtract 10, or add negative 10) and three earth beads up in the ones column (add 3) in the last equation?

10 – 4 = 6

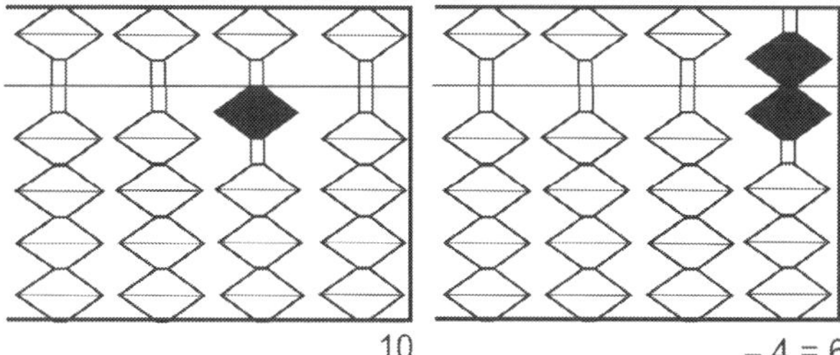

Figure 7.

We move one earth bead up on the tens column (Figure 7). Subtracting 4 is the same as subtracting 10 and adding 6. We do a triple movement: In the tens column, we move one earth bead down (subtract 10) and in the ones column we move one heaven bead down (add 5) and one earth bead up (add 1). The difference: 6.

We did this:

10 – 4 = 10 – (10 – 6) = 10 – 10 + (5 + 1) = 6

87 – 24 = 63

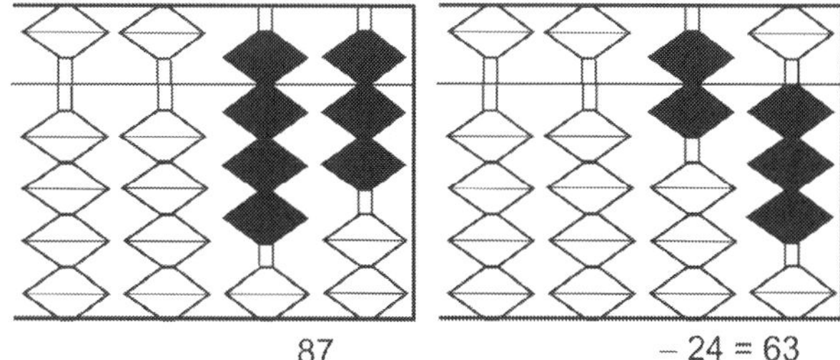

Figure 8.

In the tens column, we move one heaven bead down, and three earth beads up. Then, in the ones column, we move one heaven bead down, and two earth beads up (Figure 8).

To subtract 24, we proceed left to right, from tens to ones noting 24 = 20 + 4. To subtract two tens (20), we move two earth beads down in the tens column. Subtracting 4 is the same as subtracting 5 and adding 1. In the ones column, we move the heaven bead up (subtract 5) and one earth bead up (add 1). The difference: 63. We could also rewrite subtraction as addition, or:

87 – 24 = 87 + (-24)

Then, we note:

-24 = -20 – 4, or:
-24 = -20 + (-4), where:
-4 = 1 – 5, or:
-4 = 1 + (-5)

Do you see the movement of two earth bead downs in the tens column (subtract 20, or add negative 20) and the heaven bead up and one earth bead up in the ones column (subtract 4, or add negative 4)?

71 – 13 = 58

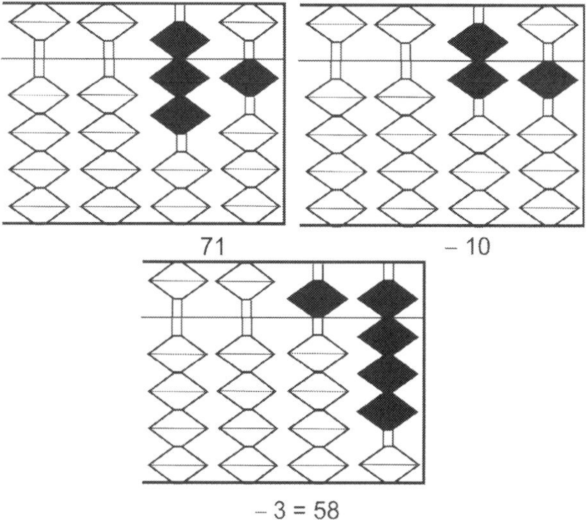

Figure 9.

First, on the tens column, we move the heaven bead down, and three earth beads up. Then, on the ones column, we move one heaven bead down, and two earth beads up (Figure 9).

To subtract 13, we first note that 13 = 10 + 3. We proceed left to right. In the tens column, we subtract 1 (subtract 10) by moving one earth bead down.

To subtract 3, we note that -3 = -10 + 7. We move an earth bead in the tens column down (-10). In the ones column, we move the heaven bead down, and two earth beads up (+7). The result, as seen in the figure, is 58.

In the last problem, note how we changed subtraction on the ones column to subtracting one ten (easy!) and adding seven ones (also easy!).

In our next lesson, we will learn how to apply this method, based on complements of 10, to make any subtracting problem easy to compute.

EXERCISES

Using the Japanese Soroban, calculate the following differences:

1. 3 – 1
2. 7 – 2
3. 8 – 2
4. 5 – 4
5. 7 – 4
6. 10 – 3
7. 10 – 6
8. 12 – 6
9. 13 – 9
10. 14 – 7
11. 16 – 8
12. 16 – 9
13. 27 – 25
14. 87 – 63
15. 71 – 58

Key words indicating subtraction:

- Decreased by
- Deduct
- Depreciate
- Difference
- Diminish
- Drop
- Fewer
- Left
- Less than
- Lose
- Loss
- Lower
- Minus
- Remain
- Remove
- Shorten
- Smaller than
- Subtract
- Take away

3.5 Complementary Removal

Concept Introduced
1. Complementary Subtraction

To ensure mastery of subtraction, these facts, consisting of a single digit minuend and subtrahend, must be memorized:

9 – 9, 9 – 8, 9 – 7, 9 – 6, 9 – 5, 9 – 4, 9 – 3, 9 – 2, 9 – 1, 9 – 0
8 – 8, 8 – 7, 8 – 6, 8 – 5, 8 – 4, 8 – 3, 8 – 2, 8 – 1, 8 – 0
7 – 7, 7 – 6, 7 – 5, 7 – 4, 7 – 3, 7 – 2, 7 – 1, 7 – 0
6 – 6, 6 – 5, 6 – 4, 6 – 3, 6 – 2, 6 – 1, 6 – 0
5 – 5, 5 – 4, 5 – 3, 5 – 2, 5 – 1, 5 – 0
4 – 4, 4 – 3, 4 – 2, 4 – 1, 4 – 0
3 – 3, 3 – 2, 3 – 1, 3 – 0
2 – 2, 2 – 1, 2 – 0
1 – 1, 1 – 0

Next, as we have noted with Japanese Soroban subtraction, we use complements. In this lesson, we will only work with complements of 10, not 5.

Table 1		
Number	Complement	Sum
1	9	10
2	8	10
3	7	10
4	6	10
5	5	10

How do we master the subtraction facts associated with all possible addition facts with single digit addends whose sum is greater than 10? Here are the associated subtraction facts for double-digit minuends greater than 10:

Table 2							
11 – 2 = 9	11 – 3 = 8	11 – 4 = 7	11 – 5 = 6	11 – 6 = 5	11 – 7 = 4	11 – 8 = 3	11 – 9 = 2
	12 – 3 = 9	12 – 4 = 8	12 – 5 = 7	12 – 6 = 6	12 – 7 = 5	12 – 8 = 4	12 – 9 = 3
		13 – 4 = 9	13 – 5 = 8	13 – 6 = 7	13 – 7 = 6	13 – 8 = 5	13 – 9 = 4
			14 – 5 = 9	14 – 6 = 8	14 – 7 = 7	14 – 8 = 6	14 – 9 = 5
				15 – 6 = 9	15 – 7 = 8	15 – 8 = 7	15 – 9 = 6
					16 – 7 = 9	16 – 8 = 8	16 – 9 = 7
						17 – 8 = 9	17 – 9 = 8
							18 – 9 = 9

Two-Digit Minuends and One-Digit Subtrahends

How can we easily compute these differences? We invoke the lessons learned from the Japanese Soroban. For example, let's compute this difference:

$$\begin{array}{r} 13 \\ \underline{-\ 4} \end{array}$$

Notice that we set up the problem vertically, not horizontally. Finding this difference is the same as (1) subtracting 10 from the minuend and then (2) adding the complement of 4 to the difference obtained in (1). We do this: (1) 13 – 10 = 3, (2) 3 + 6 = 9. Therefore, 13 – 4 = 9.

To check our work, we add the difference to the subtrahend. We should get the minuend. In symbols:

$a - b = c \Leftrightarrow c + b = a$

$$\begin{array}{r} 9 \\ +4 \\ \hline 13 \end{array}$$

Check: (above)

Let's try another:

$$\begin{array}{r} 18 \\ -\ 9 \\ \hline \end{array}$$

We do this: (1) 18 – 10 = 8, (2) 8 + 1 = 9. Therefore, 18 – 9 = 9. Check this work in your notebook. Here is another:

$$\begin{array}{r} 15 \\ -\ 7 \\ \hline \end{array}$$

We do this: (1) 15 – 10 = 5, (2) 5 + 3 = 8. Therefore, 15 – 7 = 8. Check this work in your notebook. Isn't this easy?
How would we do this, where the minuend is 10?

$$\begin{array}{r} 10 \\ -\ 8 \\ \hline \end{array}$$

We do this: (1) 10 – 10 = 0, (2) 0 + 2 = 2. Therefore, 10 – 8 = 2. Check this work in your notebook.

TWO-DIGIT MINUENDS AND SUBTRAHENDS

Let's try some differences where we have two-digit minuends and subtrahends.

$$\begin{array}{r} 85 \\ -42 \\ \hline \end{array}$$

We proceed left to right. We subtract the tens position first, and then the ones position. Tens:

$$\begin{array}{r} 85 \\ -42 \\ \hline 4 \end{array}$$

40 is our first estimate. Now the ones position:

$$\begin{array}{r} 85 \\ -42 \\ \hline 43 \end{array}$$

Knowing our basic single digit subtraction facts makes this easy. What about a problem like this?

$$\begin{array}{r} 85 \\ -\,48 \\ \hline \end{array}$$

We subtract tens first:

$$\begin{array}{r} 85 \\ -\,48 \\ \hline 4 \end{array}$$

Our first estimate: 40.

In the ones position, we note that the digit in the ones place of the minuend, 5, is less than the digit in the ones place of the subtrahend, 8. When this is the case, we invoke complements. We subtract 10 and add the complement of 8. How? Since our first estimate is 40, we subtract 10 from 40 to give us 30. Since the complement of 8 is 2, we add 2 to 5 to get 7. Our answer: 37.

$$\begin{array}{r} 85 \\ -\,48 \\ \hline \not{4}7 \end{array} \quad \text{Difference: 37}$$

$$\text{Check: } \begin{array}{r} 37 \\ +\,48 \\ \hline \overset{1}{7}5 \end{array} \quad \text{Sum: 85}$$

THE SLASH PRINCIPLE

We use slash to indicate that we are subtracting one from the digit in the position indicated. Therefore, $\not{4}$ means "4 – 1 = 3" or, recognizing place value, "40 – 10 = 30."

MULTIPLE-DIGIT MINUENDS AND SUBTRAHENDS

Let's explore more:

$$\begin{array}{r} 825 \\ -\,433 \\ \hline \end{array}$$

Invoking our left to right rule, we subtract the hundreds digits first, and then the tens digits, then the ones digits. Hundreds:

$$\begin{array}{r} 825 \\ -\,433 \\ \hline 4 \end{array}$$

Our first estimate: 400.

For the tens position work, we note that the digit in the tens place of the minuend, 2, is less than the digit in the tens place of the subtrahend, 3. We invoke complements, complements of 100. We subtract 100 and add the complement of 30. How? Since our first estimate is 400, we subtract 100 from 400 to give us 300. Since the complement of 30 is 70, we add 70 to 20 to get 90.

We write:

$$\begin{array}{r} 825 \\ -\ 433 \\ \hline \not{4}9 \end{array}$$

Our second estimate is 390.
For the ones position work, we subtract 3 from 5 to get 2. We write:

$$\begin{array}{r} 825 \\ -\ 433 \\ \hline \not{4}92 \end{array}$$

The difference is 392.

$$\text{Check: } \begin{array}{r} 392 \\ +\ 4\underline{3}3 \\ \hline \overset{1}{7}25 \end{array} \text{ Sum: } 825$$

Notice, how, in our check, we invoked complements in the tens position the way we used complements in the tens position of our original subtraction problem. Such is the dance of complements!

Figure 1. Pascaline Calculator. Source: James D. Nickel.

COMPLEMENTARY SUBTRACTION

What we have been demonstrating is, in a nutshell, left to right **complementary subtraction**. We use the same method of complements in addition for subtraction. Thus, we build upon what we already know. The benefit of complementary subtraction is that there is no need to subtract a greater digit from a lesser digit again. In left-to-right subtraction, if the digit in the minuend is less than the digit in the subtrahend, we can turn subtraction to (1) an addition problem and then (2) a subtraction problem where you just take away 1, 10, 100, 1000, etc. This left-to-right technique is simpler and faster than the right-to-left borrowing method that has been taught for generations to nearly every student of arithemtic.

Figure 2. Addiator. Source: James D. Nickel.

If you happen to have a mechanical *Pascaline* (Figure 1), a gear-based addition and subtraction calculator, you can visualize the addition and subtraction of complements.[1] The German made *Addiator*, popular in the 1960s, works on the same principle but without circular gears (Figure 2).[2]

A FEW PROBLEMS

Work out these problems in your notebook. Check your work. Check your solutions by addition and then check your work with solutions given in the second column.

Problem	Solution
1. 3418 − 1809	1. 3418 − 1809 = 2619 Difference: 1609
2. 912 − 789	2. 912 − 789 = 233 Difference: 123
3. 81,629 − 63,734	3. 81,629 − 63,734 = 28,995 Difference: 17,895
4. 123,119 − 18,804	4. 123,119 − 18,804 = 115,315 Difference: 104,315
5. 456,113 − 99,888	5. 456,113 − 99,888 = 467,335 Difference: 356,225

TWO-DIGIT MINUENDS AND ONE-DIGIT SUBTRAHENDS REVISITED

Given what we know, we can rewrite $\begin{array}{r}13\\-7\\\hline\end{array}$ as $\begin{array}{r}13\\-07\\\hline\end{array}$. Our left to right work:

$$\begin{array}{r}13\\-07\\\hline 16\end{array}\quad \text{Difference: 6}$$

[1] Check online – these calculators often appear there. They were popular from the 1950s to 1980s. With the advent of the electronic calculator in the 1980s, production of these educational gems ceased.

[2] In their construction, both the Pascaline and the Addiator show the inverse nature of subtraction. With the Pascaline, we add by turning the gears clockwise with a stylus and we subtract by turning the gears counterclockwise. Note that the engineers numbered the circular white addition digits, clockwise starting from 9, from 9 to 0, while they numbered the black circular subtraction digits, clockwise starting from 9, from 0 to 9. With the Addiator, we add by moving the digits down with a stylus and we subtract by moving the digits up. Note that the designers numbered the addition columns, reading from top to bottom, from 9 to 0, while they numbered the subtraction columns, reading from top to bottom, from 0 to 9. There is, therefore, a perichoretic interpenetration in the two designs.

$$\begin{array}{r} 06 \\ \text{Check: } +\underline{07} \\ \overset{1}{0}3 \end{array} \quad \text{Sum: } 13$$

WHAT IS HAPPENING?

Explain what is happening with this subtraction problem:

$$\begin{array}{r} 67,462 \\ -\underline{28,625} \end{array}$$

Completed work:

$$\begin{array}{r} 67,462 \\ -\underline{28,625} \\ 49,847 \end{array} \quad \text{Difference: } 38,837$$

$$\begin{array}{r} 38,837 \\ \text{Check: } +\underline{28,625} \\ \overset{1}{5}\overset{1}{6},4\overset{1}{5}2 \end{array} \quad \text{Sum: } 67,462$$

Most errors in subtraction using the traditional method involve mistakes in borrowing. Since complementary subtraction does not utilize borrowing, your answers will be more accurate, and, as a bonus, faster to compute! In mathematics, efficiency and speed are the name of the game.

A FINE NUANCE

Now, let's consider this problem because it will teach us a good lesson.

$$\begin{array}{r} 234,823 \\ -\underline{\ \ 34,824} \end{array}$$

Applying our methods, we get:

$$\begin{array}{r} 234,823 \\ -\underline{\ \ 34,824} \\ 200,0\emptyset9 \end{array}$$

What does $\emptyset$ mean in the tens place? It means take away one from 0 in the tens place. Since our difference, 234,823 – 34,824, is positive, we cannot write -1 in the tens place (0 – 1 = -1). We keep moving left until we find a nonzero digit, 2 in the hundred thousands place, and we then subtract one from 20000. The $\emptyset$ in the tens position means that you subtract one from the entire group of digits to the left of, and including the $\emptyset$: 20000 – 1 = 19999. Our work looks like this:

$$\begin{array}{r} 234,823 \\ -\underline{\ \ 34,824} \\ \cancel{200,0\emptyset}9 \end{array} \quad \text{Difference: } 199,999$$

Check:

```
  199,999
+  34,824   Sum: 234,823
  1 1 1  1 1
  123,713
```

Let's try another:

```
 5555
−1556   Difference: 3999
 4009
```

```
         3999
Check:  +1556   Sum: 5555
         1 1 1
         4445
```

For practice, work out these problems in your notebook. Check your work. Check your solutions with the ones given in the second column.

Problem	**Solution**
1. 3418 − 2419	1. 3418 − 2419 Difference: 999 1009
2. 1912 − 919	2. 1912 − 919 Difference: 993 1003
3. 81,623 − 61,628	3. 81,623 − 61,628 Difference: 19,995 20,005
4. 233,113 − 23,117	4. 233,113 − 23,117 Difference: 209,996 210,006
5. 456,113 − 96,119	5. 456,113 − 96,119 Difference: 359,994 460,004

EXERCISES

1. Using an example, explain the rule "add the complement and subtract one in the place value to the left."

Using complementary subtraction (where needed), find the difference (rewrite vertically). Check your answers by addition noting that $a - b = c \Leftrightarrow c + b = a$:

2. 24 – 9
3. 13 – 4
4. 55 – 8
5. 26 – 17
6. 14 – 7
7. 67 – 15
8. 14 – 6
9. 52 – 23
10. 116 – 19
11. 42 – 27
12. 39 – 37
13. 131 – 29
14. 526 – 17
15. 50 – 23
16. 48 – 17
17. 92 – 27

To gain mastery of subtraction sense, subtract verbally by:

18. 2s from 48 to 24 (Ans. 48, 46, 44, 42, 40, 38, 36, 34, 32, 30, 28, 26, 24)
19. 2s from 101 to 87
20. 3s from 60 to 30
21. 3s from 89 to 50
22. 4s from 88 to 72
23. 4s from 53 to 21
24. 5s from 600 to 570
25. 5s from 613 to 583
26. 6s from 99 to 39
27. 6s from 126 to 90

Copy vertically, subtract, and check, answering with the proper denominations where indicated:

28. 583 – 472
29. 583 – 399
30. 720 – 345
31. 837 – 349
32. 1003′ – 819′
33. 7215 tons – 927 tons
34. Subtract $83.25 from $119.15
35. Subtract $415.99 from $10,000.25

Copy vertically, subtract using the Algebraic Definition of Subtraction, and check:

36. 7 – 16
37. 2 – 29
38. 66 – (-95)
39. 224 – 926

40. 1008 – 1829
41. 223 – (-49)
42. 314,883 – 828,110
43. 56 – (-789)
44. 56 – 789

Answer the following problem:

45. Complete the third column of Table 3. Show all your subtraction work. Do you see a pattern in your answers?

Table 3

Order in Sequence	Fibonacci Number	Difference
1st	1	*********
2nd	1	1 – 1 = 0
3rd	2	2 – 1 – 1
4th	3	3 – 2 = 1
5th	5	
6th	8	
7th	13	
8th	21	
9th	34	
10th	55	
11th	89	
12th	144	
13th	233	
14th	377	
15th	610	
16th	987	
17th	1597	
18th	2584	
19th	4181	
20th	6765	

> You must realize that a mastery of mathematics involves both levels [computational and theoretical – JN], although your tastes may direct you more strongly to one or the other .... Be warned that deficiency at either level can ultimately hinder you in your work. Independently of need, however, it should be a source of pleasure to understand why a mathematical result is true, i.e., to understand its proof as well as to understand how to use the result in concrete circumstances.
>
> Serge Lange, *Basic Mathematics* (1971), p. xi.

3.6 Speeding Up the Backwards Dance

Term Introduced
2. Paradigm

The purpose of this lesson is to help you to increase your speed in subtraction. This quickness depends upon your sense of numbers and their complements, and your ability to automatically see the answer in your mind.

Three Possibilities

When we engage a subtraction problem involving more than two digits in the minuend or subtrahend, there are three possibilities to consider when we consider digits in their respective place values:

1. The minuend digit is greater than the subtrahend digit.
2. The minuend digit equals the subtrahend digit.
3. The minuend digit is less than the subtrahend digit.

This first example demonstrates the first two possibilities:

$$\begin{array}{r} 87 \\ \underline{-27} \end{array}$$

Moving from left to right, in the tens position, the minuend digit is greater than the subtrahend digit. In the ones position, the minuend digit equals the subtrahend digit. In both cases, we subtract:

$$\begin{array}{r} 87 \\ \underline{-27} \\ 60 \end{array}$$

Writing the "6" and 0" should be automatic, without any thought. In our second example, in the tens position, the minuend digit is greater than the subtrahend digit but, in the ones position, the minuend digit is less than the subtrahend digit:

$$\begin{array}{r} 45 \\ \underline{-19} \end{array}$$

Moving from left to right, in the tens position, we should see "3." In the ones position, we invoke complements and write "6", and then a slash left. We write:

$$\begin{array}{r} 45 \\ \underline{-19} \\ \not{3}6 \end{array} \quad \text{Difference: 26}$$

Mastering the Third Possibility

To help us master the third possibility, work the following problems. Understand that the digits are in specific places: either tens, hundreds, thousands, etc. We want to gain speed in using both complements and taking away one from the position to the left by slashing left. For example:

$$\begin{array}{r} 4 \\ \underline{-9} \\ /5 \end{array}$$

The slash / indicates we subtract 1 from the value of the digit in the position immediately to the left. If we are subtracting 9 from 44, we write:

```
  44
 – 9
 ——
 4̸5
```

The slash 4 (4̸) indicates subtract 1 from 4. The difference is 35.

Make sure that you cover the answers as you mentally work through these problems.

4	4	1	3	3	2	1	8
–9	–7	–2	–4	–5	–9	–6	–9
3	4	5	6	3	2	7	6
–8	–6	–9	–7	–6	–8	–8	–9
2	1	4	5	1	5	2	1
–5	–8	–5	–8	–4	–7	–7	–9

Answers:

4	4	1	3	3	2	1	8
–9	–7	–2	–4	–5	–9	–6	–9
/5	/7	/9	/9	/8	/3	/5	/9
3	4	5	6	3	2	7	6
–8	–6	–9	–7	–6	–8	–8	–9
/5	/8	/6	/9	/7	/4	/9	/7
2	1	4	5	1	5	2	1
–5	–8	–5	–8	–4	–7	–7	–9
/7	/3	/9	/7	/7	/8	/5	/2

Here is a final group to master (cover the answers):

7	6	5	4	3	2
–9	–8	–6	–8	–9	–3
2	1	2	1	1	3
–4	–7	–6	–5	–3	–7

Answers:

7	6	5	4	3	2
–9	–8	–6	–8	–9	–3
/8	/8	/9	/6	/4	/9

2	1	2	1	1	3
−4	−7	−6	−5	−3	−7
/8	/4	/6	/6	/8	/6

These 36 problems include all the possibilities that you will encounter with subtraction where the minuend digit is less than the subtrahend digit. Write each problem on 3″ by 5″ if you cannot immediately see the answer. Master these facts before you continue your work.

Let's continue to practice. Work these problems in your notebook and check your answer by addition.

4318	83,118	88	2319	100,678
− 2382	− 2,802	− 48	− 5199	− 829
			Remember: 2319 − 5199 = 2319 + (-5199) Then, use your addition rules from Lesson 3.3: Subtract 2319 From 5199, noting the difference is negative.	

Answers (from left to right):

1936	80,316	40	-2880	99,849

More:

5639	53,006	44,118	828
− 4662	−39,123	− 987	− 717

Answers (from left to right):

977	13,883	43,131	111

A few more:

8,816	5625	7189	418,009
− 10,199	− 596	− 6993	− 19,111
Hint: 8,816 − 10,199 = 8,816 + (-10,199)			

Answers (from left to right):

-1383	5029	196	398,898

For a challenge, try this subtraction problem. Use your notebook, check your answer, and then come back and check your solution:

$$\begin{array}{r} 86{,}118{,}496{,}003 \\ -\ \underline{44{,}232{,}818{,}007} \end{array}$$

Here is the solution:

$$\begin{array}{r} 86{,}118{,}496{,}003 \\ -\ \underline{44{,}232{,}818{,}007} \\ 42{,}986{,}688{,}006 \end{array}$$

Difference: 41,885,677,996

Check:

$$\begin{array}{r} 41{,}885{,}677{,}996 \\ +\ \underline{44{,}232{,}818{,}007} \\ {\scriptstyle 1\ 1\ 1\ \ 1\ \ 1} \\ 85{,}017{,}485{,}993 \end{array}$$

Sum: 86,118,496,003

Notice especially what is happening in the ones position: 6 + 7 = 3, record 10. In the tens position, 9 + 1 = 10, record 100. Then, in the hundreds position, 9 + 1 = 10, record 1000. In the thousands position, 5 + 1 = 6. In effect, recording a 10 in the ones position is the same as adding 1 to 599 as indicated by $\overset{1}{\overline{5{,}99}}3$. For review of this procedure, see Lesson 2.8.

THE ROLE OF PARENTHESES

Consider this problem: 225 – ((88 + 23) – (29 + 33)) = ?

The role of parentheses is to indicate priorities of order in performing the calculations; i.e., we calculate what is inside the parentheses first. Consider parentheses like a box. First, we calculate the contents of the box and then use that answer as a unit.

The problem has two sets of parentheses or two boxes within a box. First, you add 88 + 23 and 29 + 33:

$$\begin{array}{r} 88 \\ +\ \underline{23} \\ {\scriptstyle 1} \\ 101 \end{array} \qquad \begin{array}{r} 29 \\ +\ \underline{33} \\ {\scriptstyle 1} \\ 52 \end{array}$$

The problem becomes 225 – (111 – 62). Second, you subtract: 111 – 62:

$$\begin{array}{r} 111 \\ -\ \underline{62} \\ 159 \end{array}$$

The problem becomes 225 – 49:

```
 225
– 49
----
```
$\not{2}\not{8}6$

Answer: 176

THE NEW PARADIGM IN REVIEW

We have taken considerable time working through a new **paradigm**[1] of complementary addition and complementary subtraction. What is so wonderful about mathematics is that principles build upon principles. So, not only do I hope you have enjoyed this excursion into addition and subtraction, I hope, also, for the sake of your future enjoyment, that you have mastered these principles.

Let's review why complements work when adding and subtracting. Consider:

```
 6
+7
--
```

We can rewrite this addition problem by adding 10 to 6 and then subtracting the complement of 7 from the result:

```
  6
+10
---
 16
– 3
---
 13
```

Or, we can subtract the complement of 7 first and then add 10 to that result:

```
  6
– 3
---
  3
+10
---
 13
```

Now consider this subtraction problem:

```
12
–7
--
```

[1] *Paradigm* is taken from Latin roots meaning "pattern." It literally means "show beside."

We can rewrite this subtraction problem by subtracting 10 from 12 and then adding the complement of 7 to the result:

12
− 10
2
+ 3
5

Or, we can add the complement of 7 to 12 and then subtract 10 from the result:

12
+ 3
15
− 10
5

Complementary addition and complementary subtraction make use of the decimal number system regarding the ease of its use when adding or subtracting 10.

Let's try two final problems. Do the work in your notebook.

Problem 1.

5,199
289
16,123
489
10,001
+ 8,876

Answer:

5,199
289
16,123
489
10,001
+ 8,876
1 1 3 3
29,647
40,977

Problem 2. Check your answer by addition.

7,113,578
− 3,224,189

Answer:

7,113,578
− 3,224,189
4,999,499
3,889,389

Check:

3,889,389
+ 3,224,189
1 1 1 1 1
6,003,468
7,113,578

How quickly and how accurately did you do these? If you are still having some trouble, make up problems on your own until you are very comfortable with this complementary process.

EXERCISES

1. What is the meaning of paradigm?
2. What are the complements of these numbers? 8, 2, 7, 3, 6, 5, 9, 4, 1.

Verbally count by:

3. 4s from 7 to 53 and back from 53 to 7
4. 5s from 23 to 53 and back from 53 to 23
5. 6s from 27 to 63 and back from 63 to 27
6. 7s from 12 to 75 and back from 75 to 12
7. 8s from 15 to 79 and back from 79 to 15
8. 9s from 12 to 102 and back from 102 to 12

Answer the following questions and check your work:

9. From a cask containing 512 gallons, 279 gallons were drawn. How many gallons remain?
10. A section of road is 469 yards in length. 181 yards must be demolished to make room for a housing development. How many yards of road are left?
11. A man bought a computer for $1625 and a printer for $425. He sold both for $2235. How much did he gain?
12. How many are 62 and 298, less 215 and 28?
13. Eight hundred twenty-nine plus 225 equals 829 minus how many?
14. If a company published a book in 1869, how many years are there between its date of publication and the year 2010?

Determine the approximate ages of the following people based only upon the year of their birth and the year of their death. We will disregard the month and day upon which each person was born and died; these dates would, of course, be needed to determine precisely the age at death. Check your work.

15. Johannes Kepler (1571-1630), German astronomer and mathematician
16. Galileo Galilei (1564-1642), Italian mathematician
17. Isaac Newton (1642-1727), English scientist and mathematician
18. Michael Faraday (1791-1867), English scientist
19. Leonhard Euler (1707-1783), Swiss mathematician
20. Carl Friedrich Gauss (1777-1855), German mathematician
21. Sophie Germain (1776-1831), French mathematician
22. Blaise Pascal (1623-1662), French theologian and mathematician
23. David Hilbert (1862-1943), German mathematician
24. Alfred North Whitehead (1861-1947), English philosopher and mathematician
25. James Clerk Maxwell (1831-1879), Scottish mathematician and physicist
26. Bertrand Russell (1872-1970), English philosopher and mathematician
27. Albert Einstein (1879-1955), German physicist
28. Bernard Riemann (1826-1866), German mathematician
29. Howard Eves (1911-2004), American mathematics historian
30. Walter Warwick Sawyer (1911-2008), English mathematician

Replace the x in each of the following problems with a number that makes the equations true. Calculate from left to right unless you see parentheses:

31. $23 + 18 - (16 + 15) = x$
32. $212 + 88 - 25 = x$
33. $155 + 45 - 50 = x$
34. $119 - 19 + 83 = x$
35. $625 - (425 + 175) = x$
36. $35 + 15 = 50 - x$
37. $87 - 62 = 20 + x$
38. $55 + 45 = x + 11$
39. $99 - 19 = 119 - x$
40. $49 + 25 = 1000 - x$
41. $99 - (27 + 32) = x + 25$
42. $65 + (93 - 37) = 145 - x$
43. $125 + 89 = 123 + x$
44. $256 - 128 = 160 - x$
45. $345 - 337 = x + 2$
46. $225 + 378 = 1315 - x$
47. $87 - x = 55 + 29$
48. $825 - x = 925 - (280 - 180)$
49. $415 + 983 = (785 - 289) + x$

3.7 Properties of Addition and Subtraction

Terms, Symbols & Concepts Introduced
1. $\notin$
2. $\in$
3. Associative Law of Addition
4. Axiom
5. Balance Beam Principle
6. Commutative Law of Addition
7. Counterexample
8. Identity
9. Identity Element
10. Identity Element in Addition and Subtraction
11. Law of Change in Addition
12. Law of Decrease in Addition
13. Law of Increase in Addition
14. Law of Minuend Change in Subtraction
15. Law of Subtrahend Change in Subtraction
16. Proposition
17. Theorem

In this lesson, we shall explore and explain the basic properties of addition and subtraction; i.e., how numbers interact with each other when we add or subtract.

The Property of Order: The Commutative Law of Addition

So far, we discovered that the addends can change places in an addition problem. To illustrate (Figure 1), pick up three coins and put them in your left hand. Next, pick up two coins with your right hand. How many coins in both hands, starting from left to right? 3 + 2 = 5. Now, put the coins down and put two in your left hand and three in your right hand. How many coins in both hands, starting from left to right? 2 + 3 = 5.

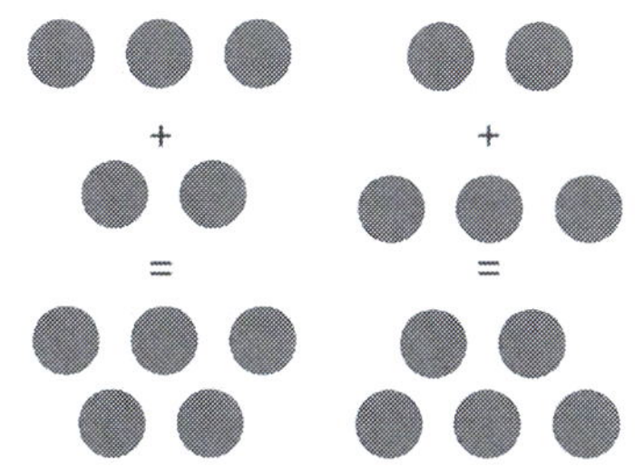

Figure 1.

What we have done reveals a very important property in arithmetic, the **Commutative Law of Addition.**[1] This law is foundational to the way numbers can be combined. This property is inherent in every pair of addends and embodied in the foundation of the structure of reality. This reality reveals this order because it is grounded in the person of Jesus Christ, the creator. Therefore, the commutative law of addition holds together or is sustained by the person of God in Christ (Colossians 1:15-17).

In symbols, if a and b are letters that represent any number, as a generalization (Lesson 2.3), we write:

$a + b = b + a$

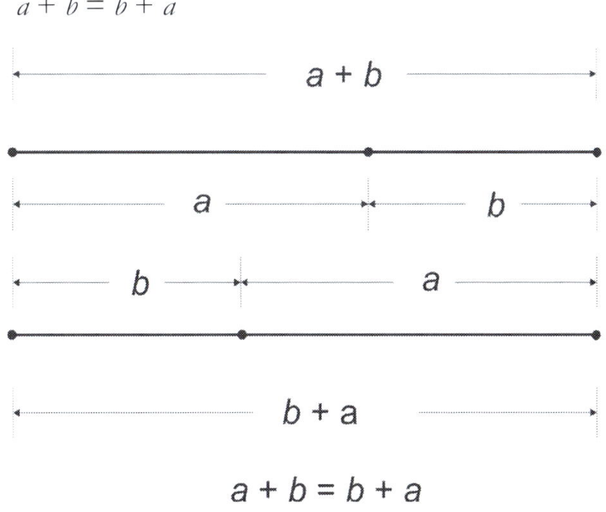

Figure 2.

Think carefully about this equation. Make sure that it makes sense in your understanding, especially the use of the letters, addition sign, and the equals sign. Compare it with Figure 1 and Figure 2, grouping dots and the number line.

[1] *Commutative*, from the Latin *commutare*, means "exchange" or "interchange."

Writing a statement, or **proposition**,[2] using algebraic symbols is a way of stating a general law governing the dance of number. It is the preference of mathematicians to state truths in concise ways, using symbols. It is the bread and butter of mathematics.

THE PROPERTY OF GROUPING: THE ASSOCIATIVE LAW OF ADDITION

We have already seen this law of grouping when we added a list of single digit addends and looked for pairs of addends that summed to 10. Consider this equation (Figure 3):

$3 + 2 = 5$

Adding 4 to the sum, we get:

$5 + 4 = 9$

Now, consider this equation:

$2 + 4 = 6$

Adding 3 to the sum, we get:

$6 + 3 = 9$

"5 + 4" and "6 + 3" have the same sum: 9. What we have done is to show this:

$3 + 2 + 4 = 2 + 4 + 3$

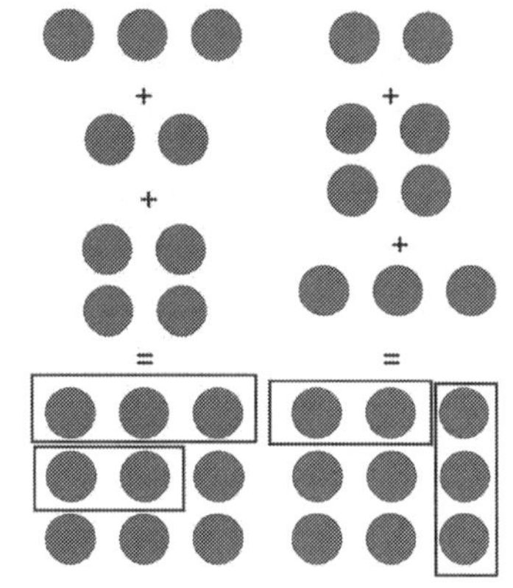

Figure 3.

When we have an addition problem with three or more addends, we can group or associate[3] the addends in pairs in any way we choose.[4] This choice will not change the sum, if we add correctly.

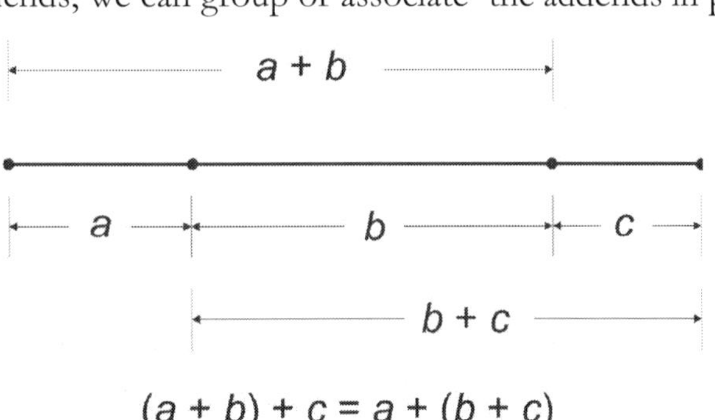

Figure 4.

We encountered an application of this law when we tried to sum two or more addends. For example, 12 + 72 + 88 can be rearranged as 12 + 88 + 72 = 100 + 72 = 172. In this example, changing the order of the addends can help us identify two addends that sum to 100.

In symbols, if a, b, and c are letters that represent any number, we state the **Associative Law of Addition** as follows:

$$(a + b) + c = a + (b + c)$$

Again, think carefully about this equation. Make sure that it makes sense in your understanding, especially the use of parentheses that tell us to do what is inside them first (Figure 4).

---

[2] A proposition is a mathematical statement that is either true or false. Sometimes propositions are statements that are assumed to be true, named **axioms** (from the Greek meaning to "reckon as worthy"). Axioms are self-evident truths in a system of thought. At other times, propositions are statements that must be proven using the rules of logic (see Step 13), named **theorems** (from the Latin meaning "to see"). Sometimes propositions are named laws or principles.

[3] *Associative*, from the Latin *associatus,* means "to join to" or "unite."

[4] This pairing reflects on the binary nature of addition; we add two addends at a time.

THE BALANCE BEAM PRINCIPLE

Now, consider this equation:

$13 + 9 = 22$

What happens if we change one of the addends on the left side? What happens to the right side when we add five to the first addend on the left side? Since $13 + 5 = 18$, then:

$18 + 9 = 27$

What happened to the sum? It also increased by five. Adding five to both sides of an equation is an example of a powerful principle in mathematics, the **Balance Beam**[5] **Principle**: When you add a quantity to the left side of an equation, to keep the equation balanced you must also add that same quantity to the right side of the given equation. Since we can understand subtraction can as addition by the Algebraic Definition of Subtraction, the Balance Beam principle also holds when we subtract a quantity from both sides of an equation.

Figure 5. Scientific balance beam. Source: James D. Nickel.

In our example, we added five to both sides of the equation. When we did so, the equation remained balanced, or true; i.e., $18 + 9 = 27$ is a true statement.

In summary, this is what we have done (note the equivalence sign):

$13 + 9 = 22 \Leftrightarrow 18 + 9 = 27$

We express the Balance Beam Principle as a law, the law of increase in addition.

LAW OF INCREASE IN ADDITION

When one of the addends increases by a quantity, then the sum increases by the same quantity.

This property originates from the nature of addition; i.e., the sum depends upon each of the addends, or the sum contains the addends, and the addends contain the sum. In this way, the addends mutually indwell the sum, a relation reflecting the rationality of the perichoretic nature of number. When we change one of the addends, the same change happens to the sum.

Hidden behind this principle is another law. Let's see if we can uncover it by some analysis. Since $15 = 10 + 5$, we can rewrite $15 + 3$ as follows:

$15 + 3 = (10 + 5) + 3$

The use of parentheses indicates that we are grouping the 10 and the 5. Because of this, 15 is partitioned into two parts. We calculate the sum of $(10 + 5) + 3$ in two steps.

Step 1. $10 + 5 = 15$ (We have performed the sum within the parentheses.)
Step 2. $15 + 3 = 18$

This two-step addition process is very useful in calculating sums. For example, $40 + 52$ can be rewritten as:

---

[5] In science, a balance is a lever (Lesson 8.3) with two equal arms and a pan suspended from each arm.

$40 + (50 + 2)$

Next, we reorder by rearranging the parentheses (applying the Associative Law of Addition):

$(40 + 50) + 2$

Next, we add $40 + 50$ since we sum the two addends within the parentheses first. We get 90. Finally, we calculate $90 + 2 = 92$.

Let's consider something different. Would you think the law of change in addition would also work for removing, or subtracting, a given quantity? Let's again consider the equation:

$9 + 7 = 16$

Let's subtract two from seven:

$9 + 5 = 14$

When we change one the addends, it results in the same change to the sum. When we decrease the left side of the equation by a quantity, then the right side of the equation decreases by the same amount.

LAW OF DECREASE IN ADDITION

> When one of the addends decreases by a quantity, the sum decreases by the same quantity.

Combining both laws, we can state:

LAW OF CHANGE IN ADDITION

> Changing the addend engenders the same change in the sum.

How can we make use of this law? Is it easier to add 6 to 9 or 6 to 10? Consider $352 + 999$. Note that it is much easier to calculate $352 + 1000 = 1352$. But, 999 is 1 less than 1000, or $1000 - 1 = 999$. By the law of change in addition, $352 + 999$ is also one less than $352 + 1000$. We write:

$(352 + 1000) - 1 = 1352 - 1 = 1351$
$352 + (1000 - 1) = 352 + 999 = 1351$

Explain how would you use the law of decrease in addition to add:

Problem 1. $367 + 999$
Problem 2. $4215 + 9997$
Problem 3. $24{,}812 + 99{,}998$

Answers:

Problem 1. $367 + 999 = 367 + (1000 - 1) = (367 + 1000) - 1 = 1367 - 1 = 1366$
Problem 2. $4215 + 9997 = 4215 + (10{,}000 - 3) = (4215 + 10{,}000) - 3 = 14{,}215 - 3 = 14{,}212$
Problem 3. $24{,}812 + 99{,}998 = 24{,}812 + (100{,}000 - 2) = (24{,}812 + 100{,}000) - 2 = 124{,}812 - 2 = 124{,}810$

Like addition, the operation of subtraction is undergirded by laws that are universal. Let's first consider the laws of change in subtraction. Consider the difference in $16 - 7 = 9$. What happens to the difference if we increase the minuend by two? Our difference becomes $18 - 7 = 11$.

We see that when we increase the minuend by two, then the difference increases by two. Let's state what we have done, using parentheses:

$16 - 7 = 9$

Add two to the minuend:

$(16 + 2) - 7 = 9 + 2 = 11$

LAW OF MINUEND CHANGE IN SUBTRACTION

> When the minuend increases by a quantity, then the difference increases by the same quantity.
> Or, when the minuend decreases by a quantity, then the difference decreases by the same quantity.
> Changing the minuend engenders the same change in the difference.

This property originates from the nature of subtraction, i.e., the difference depends the minuend and the subtrahend, or the difference contains the minuend and subtrahend and the minuend and subtrahend contain the difference. As with addition, the minuend and subtrahend mutually indwell the difference, a relation reflecting the rationality of the perichoretic nature of number. Changing the minuend results in the same change to the difference.

Now, let's increase the subtrahend by two. We start with 16 – 7 = 9. When we add 2 to the subtrahend, the subtrahend is then 9. Our difference becomes 16 – 9 = 7. Here we discover that when we increase the subtrahend by two, then the difference decreases by two. Let's state what we have done, using parentheses, starting from 16 – 7 = 9 and adding two to the subtrahend:

$16 - (7 + 2) = 9 - 2 = 7$

LAW OF SUBTRAHEND CHANGE IN SUBTRACTION

> When the subtrahend increases by a quantity, then the difference decreases by the same quantity.
> Or, when the subtrahend decreases by a quantity, then the difference increases by the same quantity.
> Changing the subtrahend engenders the opposite change in the difference.

We can employ these rules of change in creative ways to help simplify a subtraction problem. Consider 32 subtracted from 80. Using complementary subtraction, we calculate the difference:

```
 80
-32   Difference: 48
---
 58
```

(the 5 in 58 is struck through)

Another way to do this problem is first to observe that 32 = 30 + 2. If we subtract 30 from 80 we get 50; i.e., 80 – 30 = 50. We want to subtract 32, and we note that to do so, we must add 2 to the subtrahend. Adding 2 to the subtrahend means, by our law of change in subtraction, that we must subtract 2 from the difference:

$50 - 2 = 48$

Using complementary subtraction:

$$\begin{array}{r} 50 \\ -\ 2 \\ \hline \not{5}8 \end{array}$$ Difference: 48

Let's try another problem. Consider the difference of 83 – 55. We know that 55 = 53 + 2. We can easily subtract 53 from 83: 83 – 53 = 30. We want to subtract 55 so we must add 2 to the subtrahend. To calculate our answer, by our law of change in subtraction, we must subtract 2 from the difference: 30 – 2 = 28.

Therefore:

83 – 55 = 28

Using complementary subtraction:

$$\begin{array}{r} 83 \\ -\ 55 \\ \hline \not{3}8 \end{array}$$ Difference: 28

SUBTRACTION AND THE COMMUTATIVE LAW

Does the Commutative law hold for subtraction? If we exchange the minuend with the subtraction, will the difference remain unchanged? We have already noted that the difference does change. For example:

7 – 3 = 4

Exchanging the minuend with the subtrahend, we get:

3 – 7 = 3 + (-7) = -4

With this **counterexample**,[6] we know that the Commutative law does not hold for subtraction.

Does the Associative Law hold for subtraction? Can we create a string of subtractions and then rearrange groups with parentheses and get the same final answer? Consider 8 – 3 – 4. We subtract 3 from 8 first. Using parentheses, we write (8 – 3) – 4. Performing the first subtraction, we get:

8 – 3 = 5

Now, subtract 4 from 5. We get:

5 – 4 = 1

To test this question, regroup the string by changing the parentheses, or 8 – (3 – 4). Subtracting 4 from 3 first, we get:

3 – 4 = 3 + (-4) = -1

Now, we subtract -1 from 8. We get:

8 – (-1) = 8 + 1 = 9

We note that -1 ≠ 9. Again, with this counterexample, we know that the Associative law does not hold for subtraction.

[6] In mathematics, one way to show that a given proposition is not true is to find one counterexample, thus proving that the proposition fails the "truth test."

THE ROLE OF ZERO IN ADDITION

How does zero interact with the operations of addition and subtraction? Can we state this interaction as a property? In the mathematical world, these properties are sometimes named laws. At other times, these properties/laws are named identities. These three words are interchangeable. Remember, a numerical law is a general statement that holds true for all numbers that we are considering.

> An **identity** in arithmetic is a relationship, a law order, that is true for all numbers that we are considering.

Let's suppose that Steve has three marbles, and Mary, not being a marble player, has none. How many marbles do Steve and Mary have altogether? We write this equation:

$3 + 0 = 3$

It should not take you much thought to conclude that if we add zero to any number, the sum will always be that number. Mathematicians call numbers that act in this way **identity elements**.[7] In our example, 0 is an element or a part of a whole. We let the whole in this case be the set of integers, or $\mathbb{Z}$. In review, using set-building notation:

$\mathbb{N} = \{1, 2, 3, \ldots\}$

$\mathbb{Z} = \{\ldots, -3, -2, -1, 0, 1, 2, 3, \ldots\}$

Note the zero is an element or a member of $\mathbb{Z}$, but not an element or member of $\mathbb{N}$. We note that $\in$ means "belongs to," "belonging to," or "is."[8] In symbols, we write:

$0 \in \mathbb{Z}$ and $0 \notin \mathbb{N}$

Since $\neq$ means not equal to, then what do you think $\notin$ means? You are right, $\notin$ means "does not belong to."

If we consider $\mathbb{Z}$ and the arithmetical operation of addition, then we have discovered that adding zero to any element, or number, in $\mathbb{Z}$ leaves that element, or number, unchanged. The identity element of a set of numbers is a number in that set such that, when an arithmetic operation (i.e., addition, subtraction, multiplication, division) using this number is performed on any other number of the set, this operation leaves this second number unchanged. Given $\mathbb{Z}$, the identity element of addition is zero.

THE IDENTITY ELEMENT OF ADDITION AND SUBTRACTION

Using symbols, letting a stand for any number in $\mathbb{Z}$, we write the **Identity Element for Addition**:

For every $a \in \mathbb{Z}$, $a + 0 = a$

Note, again, the conciseness of this symbolic representation. Note also that if $a = 0$, then we have:

$0 + 0 = 0$

[7] An *identity element* is the element of a set of numbers that when combined with another number in an arithmetic operation leaves that number unchanged.

[8] The Italian mathematician Giuseppe Peano (1858-1932) first introduced this symbol in 1889. $\in$ is a stylized form of the Greek letter *epsilon*, ε, the first letter of the Greek word εστι, which means "is."

Is zero the identity element for subtraction? Again, let's suppose that Steve has three marbles, and Mary, not being a marble player, has none. How many more marbles does Steve have than Mary? Write an equation:

$3 - 0 = 3$

Yes, zero is the **Identity Element for Subtraction**. Letting a stand for any number in $\mathbb{Z}$, we write:

For every $a \in \mathbb{Z}$, $a - 0 = a$

Finally, make sure you understand these three observations:

1. When the minuend is non-zero, and the subtrahend is zero, the difference is the minuend. For example:

   $7 - 0 = 7$

2. When the subtrahend is nonzero, and the minuend is zero, the difference is the opposite of the subtrahend. For example:

   $0 - 7 = 0 + (-7) = -7$
   $0 - (-7) = 0 + 7 = 7$

3. When the minuend and subtrahend are equal, the difference is zero. For example:

   $7 - 7 = 0$ (Note: Both the minuend and subtrahend are the same; i.e., 7.)
   $-7 - (-7) = -7 + 7 = 0$ [Both the minuend and subtrahend are the same; i.e., -7.]

EXERCISES

Define the following terms:

1. Commutative Law of Addition
2. Proposition
3. Axiom
4. Theorem
5. Associative Law of Addition
6. Balance Beam Principle
7. Law of Increase in Addition
8. Law of Decrease in Addition
9. Law of Change in Addition
10. Law of Minuend Change in Subtraction
11. Law of Subtrahend Change in Subtraction
12. Counterexample
13. Identity
14. Identity Element
15. $\in$
16. $\notin$
17. Identity Element in Addition and Subtraction

Table 1: Mel Ott

Year	Home Runs	Year	Home Runs
1926	0	1937	31
1927	1	1938	36
1928	18	1939	27
1929	42	1940	19
1930	25	1941	27
1931	29	1942	30
1932	38	1943	18
1933	23	1944	26
1934	35	1945	21
1935	31	1946	1
1936	33	1947	0

Answer the following problems:

18. Mel Ott (1909-1958) was an American baseball player. He played baseball for the New York Giants from 1926 to 1947, performing as an outfielder and infielder. In Table 1, calculate the number of home runs that he hit during this period.

19. Willie Mays (1931) hit 660 home runs during his 22 year major league baseball career. How many more home runs did Mays hit than Ott?
20. If Mel Ott hit 36 home runs in 1941 and 15 home runs in 1945, what are the answers to Questions 18 and 19?

21. Sandy Koufax (1935-) was an American baseball pitcher. He played baseball for the Brooklyn and Los Angeles Dodgers from 1955 to 1966. In Table 2, calculate the number of batters that he struck out during this period.

Table 2: Sandy Koufax

Year	Strikeouts	Year	Strikeouts
1955	30	1961	269
1956	30	1962	216
1957	122	1963	306
1958	131	1964	223
1959	173	1965	382
1960	197	1966	317

22. Nolan Ryan (1947-) struck out 5714 batters in his 27-year major league baseball career. How many more strikeouts did Ryan have than Koufax?
23. If Sandy Koufax struck out 297 batters in 1963 and 250 batters in 1964, what are the answers to Questions 21 and 22?

Answer the following questions:
24. Show by a counterexample that the commutative law does not hold for subtraction.
25. Show by a counterexample that the associative law does not hold for subtraction.
26. Show by a counterexample that $a = -(-(-a))$ is false.
27. State the identity element for addition and subtraction using an equation with symbols.

Use the Associative Law of Addition to find x:
28. $19 + 18 + 82 = x$
29. $245 + 68 + 55 + 32 + 25 = x$
30. $9918 + 84 + 82 + 16 + 3 = x$
31. $27 + 53 + 8 + 47 + 92 + 73 = x$

Given $93 - 74 = 19$:
32. If we add 6 to 93, what happens to the difference?
33. If we add 6 to 74, what happens to the difference?

Given $456 + 128 = 584$:
34. If we add 6 to 456, what happens to the sum?
35. If we subtract 6 from 456, what happens to the sum?

Given $-82 + 93 = 11$:
36. If we add 6 to -82, what happens to the sum?
37. If we subtract 6 from -82, what happens to the sum?

Given $16 - (-9) = 25$:
38. If we add 6 to 16, what happens to the difference?
39. If we subtract 6 from 16, what happens to the difference?
40. If we add 6 to -9, what happens to the difference?
41. If we subtract 6 from -9, what happens to the difference?

Given $-16 - (-9) = -7$:
42. If we add 6 to 16, what happens to the difference?
43. If we subtract 6 from 16, what happens to the difference?
44. If we add 6 to -9, what happens to the difference?
45. If we subtract 6 from -9, what happens to the difference?

Show that the following equations are true:

46. $-(8 + 7) = (-8) + (-7)$
47. $-(8 + 7) = -8 - 7$

State the following relationships in words:

48. $a, b \in +\mathbb{Z} \Rightarrow a + b \in +\mathbb{Z}$

49. $a, b \in -\mathbb{Z} \Rightarrow a + b \in -\mathbb{Z}$

Assume that $a, b, c \in \mathbb{Z}$. We have these two relationships:

$a + b = c \Leftrightarrow a = c - b \qquad a + b = c \Leftrightarrow b = c - a$

Examples:

$4 + 5 = 9 \Leftrightarrow 4 = 9 - 5$

$6 + 5 = 11 \Leftrightarrow 5 = 11 - 6$

$-2 + 3 = 1 \Leftrightarrow -2 = 1 - 3$

$-2 + 3 = 1 \Leftrightarrow 3 = 1 - (-2)$

$-4 - 6 = -10 \Leftrightarrow -4 + (-6) = -10$

$-4 + (-6) = -10 \Leftrightarrow -4 = -10 - (-6) \Leftrightarrow -4 = -10 + 6$

$-4 + (-6) = -10 \Leftrightarrow -6 = -10 - (-4) \Leftrightarrow -6 = -10 + 4$

Given these relationships and the Algebraic Definition of Subtraction (where needed), find the value of x that makes the following equations true (Check your solution with the given equation):

50. $6 + x = 4$

    Ans. $6 + x = 4 \Leftrightarrow x = 4 - 6 \Leftrightarrow x = -2$

    Check: $6 + (-2) = 4$

51. $x + 8 = 12$
52. $-2 + x = 4$
53. $2 - x = 5$
54. $x - 3 = 7$
55. $-x + 4 = -1$
56. $4 - x = 8$
57. $-5 - x = -2$
58. $-7 + x = -10$
59. $-3 + x = 4$

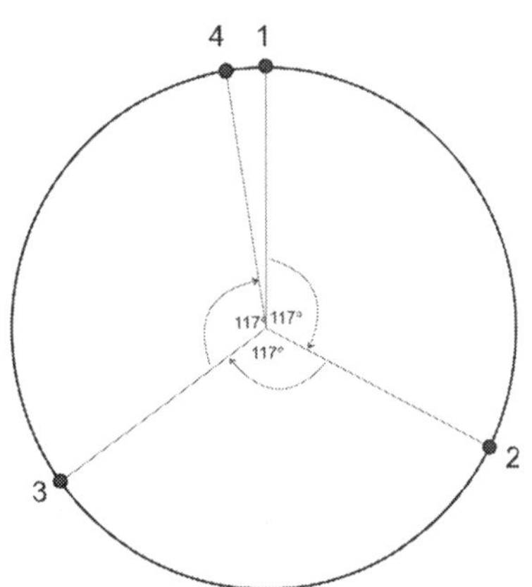

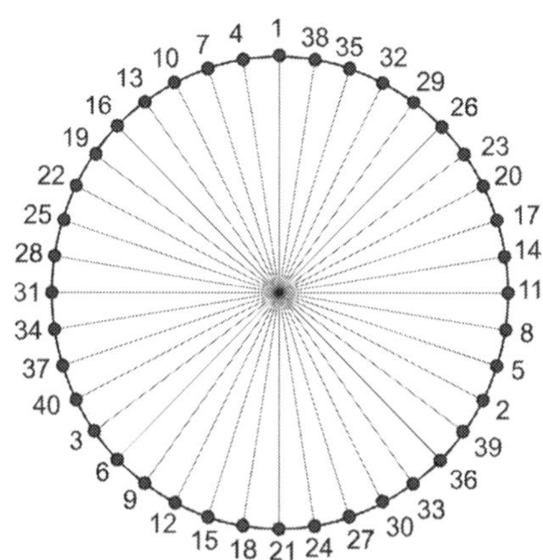

Figure 6. In the 17th century, the astronomer/mathematician Johannes Kepler (1571-1630) developed a geometric method to plot the places in the heavens when Jupiter and Saturn passed near each other, a conjunction (See Lesson 9.4). He knew, by observation, that if they pass each other at a specific point, in twenty years they would meet again, at a point 117° away, measured clockwise. The figure on the left shows the positions for a 60-year period. The figure at the right shows 800 years of positions (40 times 20, a harbinger of Step 4).

3.8 Counting Change

Counting backwards is another way to help you solve a subtraction problem. Let's consider the problem 50 – 3. By counting backwards, we start from 3 and count to 50:

3 is your starting point, and we ask, what must we add to 3 to get 10? 7
Then, what must we add to 10 to get 50? 40

I've counted backwards:

7 + 40 = 47

Note that this method is a way of using addition, the inverse of subtraction, to find the missing addend:

3 + 47 = 50

Let's try calculating the difference between 1000 and 182. Count from 182 to 1000:

182 + 18 = 200
200 + 800 = 1000
Answer: 18 + 800 = 818

We have just shown that 182 + 818 = 1000.

American currency	Value in cents	Value in dollars
Penny	1 cent	$.01
Nickel	5 cents	$.05
Dime	10 cents	$.10
Quarter	25 cents	$.25
Half-dollar	50 cents	$.50
Dollar	100 cents	$1.00

We can always check a subtraction problem such as 44 – 38 = 6 by adding 6 to 38 to get 44. Before the advent of electronic calculators and cash registers, sales clerks, when making change, used addition to check subtraction. Suppose you bought something for $3.27 and used a $5 bill for payment. Subtraction will show you that $5.00 – $3.27 = $1.73 (your change). Clerks used to calculate the change in this manner:

1. Customer gives the clerk $5.00 for the merchandise.
2. The clerk gives the customer the merchandise saying, "$3.27."
3. Next, the clerk gives three pennies to the customer saying "28, 29, 30" …
4. Next, the clerk gives the customer two dimes saying "3.40, 3.50" …
5. Next, the clerk gives the customer two quarters saying "3.75, 4 dollars" …[1]
6. Finally, the clerk gives the customer a dollar bill saying "5 dollars."

What the clerk did was check the change by adding it to the cost of what the customer purchased to come to the amount the customer tendered in payment. The change was 3 cents plus 20 cents plus 50 cents plus $1 (i.e., $1.73).

Let's try a few more examples. Cost of merchandise: $3.42. Paid: $5.00. How much change?

1. + 3 cents – say "$3.45."
2. + 5 cents – say "$3.50."
3. + 50 cents – say "$4.00."

---

[1] The clerk could give the customer one fifty-cent piece, a half-dollar, saying "4 dollars." Today, half-dollars are rare coins. If they are are obtained, many people save them instead of using them.

4. + 1 dollar – say "$5.00."
5. Change: $1.58.

Cost of merchandise: $3.75. Paid: $10.00. How much change?

1. + 5 cents – say "$3.80."
2. + 10 cents – say "$3.90."
3. + 10 cents – say "$4.00."[2]
4. + 1 dollar – say "$5.00."
5. + 5 dollars – say "$10.00."
6. Change: $6.25

Cost of merchandise: $1.79. Paid: $5.00. How much change? (Cover the sequence of steps below and see if you can make change in your head. Then, check your answer.):

1. + 1 cent – say "$1.80."
2. + 10 cents – say "$1.90."
3. + 10 cents – say "$2.00."
4. + 3 dollars – say "$5.00."
5. Change: $3.21

Cost of merchandise: $1.39. Paid: $2.00. How much change? (Cover, make change, and check.):

1. + 1 cent – say "$1.40."
2. + 10 cents – say "$1.50."
3. + 50 cents – say "$2.00."
4. Change: 61 cents

Cost of merchandise: $9.23. Paid: $10.00. How much change? (Cover, make change, and check.):

1. + 2 cents – say "$9.25."
2. + 25 cents – say "$9.50."
3. + 50 cents – say "$10.00."
4. Change: 77 cents.

Cost of merchandise: $13.32. Paid $20. How much change? (Cover, make change, and check.):

1. + 3 cents – say "$13.35."
2. + 5 cents – say "$13.40."
3. + 10 cents – say "$13.50."
4. + 50 cents – say "$14.00."
5. + $1 – say "$15.00."
6. + $5 – say "$20.00."
7. Change: $6.68

Cost of merchandise: $6.51. Paid: $7. How much change? (Cover, make change, and check.):

1. + 4 cents – say "$6.55."
2. + 5 cents – say "$6.60."
3. + 10 cents – say "$6.70."
4. + 10 cents – say "$6.80."
5. + 10 cents – say "$6.90."
6. + 10 cents – say "$7.00."

---

[2] The clerk could give the customer one quarter, saying "4 dollars," thus bypassing two steps.

7. Change: 19 cents.

I make a game of this when I buy merchandise with cash. I try to beat the cashier to the draw by stating the change I should get before the clerk's computer screen gives the information. Try doing the same. By doing so, you will increase your ability to add sums quickly in your head.

In conclusion, you are free to use complementary subtraction or count backwards. Always look at a subtraction problem first and then determine the most efficient way to calculate the answer.

EXERCISES

Subtract the following numbers: (Try using shortcuts and check each answer.)
1. 68 – 29 (Hint: Consider 68 – 30.)
2. 157 – 38 (Hint: Consider 157 – 40.)
3. 97 – 59 (Hint: Consider 97 – 60.)
4. 85 – 46 (Hint: Consider 85 – 45.)
5. 625 – 98
6. 3898 – 993
7. 8225 – 6126
8. 13,687 – 8199
9. 53,452 – 9998
10. 729,123 – 426,099

Do the following set of exercises in your head because you should be doing this mentally every time you receive change from a cashier: (Assume that the cashier has sufficient groups of these coins (1 cent, 5 cents, 10 cents, 25 cents) and these bills [$1, $5, $10, and $20].)

If you give a cashier $1, state the change you should receive from a purchase of:
11. $0.28
12. $0.57
13. $0.76
14. $0.91

If you give a cashier $5, state the change you should receive from a purchase of:
15. $3.18
16. $1.09
17. $4.85
18. $2.25

If you give a cashier $10, state the change you should receive from a purchase of:
19. $1.25
20. $5.75
21. $6.19
22. $9.33

If you give a cashier $20, state the change you should receive from a purchase of:
23. $1.09
24. $3.25
25. $15.12
26. $9.75

If you give a cashier $50, state the change you should receive from a purchase of:
27. $3.10
28. $0.99
29. $21.99
30. $35.27

Calculate the number of years from the date of each of the following until the year 2010 and check each answer: (You can either count backwards or use complementary subtraction.)
31. The Arabs first used the decimal system in 890.
32. Christopher Columbus discovered the New Word in 1492.
33. Martin Luther posted his 95 theses objecting to abuses within the Roman Catholic Church in 1517.
34. Galileo Galilei first used the telescope to see the moons of Jupiter in 1610.

35. Isaac Newton's first published *Principia Mathematica* (Mathematical Principles of Natural Philosophy) in 1687.
36. The first crossing of the Atlantic Ocean by steam was in 1839.
37. The first use of the electric telegraph in the United States was in 1844.
38. The American War Between the States ended in 1865.
39. The Wright brothers first flew in a fixed-wing, engine-powered, heavier-than-air machine in 1903.
40. World War I ended in 1918.
41. World War II ended in 1945.
42. A human being first set foot on the Moon in 1969.

43. Choose any five of the above dates and calculate the difference between that year and the year of your birth. Check each answer.

Archimedes (ca. 287-212 BC), an ancient Greek mathematician, used the balance-beam principle for many applications. One of these is the hanging mobile. In the following problems (44-46), find the value of each weight, in grams (abbreviated g), in these balanced mobiles. Disregard the weight of the connecting rods.

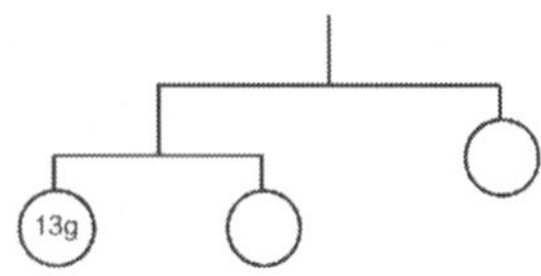

Figure 1.

44. (Hint: 13 g will balance with 13g.)

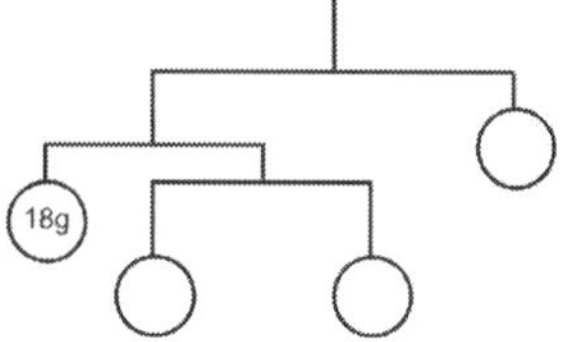

Figure 2.

45. (Hint: 18 g will balance with two equal weights that total 18 g.)

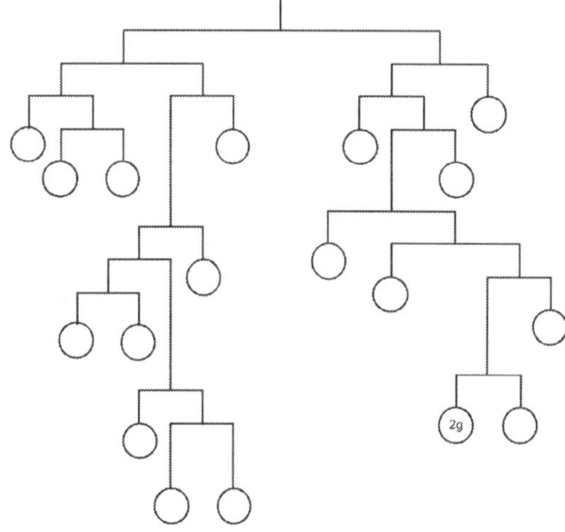

Figure 3.

46.

Lifetime Field Project: Begin to calculate change when you buy anything in a store; i.e., buy an object using cash and learn to figure the change in your head before you receive it from the store clerk. If you are a store clerk, practice this method. Your customers might be astounded!

A Brief History of Mathematics per to Archaeological and Historical Records: BC		
Date	**Location**	**Development**
ca. 2400 BC	Sumeria	Place Value System.
ca. 1750 BC	Babylonia (Southern Iraq)	Cuneiform writing to record numbers.
ca. 1650 BC	Egypt	Hieroglyphics first used to record numbers.
ca. 1550 BC	China	Decimal numbers first used; bamboo rods used to record numbers.
ca. 900 BC	India	First use of zero.
ca. 300 BC	Greece	Euclid writes *The Elements*, a thirteen-volume work on geometry and *logistica*. These volumes become the standard text for the next 2000 years.
ca. 100 BC	China	First use of negative numbers.

From ca. 4 BC–30 AD, in Palestine: The life, death, resurrection, and ascension of the Lord Jesus Christ, the Creator, Sustainer, and Light of the Cosmos, the God-man titled the Savior of the World, whose impact on every branch of learning and culture has been immeasurable.

A Brief History of Mathematics per Archaeological and Historical Records: AD		
Date	**Location**	**Development**
800	Arabia	The beginning of symbolic algebra.
1000	Europe	Decimal number system appears because of trade with Muslims.
1000-1500	England and Europe	A host of seminal mathematical and scientific concepts were developed by a line of medieval theologians whose starting points were three-fold: (1) The Biblical God was fully rational, (2) His creation was also fully rational and (3) Man, made in God's image, can explore and understand this rational and orderly creation. These three starting points proved to be the necessary ingredients for the birth of modern science, along with its associated mathematical developments, which could nurture itself into a self-sustaining enterprise.
1514	Holland	First use, in the modern sense, of the symbols for addition and subtraction: + and –.
1614	Scotland	John Napier introduces logarithms, a tool for quickly multiplying and dividing by adding and subtracting.
1630s	France	Development of analytic or coordinate geometry based upon seminal work done by scholastics in the 14th century.
1660s	England	Beginnings of a new branch of mathematics, Statistics.
1660-1670s	England and Germany	Isaac Newton and Gottfried Wilhelm Leibniz simultaneously developed Calculus, a singular mathematical and scientific gift to the world by Western civilization.
1830s	Germany	Explorations in geometries that are different than the geometry developed by Euclid.
1960	United States	Development of fractal geometry.
1980s	United States	Development of Chaos Theory to understand complex systems like the weather.

Figure 4. Published in *Nature* (1957) by the British information theorist Donald M. MacKay (1922-1987). The eye is stimulated to see a scintillating pattern of fine lines that appear to run at right angles to the rays.

STEP 4. THE PULSE OF MULTIPLICATION

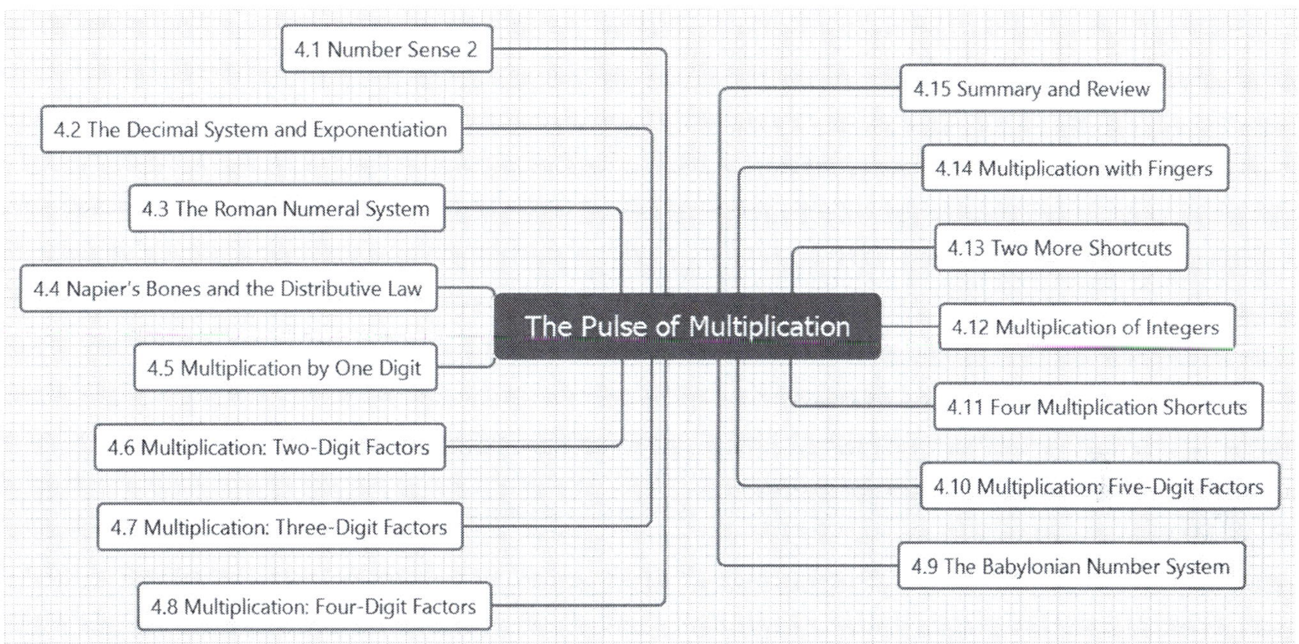

> Mathematics, like the Nile, begins in minuteness, but ends in magnificence.
>
> Charles Caleb Colton (1780-1832).

4.1 Number Sense 2

Terms & Concepts Introduced
1. Algebraic expression
2. Factor
3. Multiple
4. Multiplicand
5. Multiplication
6. Multiplier
7. Product
8. Signs of multiplication

In Step 4 we will explore the operation of multiplication. For the operations of addition and subtraction, we used the Japanese Soroban to guide us into a fast and easy way of computing with complements. Multiplication, along with its inverse, division, uses addition and subtraction as intermediate operations so we will be able to use complements to make the work easier.

Figure 1. Three-leaf clover. Source: iStockPhoto.

We define multiplication as repeated addition of the same number, or more briefly as multiple addition. Multiplication builds on addition. In addition, we count forward. In multiplication, we count forward by groups of units instead of by "one and one and one, etc." We repeat addition by repeating a unit several times. If we note that a clover has three leaves, and we have a collection of four clovers, the total number of leaves in our collection is twelve. We do this addition:

$3 + 3 + 3 + 3 = 12$

Three is repeated as four addends to get the sum of 12. Multiplication, therefore, is a higher-order operation when compared to addition. It is counting forward faster than counting forward by addition. Remember this.

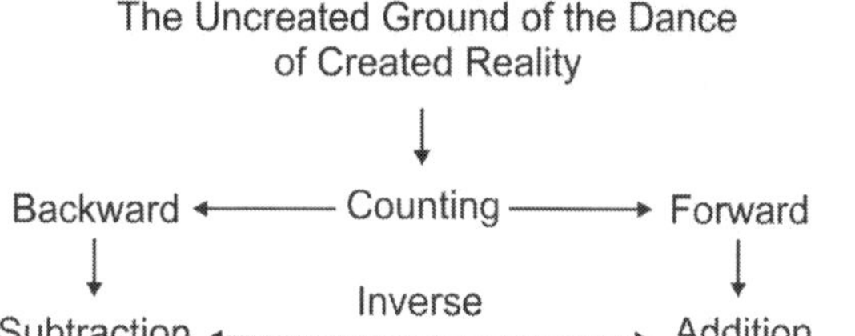

Figure 2.

Multiplication Signs

Using symbols, we write: $3 \times 4 = 12$. The English mathematician William Oughtred (1575-1660) introduced the "×" sign of multiplication in 1631 in his book *Clavis mathematicae* (*Key to Mathematics*). We read this sign as "times," or "multiplied by." When placed between two numbers, it indicates that we multiply the two numbers together.

There are four symbols commonly used for multiplication:

Cross:[1] ×
Dot:[2] ·
Star:[3] *
Parentheses: ()

[1] The cross, based on the St. Andrew's Cross, originates from the cross-multiplication of proportions (Lesson 6.7 and Lesson 8.1). See Lesson 4.14 for another possible origin of this symbol.
[2] In some countries (e.g., Ireland, Korea, Malaysia, New Zealand, Philippines, Singapore, Taiwan, Thailand, United Kingdom), the dot is used as a decimal point; i.e., 3.45 can be written as 3·45.
[3] The star (*) is commonly used in computer code to signify multiplication.

Therefore, we can also write:

$3 \cdot 4 = 12$
$3 * 4 = 12$
$(3)(4) = 12$

Figure 3. The German mathematician Gottfried Wilhelm Leibniz (1646-1716) first employed this symbol to avoid confusion with the normal use of the algebraic variable x. In 1686, he started using the dot symbol for multiplication.

The general form of an operation of multiplication includes three components:

In Latin (15th century),	In English,
Numerus multiplicandus	Multiplicand
Numerus multiplicans	× Multiplier
Numerus productus	= Product

Multiplication is the process of taking one of two numbers as many times as there are units in the other. Or, it is a short-cut method of adding equal numbers.[4] This economy is not that recognizable when multiplying a single digit number by a single digit number (e.g., 4 × 6), but it comes in handy for large numbers. If a shepherd had a pasture of 3000 sheep, to count the number of legs by addition he must do this:

$$\underbrace{4 + 4 + \ldots + 4}_{3000 \text{ addends of } 4}$$

Using multiplication, just one operation is needed:

4×3000

The number to be multiplied is the **multiplicand**. The **multiplier** is the number by which to multiply. The multiplier shows how many times we add the multiplicand. The **product** means "something produced." It is the result obtained by multiplication.

THE COMMUTATIVE LAW OF MULTIPLICATION

If we represent multiplication horizontally, 4 × 5 (read "four times five") means 4 + 4 + 4 + 4 + 4 (the multiplicand appears as an addend five times). Four is the multiplicand, and five is the multiplier. 5 × 4 (read "five times four") means 5 + 5 + 5 + 5 (the multiplicand appears as an addend four times). Five is the multiplicand, and four is the multiplier. Notice that 4 × 5 and 5 × 4 generate the same product, i.e., 20. The commutative law holds for multiplication as well as addition. (See Lesson 5.15 for a formal definition.)

Figure 4. Honeycomb. Source: iStockPhoto.

The multiplier is in most cases regarded as an abstract number while the multiplicand and the product

[4] *Multiplication* comes from two Latin words, *multus*, meaning "many," and *plicare*, meaning "to fold." Multiplication is like the English word *manifold* meaning "of many kinds." English words like *three-ply* and *four-ply* illustrate the use of *plicare*. From these meanings, we get the idea of the "addition or folding of many numbers."

are like numbers (Lesson 2.2), either both abstract or both concrete. For example, consider a regular hexagon (Lesson 2.3), the honeycomb, with each of its six sides measuring 3 cm. Since all its sides are equal because it is regular. To find the perimeter, we multiply:

$(3 \text{ cm})(6) = 3 \text{ cm} + 3 \text{ cm} + 3 \text{ cm} + 3 \text{ cm} + 3 \text{ cm} + 3 \text{ cm} = 18 \text{ cm}$

The multiplicand is the measure of each side, 3 cm, a concrete number. The multiplier is the number of sides, 6, an abstract number. The perimeter, 18 cm, is a concrete number. When a multiplication problem contains a concrete number and an abstract number, the concrete number is the multiplicand and the abstract number is the multiplier. We write:

multiplicand × multiplier = product

$3 \text{ cm} \times 6 = 18 \text{ cm}$

When either the multiplicand or the multiplier has more than one digit or both have more than one digit, we usually assign the multiplier to be the number with the lesser number of digits. You will discover, if you haven't already, that when writing a multiplication problem vertically, it is more efficient to calculate a product when the number of digits of the multiplicand is greater than the number of digits of the multiplier. For example, it is much easier to calculate 876 times 9 like this:

$$\begin{array}{r} 876 \\ \times\ 9 \\ \hline \end{array}$$

The multiplication is a little more cumbersome when you set it up this way:

$$\begin{array}{r} 9 \\ \times\ 876 \\ \hline \end{array}$$ (It still can be done, though.)

FACTORS AND MULTIPLES

When looking at the multiplicative structure of a given number, the terms multiplicand and multiplier are substituted by the word **factors**. For example, 3 and 6 are factors of 18 because $3 \times 6 = 18$.[5] Also, 18 is a **multiple** of both 3 and 6.[6]

If, during the summer, an ice-cream truck visits a neighborhood every three days, the truck will visit on day 3, 6, 9, 12, 15, 18, 21, 24, etc. By this process (counting by three), we have found the multiples of three. The fifth multiple of six is 5×6, or 30. The sixth multiple of five is also 30 (6×5). The eighth multiple of nine is 8×9, or 72. The ninth multiple of eight is also 72 (9×8).

Remember our previous discussion (Lesson 2.4) about number sense where we had arranged six pennies in two rows?

This arrangement of pennies could also represent a group of trees in an orchard, or people in chairs. We draw pictures like this to represent arithmetical operations. With addition, instead of counting from one to six, you could add three to three. Instead of adding, we have joined two groups of three together to get six. Or, you could say that there are two rows of three columns and invoke multiplication. Using symbols:

$2 \times 3 = 6$

---

[5] When two numbers are multiplied, both numbers are *factors* of the product.

[6] A *multiple* is a number that is the product of another number an integral, or whole, number of times. For example, 18 is a multiple of 6 because $18 = 6 \times 3$.

The product 6 is contained in the two factors, 2 and 3, and the two factors are contained in the product. Like the addends and the sum, the factors and the product mutually indwell each other.

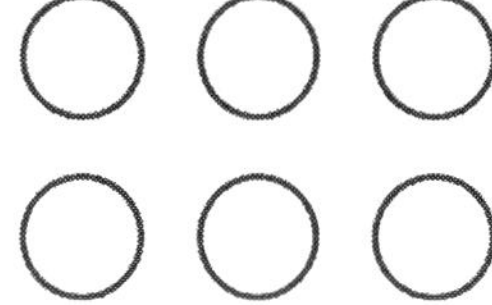

Figure 5. $2 \times 3 = 6$

By experimenting with groups of coins, we gain a rudimentary understanding of the perichoretic relationships that exist between numbers.

The numbers 2 and 5 for a group of 10 pennies mutually indwell each other; they inhere in each other under the operation of multiplication.

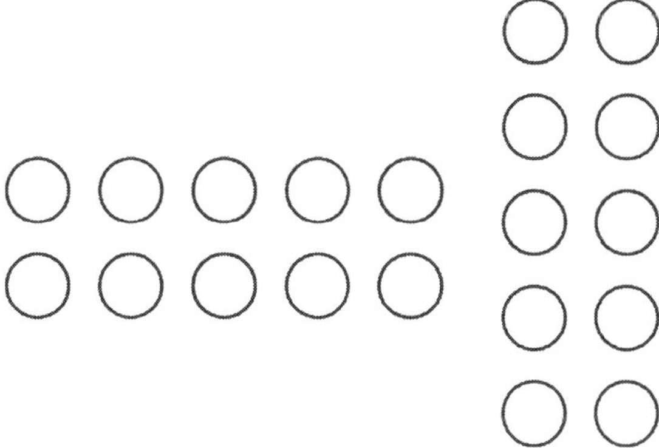

Figure 6. $2 \times 5 = 5 \times 2 = 10$

We see the same perichoresis in 3 and 4, and 2 and 6, for a group of 12 pennies.

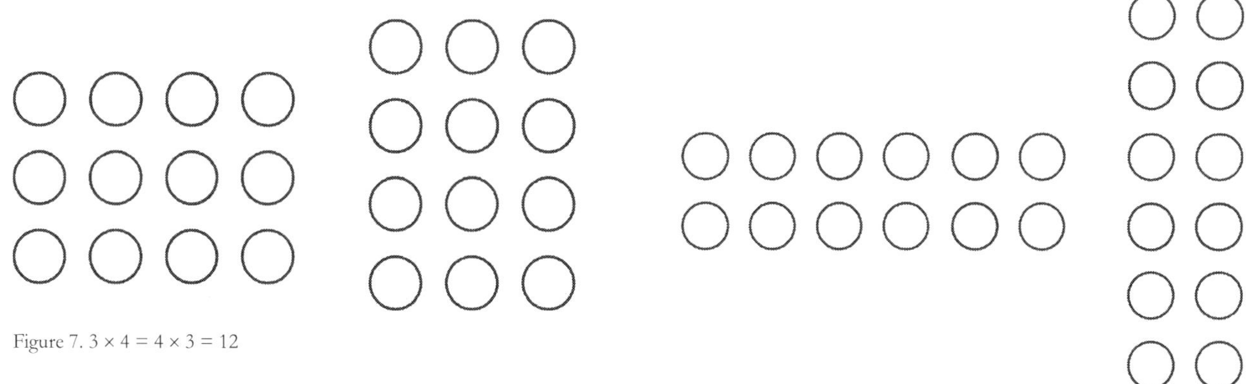

Figure 7. $3 \times 4 = 4 \times 3 = 12$

Figure 8. $2 \times 6 = 6 \times 2 = 12$

2 and 10, and 4 and 5, generate relational row/column arrangements for a group of 20 pennies.

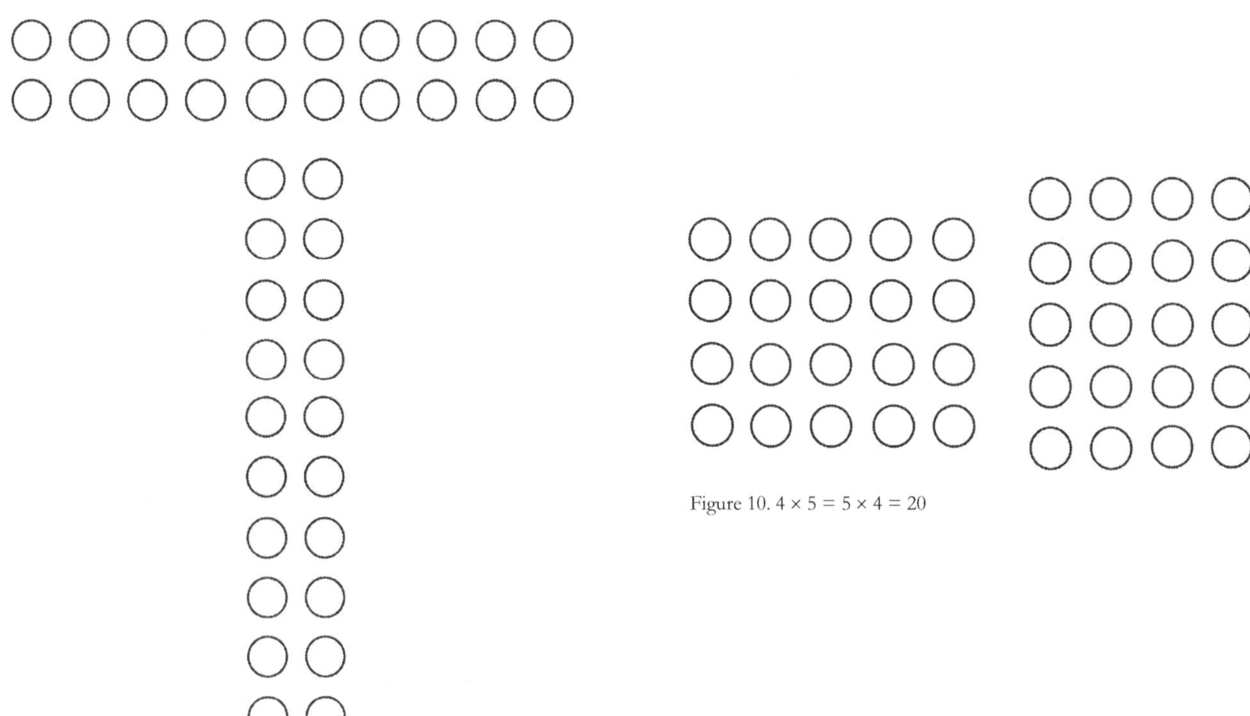

Figure 9. $2 \times 10 = 10 \times 2 = 20$

Figure 10. $4 \times 5 = 5 \times 4 = 20$

From these observations, we see that it is often easier to see many objects by arranging them in ordered groups. We also discover that there is a connection between addition and multiplication. What is it?

$5 + 5 = 5 \times 2 = 10$
$2 + 2 + 2 + 2 + 2 = 2 \times 5 = 10$

(Note: $5 \times 2 = 2 \times 5$)

$3 + 3 + 3 + 3 = 3 \times 4 = 12$
$4 + 4 + 4 = 4 \times 3 = 12$

(Note: $3 \times 4 = 4 \times 3$)

$6 + 6 = 6 \times 2 = 12$
$2 + 2 + 2 + 2 + 2 + 2 = 2 \times 6 = 12$

(Note: $6 \times 2 = 2 \times 6$)

$4 + 4 + 4 + 4 + 4 = 4 \times 5 = 20$
$5 + 5 + 5 + 5 = 5 \times 4 = 20$

(Note: $4 \times 5 = 5 \times 4$)

With natural numbers, multiplication and addition unfold a beautiful coinherence because we can understand multiplication as repeated addition of the same number.

For example, 6×3 means that we are adding three groups of six:

$6 + 6 + 6 = 18$

3×6 means that we are adding six groups of three:

$3 + 3 + 3 + 3 + 3 + 3 = 18$

As we have noted, since we can understand addition as counting forward, multiplication is an extension of this type of counting.

ALGEBRA AND MULTIPLICATION

If letters are used to represent numbers, as is common in Algebra and in defining general rules or properties, we drop the ×, *, ·, and ()(). If the letters a and b represent numbers, then we represent a "times" b as:

ab (Note: We say "ab." We do not say "a times b.")

But, not like this:

$a \times b$

$a \cdot b$

$a * b$

$(a)(b)$

$(a)b$

$a(b)$

> The symbol × for multiplication is primarily used in Arithmetic but very seldom used in Algebra and beyond.

Dropping the sign makes sense because, since multiplication is like counting; "six marbles" and "six times a marble" mean the same thing. As in counting, we do not write a sign between the number and its denomination; we just say or write "six marbles." So, if we let x represents marbles, we write six marbles as $6x$. You will see a lot of this kind of abbreviation in algebraic work.[7]

THE MULTIPLICATION TABLE

Proficiency in multiplication means mastery of Table 1.[8]

Table 1

×	2	3	4	5	6	7	8	9
2	4	6	8	10	12	14	16	18
3		9	12	15	18	21	24	27
4			16	20	24	28	32	36
5				25	30	35	40	45
6					36	42	48	54
7						49	56	63
8							64	72
9								81

> I had been to school ... and could say the multiplication tables up to 6 x 7 = 35, and don't reckon I could ever get any further than if I was to live forever. I don't take stock of mathematics.
>
> Mark Twain, *Huckleberry Finn* (1884).

[7] The symbols "$6x$" represent an **algebraic expression**. An algebraic expression, cf. numeric expression in Lesson 3.1, is a combination of letters, numerals, and symbols representing arithmetic operations. We say "six x" not "six times x."

[8] The ancient mathematician Nicomachus (ca. 60-120), born in the Roman city of Gerasa, north of Jordan, gives the multiplication table precisely as we learn it today in his book *Introductio Arithmeticae* (Introduction to Arithmetic). The *Introductio* was the standard textbook in arithmetic for nearly one thousand years.

The two one-digit factors are in row one and column one, meaning that, for example, 4 and 6 are factors of 24. Of course, $12 \times 2 = 24$, but we are only considering one-digit factors.

Only half of the table is completed because if we know that $2 \times 3 = 6$, we also know that, by the Commutative Law of Multiplication, $3 \times 2 = 6$.

We do not include zero and one since they reveal two important properties in multiplication:

1. Zero multiplied by any number is zero. In symbols, $0a = 0$ for every number a.
2. One multiplied by any number is that number. In symbols, $1a = a$ for every number a. This relationship means that 1 is the identity element for multiplication.

We will formally define these properties in Lesson 5.15.

The patterns in Table 1 will assist us in the memorization of the following relationships.

MULTIPLICATION BY 2

We can understand four as a pair of pairs, or ten as a pair of fives. Therefore, if we multiply by 2, we double the other factor. For example, $2 \times 7 = 14$ because $7 + 7 = 14$. Doubling seven generates 14. Doubling 9 generates 18, $9 + 9 = 18$, or $2 \times 9 = 18$.

MULTIPLICATION BY 3

If we multiply by 3, we double the other factor and then add that factor. For example, $3 \times 7 = 21$ because doubling 7 is 14 and then adding 14 to 7 is 21.

$3 \times 7 = (7 + 7) + 7 = 14 + 7 = 21$

MULTIPLICATION BY 4

If we multiply by 4, we double the other factor and then double the result. For example, $4 \times 8 = 32$ because doubling 8 is 16 and doubling 16 is 32.

$4 \times 8 = (8 + 8) + (8 + 8) = 16 + 16 = 32$

Do you see the two doublings? See the two groups of $8 + 8$?

> When doing any serious mathematical thinking, you're going to have to multiply 6 by 8 sometimes, and if you have to reach for your calculator each time you do that, you'll never achieve the kind of mental flow that actual thinking requires. You can't write a sonnet if you have to look up the spelling of each word as you go.
>
> Jordan Ellenberg, *How Not to Be Wrong: The Power of Mathematical Thinking* (2014), p. 57.

MULTIPLICATION BY 8

If we multiply by 8, we double the other factor three times. For example, $5 \times 8 = 40$ because doubling 5 is 10, doubling 10 is 20, and doubling 20 is 40. Note:

$5 \times 8 = (5 + 5) + (5 + 5) + (5 + 5 + 5 + 5) = 10 + 10 + 20 = 40$

Do you see the three doublings? See the four groups of $5 + 5$?

MULTIPLICATION BY 10

Multiplying by 10 is easy; we just append 0 to the other factor.
For example:

$15 \times 10 = 150$

MULTIPLICATION BY 5

If we multiply by 5, we multiply the other factor by 10 and then take half of that.
For example:

$5 \times 8 = 40$ because half of 80 is 40. To find half of a number, we ask, "Since 80 is equal to the sum of two equal addends, what are those addends?" ($80 = 40 + 40$)

When we take half of a number, we enter the world of division (Step 5).

Another example:

$5 \times 9 = 45$ because half of 90 is 45. Fast, isn't it?

We note:

$90 = 45 + 45$

MULTIPLICATION BY 6

If we multiply by 6, we multiply the other factor by 5 and then add that factor to the result. For example, $6 \times 7 = 42$ because $5 \times 7 = 35$ and $35 + 7 = 42$. Or, $3 \times 6 = 18$ because $3 \times 5 = 15$ and $15 + 3 = 18$.
We are doing this:

$6 \times 7 = 7 \times 6 = 7 \times (5 + 1) = (7 \times 5) + (7 \times 1) = 35 + 7 = 42$
$3 \times 6 = 3 \times (5 + 1) = (3 \times 5) + (3 \times 1) = 15 + 3 = 18$

A very important principle is invoked for this rule. It is named the Distributive Law of Multiplication over Addition. We shall define, investigate, explore, and invoke this key principle of multiplication in Lesson 4.4. For now, ponder the two examples carefully, noting where the source of "35 + 7" and "15 + 3."

MULTIPLICATION BY 7

If we multiply by 7, we multiply the other factor by 5 and then add twice that factor to the result. For example, $4 \times 7 = 28$ because $4 \times 5 = 20$, $4 \times 2 = 8$, and $20 + 8 = 28$. Or, $9 \times 7 = 63$ because $9 \times 5 = 45$, $9 \times 2 = 18$, and $45 + 18 = 63$.
Invoking the Distributive Law, we are doing this:

$4 \times 7 = 4 \times (5 + 2) = (4 \times 5) + (4 \times 2) = 20 + 8 = 28$
$9 \times 7 = 9 \times (5 + 2) = (9 \times 5) + (9 \times 2) = 45 + 18 = 63$

MULTIPLICATION BY 9

If we multiply by 9, we subtract one from the other factor to form the tens position of the product. For the ones position, write the complement of 9. For example, $9 \times 8 = 72$. Note that the tens position is $8 - 1 = 7$. The complement of 7 is 2, the digit in the ones position. Now, that is easy! We will study this amazing property and how it relates to division in Lesson 5.4.

These multiplication principles are applicable for multiplying any number by one-digit. We will engage in this type of multiplication again in Lesson 4.5.

EXERCISES

Define the following terms:

1. Multiplication
2. Multiplicand
3. Multiplier
4. Product
5. Factor
6. Multiple
7. Algebraic expression

8. What are the signs of multiplication?

9. If m and n are factors of a number, which symbolic representation correctly represents "m times n" when used in Algebra?"

   (a) $m \times n$
   (b) mn
   (c) $m * n$
   (d) $(m)n$
   (e) $(m)(n)$
   (f) $m(n)$
   (g) $m \cdot n$

10. Master the rules for multiplying by a single digit by 2, 3, 4, 5, 6, 7, 8, 9 (Table 1), and 10.

Calculate the first ten multiples of the following numbers:

11. 2 (Hint: the first two multiples of 2 are 2 and 4.)
12. 3
13. 4
14. 5
15. 6
16. 7
17. 8
18. 9
19. 10
20. 26
21. 53

Tally marks, as represented by the barred-gate system (Lesson 1.2), still serve a purpose when the aim is to record a quantity that increases by one, such as the number of points scored in a basketball game.

Key:
2-FGM (Two-point field goals made)
2-FGA (Two-point field goals attempted)
3-FGM (Three-point field goals made)
3-FGA (Three-point field goals attempted)
FTM (Free throws made: worth one point)
FTA (Free throws attempted)
OREB (Offensive rebounds)
DREB (Defensive rebounds)
REB (Total rebounds = OREB + DREB)
A (Assists)
TO (Turnovers)
PF (Personal fouls)
TP (Total points)

Figure 11. Basketball. Source: Public Domain.

Consider this score sheet:

Player	2-FGM	2-FGA	3-FGM	3-FGA	FTM	FTA	OREB	DREB	REB	A	TO	PF	TP
Stevens	IIII	卌 卌 I	I	III	I	IIII	I	III		I		II	
Gatlin	卌 II	卌 III			IIII	IIII		III		III	卌	卌	
Norcross	卌 II	卌 卌 IIII	III	卌	卌 I	卌 I	I	III		I			
Wiebe		I		I	II	II		III		II	I	I	
Barnett	III	卌 卌 IIII	III	卌 II	I	II	II	IIII		卌		I	
Davis	II	III			I	II	IIII	I				卌	
Hauptmann	III	卌	I	II		II		III		III	II	II	
Whitson		I					I			I		I	

22. Re-create the table replacing the tally marks with numbers. Perform the arithmetic to determine total rebounds, total points per player, and totals of all columns for the team.

You meet a mother who has three daughters. She tells you, "The ages of my daughters are positive integers and the product of their ages is 36." You ask the mother, "How old are your daughters?" She responds, "If I told you the sum of their ages, it wouldn't help you, but my oldest daughter loves her pet parakeet."

23. From this information, state the ages of the three daughters. (Hint: The mother has given you three clues that will help you solve this problem.)

> In learning that two times two equals four, the child is thinking conjunctively on three levels at the same time, the word level, the number or symbolical level, and the empirical level [based on experience – JN], and has no difficulty in doing so – yet that turns out to be precisely the structure of the most developed science.
>
> Thomas F. Torrance, "The Framework of Belief," in *Belief in Science and in Christian Life* (1980), p. 23.

4.2 The Decimal System and Exponentiation

Terms & Concepts Introduced
1. Algebraic term
2. Area
3. Base of the power
4. Coefficient
5. Congruent figures
6. Exponent or degree of the power
7. Exponential form
8. Exponentiation
9. Formula
10. Googol
11. Googolplex
12. Power
13. Solid
14. Unit cube
15. Unit of area
16. Unit of volume
17. Unit square
18. Variable
19. Volume

The base of the decimal system is 10. We have already made use of 10 in our work with complementary addition and complementary subtraction. In our study of place value, we also discovered that the orders increase (from 1st to 2nd to 3rd, etc.) by a factor of 10.

Table 1: Place Value

Trillions (5th) period			Billions (4th) period			Millions (3rd) period			Thousands (2nd) period			Ones (1st) period		
15	14	13	12	11	10	9	8	7	6	5	4	3	2	1
Hundreds	Tens	Ones	Hundreds	Tens	Ones	Hundreds	Tens	Ones	Hundreds	Tens	Ones	Hundreds	Tens	Ones

We are all familiar with machines, pulleys, levers, gears, etc. that man uses to multiply his strength. When a man breaks in a horse, he passes a rope round a post. The man holds one end and fastens the other end to the
horse. For the horse to get away, it must pull many times harder than the man. The rope and post serve to multiply the muscle power of the man.

Exponentiation

Suppose that we have a rope and a post that together multiplies one's strength by ten when the rope makes one complete turn (Figure 1).

Figure 1.

What will be the effect if we have a series of such posts? After one turn, post 1, your strength is increased by 10. After a second turn, post 2, your strength is increased by 100 (10 × 10). After a third turn, post 3, your strength is increased by 1000 (100 × 10), etc.

This sequence of power increase interpenetrates the orders in the decimal place value system: 1, 10, 100, 1000, 10,000, etc. We show this in Table 2.

Table 2

Order	Power Increase Factor	Multiplication	Post
1st	1		0
2nd	10	1 × 10 = 10	1
3rd	100	10 × 10 = 100	2
4th	1000	100 × 10 = 1000	3
5th	10,000	1000 × 10 = 10,000	4
6th	100,000	10,000 × 10 = 100,000	5
7th	1,000,000	100,000 × 10 = 1,000,000	6
8th	10,000,000	1,000,000 × 10 = 10,000,000	7

This table is an open window for the exploration of one more way in which mathematicians economize their work. Consider column two, the power increase. Is there a shorter way to write large numbers like 1,000,000? Before we reveal it, consider 1,000,000. How many times must you multiply 10 by itself to get 1,000,000? Inspect Table 3.

Table 3

Power Increase Factor	Multiplication by Factors That Are 10
100	10 × 10 = 100
1000	10 × 10 × 10 = 1000
10,000	10 × 10 × 10 × 10 = 10,000
100,000	10 × 10 × 10 × 10 × 10 = 100,000
1,000,000	10 × 10 × 10 × 10 × 10 × 10 = 1,000,000

In these examples, we are repeating the same factors. When we defined multiplication as repeated addition of the same addend, we observed that multiplication is an extension of counting. Now we have an operation in which we repeatedly multiply by the same factor. This process is an extension of multiplication, which is an extension of addition, which is an extension of counting. Counting takes us a long way!

Repeatedly multiplying by the same number is **exponentiation.**[1] A **power** of a number, an appropriate name given our fence post example, is the product of factors, each of which is equal to that number. Therefore, 100 is the second power of 10 since 100 = 10 × 10.

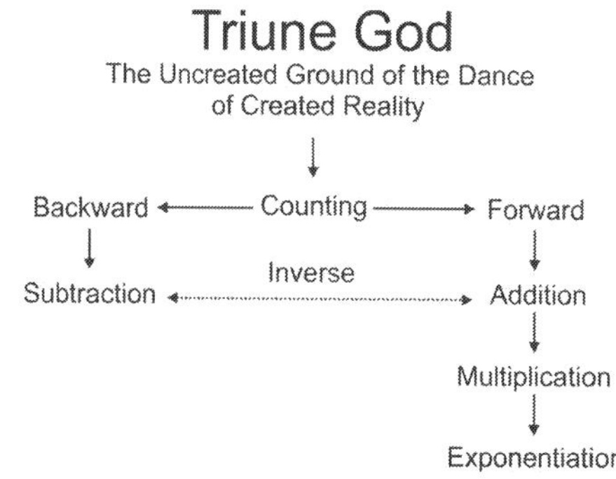

Figure 2.

[1] *Exponentiation* comes from the Latin word *exponere,* which means "expound." *Expound* means "to set forth" or to "put out." In exponentiation, the exponent is "put out" using the superscript. In older textbooks, exponentiation was named "involution," meaning, in Latin, to "roll up."

Mathematicians symbolize exponentiation or a power by using superscript notation: $10 \times 10 = 10^2$ (say "ten raised to the second power" or "ten raised to the power of two" or "ten squared"). Ten is the **base of the power** (this should make sense) and two, as a superscript,[2] is the **exponent of the power** or the **degree of the power.**[3] The degree tells us how many times the base is used as a factor to produce the power. It is the number of posts used to multiply our strength. When we see 10^2 we know we are dealing with "the second power of ten" or "ten as a factor two times" (10×10).

We now have Table 4.

Table 4			
Power of 10	**Repeated Multiplication**	**Exponential Form**	**Posts**
100	10×10	10^2	2
1000	$10 \times 10 \times 10$	10^3	3
10,000	$10 \times 10 \times 10 \times 10$	10^4	4
100,000	$10 \times 10 \times 10 \times 10 \times 10$	10^5	5
1,000,000	$10 \times 10 \times 10 \times 10 \times 10 \times 10$	10^6	6

Exponential form, that is, $\text{base}^{\text{exponent}}$, is economy extraordinaire. Do you see the pattern? The exponent not only tells us how many times we have repeatedly multiplied the base, it tells you how many zeroes are after the numeral 1 in the multiple or power of 10.

Scientists use the exponential form to write very large numbers. For example, astronomers estimate the number of stars in the universe to be between 10^{21}, 1 followed by 21 zeroes, and 10^{26}, 1 followed by 26 zeroes.[4] Quite a number, isn't it? Note the declaration of worship by the Psalmist David (in his day, he could probably count three to four thousand stars with the naked eye):

> "He counts the number of the stars; He gives names to all of them. Great is our Lord and abundant in strength; His understanding is infinite." (Psalm 147:4-5).

The prophet Isaiah echoes the same sentiments:

> "Lift up your eyes on high and see who has created these stars, the One who leads forth their host by number, He calls them all by name; Because of the greatness of His might and the strength of His power, not one of them is missing." (Isaiah 40:26).

Table 5 is an extenstion of Table 4. Based on the pattern revealed, what should go in the blanks (column three, first two rows)?

Table 5			
Power of 10	**Repeated Multiplication**	**Exponential Form**	**Posts**
1	1		0
10	10		1
100	10×10	10^2	2
1000	$10 \times 10 \times 10$	10^3	3

---

[2] We have French mathematician and philosopher, René Descartes (1596-1650), to thank for the use of superscripts to denote exponents.

[3] In England and other Commonwealth nations, index is used instead of exponent or degree.

[4] If we resize a galaxy to the shape of a pea, all the galaxies in the universe would fill a large auditorium like the Royal Albert Hall in London, England. Counting one star every second, it would take us approximately ten thousand trillion centuries to count to 10^{26}.

Table 6			
Power of 10	Repeated Multiplication	Exponential Form	Posts
10,000	10 × 10 × 10 × 10	10^4	4
100,000	10 × 10 × 10 × 10 × 10	10^5	5
1,000,000	10 × 10 × 10 × 10 × 10 × 10	10^6	6

The pattern indicates that $10 = 10^1$ (one post) and $1 = 10^0$ (no post). The meaning of repeated multiplication breaks down in these two cases, but the pattern drives the conclusion.

> Basic to arithmetic and to all of mathematics is identifying patterns and reasoning to conclusions based on patterns.
>
> James D. Nickel

Do you see a wonderful pattern revealed by this table? In Lesson 4.1, we noted that to multiply any positive integer by 10, we append 0 to the right of the positive integer. Or, to multiply a positive integer by 10, we add a 0 to the right of that positive integer. Examples:

15 × 10 = 150
63 × 10 = 630
3284 × 10 = 32,840

What happens when you multiply a given number by 100? by 1000? by 10,000? by 100,000?

32 × 100 = 3200
32 × 1000 = 32,000
32 × 10,000 = 320,000
32 × 100,000 = 3,200,000

MULTIPLYING BY POWERS OF TEN

What is the rule? When you multiply a any number by a power of 10, all you do is append to the given number as many zeroes as there are after the 1 in the multiple of 10. Woudn't it be fun if all multiplication problems are that easy? Sorry, they are not. But, by making use of a powerful law of mathematics, the Distributive Law (Lesson 4.4), we will be able to perform any multiplication problem.

EXPONENTIAL REPRESENTATION OF AN INTEGER IN BASE 10

In exponential form, we can write any positive integer as the sum of the powers of 10 times the digits of the integer. In words:

435 = 4 hundreds plus 3 tens plus 5 ones

Since the meaning of 4 hundreds is 4 times 100, the meaning of 3 tens is 3 times ten, etc., we write in symbols:

$435 = 4 \times 100 + 3 \times 10 + 5 \times 1$
$435 = 4 \times 10^2 + 3 \times 10^1 + 5 \times 10^0$

Let's see if we can generalize the representation of any positive integer in the base 10 positional system. We let $a_1a_2\ldots a_n$ be the digits of an n-digit postive integer. We say a_1 as "a sub 1," a_2 as "a sub 2," etc. (Note: Lesson 11.2 where the same form appears in a different context.) For example, in 435, $a_1 = 4$, $a_2 = 3$, and $a_3 = 5$.

For a three-digit positive integer, we write:

$a_1a_2a_3 = a_1 10^2 + a_2 10^1 + a_3 10^0$

For an n-digit positive integer, we write:

$a_1a_2a_3 \ldots a_n = a_1 10^{n-1} + a_2 10^{n-2} + a_3 10^{n-3} + \ldots + a_{n-3} 10^3 + a_{n-2} 10^2 + a_{n-1} 10^1 + a_n 10^0$
(Note: $n = 4 \Rightarrow 10^{n-1} = 10^{4-1} = 10^3$, $10^{n-2} = 10^{4-2} = 10^2$, etc.)

Since $10 = 10^1$ and $1 = 10^0$, we could write this equation:

$a_1a_2a_3 \ldots a_n = a_1 10^{n-1} + a_2 10^{n-2} + a_3 10^{n-3} + \ldots + a_{n-3} 10^3 + a_{n-2} 10^2 + a_{n-1} 10 + a_n$

Example:
$8176 = 8 \times 10^3 + 1 \times 10^2 + 7 \times 10 + 6$

For an eight-digit integer, we have:
$a_1a_2a_3a_4a_5a_6a_7a_8 = a_1 10^7 + a_2 10^6 + a_3 10^5 + a_4 10^4 + a_5 10^3 + a_6 10^2 + a_7 10 + a_8$

Example:
$24{,}708{,}017 = 2 \times 10^7 + 4 \times 10^6 + 7 \times 10^5 + 0 \times 10^4 + 8 \times 10^3 + 0 \times 10^2 + 1 \times 10 + 7$

(Note: It is not necessary to write powers of ten for place values of zero, i.e., where $a_n = 0$ for some n.) We can write this:

$24{,}708{,}017 = 2 \times 10^7 + 4 \times 10^6 + 7 \times 10^5 + 8 \times 10^3 + 1 \times 10 + 7$

For 10,000, there is no need to write the powers of ten for thousands, hundreds, tens, and one. We write:

$10{,}000 = 1 \times 10^4$

Or, more simply, since any number multiplied by 1 is that number (Lesson 4.1 and Lesson 5.15):

$10{,}000 = 10^4$

MULTIPLYING AND AREA

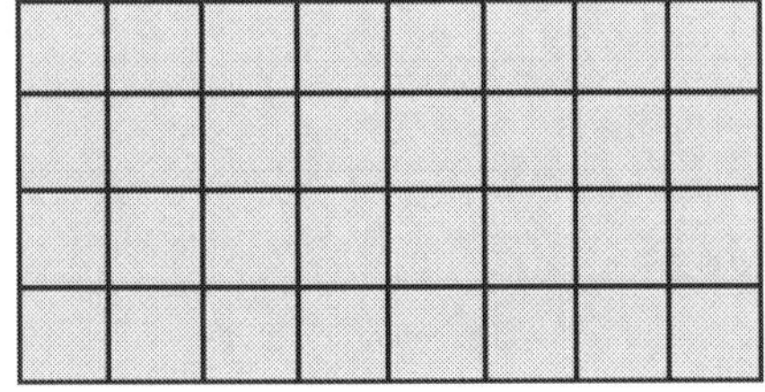

Figure 3. 4 by 8 rectangle.

The rectangle forms the basis for understanding multiplication using the concept of **area**. Area reveals the size of an object in two dimensions. In a rectangle, the area is comprised of square sections (Figure 3). How many squares? We can either count them or multiply the number of rows by the number of columns, i.e., $4 \times 8 = 32$.

The area of a geometric figure, no matter what its shape, is a quantity expressed by a number greater than zero, a number that is well-defined or measurable. This quantity will always have a denomination. For example, let's consider the area of the rectangle in Figure 4. We measure its length and its width as 4 cm and 2 cm, respectively (cm is the abbreviation for "centimeter"). When we measure, we can visualize four 1-cm sections, for the length, and two 1-cm sections for the width. We take our ruler and mark these sections and then draw straight

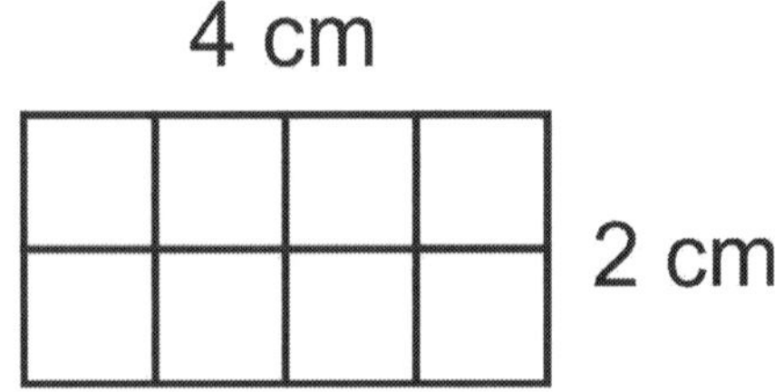

Figure 4. 4 cm by 2 cm rectangle.

lines across and down. Our result is that we have created a network of squares, each square measuring 1 cm by 1 cm. How many squares are there? There are eight such squares or:

(2 cm)(4 cm) = 8 square centimeters

Notice in this case that both factors and product are concrete numbers. The symbol for square centimeters is cm^2. The superscript 2, like our exponents in exponentiation, represents the idea that $cm \times cm = cm^2$ just like $10 \times 10 = 10^2$.

Multiplication, like addition, is perichoretic in its nature. The factors, 2 cm and 4 cm, are contained in the area, 8 cm^2, and the area is contained in the factors. The two-dimensional area of a rectangle and its one-dimensional length and width mutually indwell each other. They dance in unity; they sing in unison.

UNITS OF LENGTH AND AREA

One centimeter is a unit, or denomination, of length. A square centimeter is a unit, or denomination, of area. Every unit of length has a corresponding unit of area. The **unit square**, whose side is one unit of length (centimeters, inches, etc.), is one **unit of area**, i.e., we set the number expressing the area of such a square to one. This unit square will vary in size based upon the unit of length used.

Table 6

Unit of Length	**Unit of Area**
1 inch (abbreviated in)	1 square inch (abbreviated in^2)
1 foot (abbreviated ft)	1 square foot (abbreviated ft^2)
1 mile (abbreviated mi)	1 square mile (abbreviated mi^2)
1 millimeter (abbreviated mm)	1 square millimeter (abbreviated mm^2)
1 meter (abbreviated m)	1 square meter (abbreviated m^2)
1 kilometer (abbreviated km)	1 square kilometer (abbreviated km^2)

UNIT OF AREA AND CONGRUENCE

Congruent figures have equal areas.

There is another fundamental property revealed by this rectangle. Note that we have partitioned the area of the rectangle into eight squares. The area of the whole is equal to the area of the sum of its parts:[5]

$8\ cm^2 = 1\ cm^2 + 1\ cm^2 + 1\ cm^2 + 1\ cm^2 + 1\ cm^2 + 1\ cm^2 + 1\ cm^2 + 1\ cm^2$

Of course, multiplication streamlines this work into:

$(2\ cm)(4\ cm) = 8\ cm^2$

[5] The ancient Greek geometer Euclid (ca. 300 BC) articulated the truth that "the whole is equal to the sum of its parts" or "the whole is greater than its parts" in his renowned mathematical work *The Elements*.

Figure 5. Cube. Source: Public Domain.

Note that these square unit areas, each 1 cm^2, are all eqaual. Geometric objects that can be superimposed on each other so that both they are all identical to each other are **congruent**[6] objects. Therefore, congruent geometric objects have equal areas.

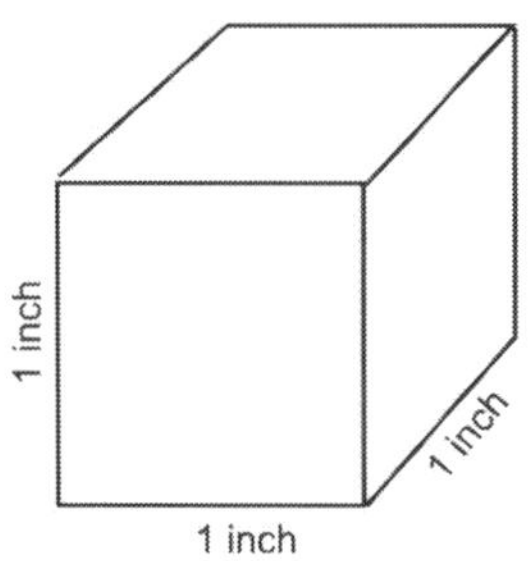

Figure 6. Unit cube, volume of 1 in^3.

SOLIDS AND VOLUME

Look at your room. It has three dimensions, length, width, and height. It is an example of a **solid**. Its unit of measure is **volume**.

As we have noted, centimeter is a unit, or denomination, of length, and square centimeter is a unit, or denomination, of area. What would be the unit of volume?

Every unit of length has a corresponding unit of volume. The **unit cube**, whose side is one unit of length (centimeters, inches, etc.), is one **unit of volume**, i.e., we set the number expressing the volume of such a cube to one. This unit cube will vary in size based upon the unit of length used.

Table 7

Unit of Length	Unit of Volume
1 inch (abbreviated in)	1 cubic inches (abbreviated in^3)
1 foot (abbreviated ft)	1 cubic foot (abbreviated ft^3)
1 mile (abbreviated mi)	1 cubic mile (abbreviated mi^3)
1 millimeter (abbreviated mm)	1 cubic millimeter (abbreviated mm^3)
1 meter (abbreviated m)	1 cubic meter (abbreviated m^3)
1 kilometer (abbreviated km)	1 cubic kilometer (abbreviated km^3)

All three dimensions mutually indwell the volume of a solid; i.e., the volume is contained in all three dimensions and all three dimensions are contained in the volume. There is a perichoretic union between the volume and the three dimensions.

THE ASSOCIATIVE LAW OF MULTIPLICATION

There is another remarkable subtlety revealed in the equation (2 cm)(4 cm) = 8 cm^2. A centimeter is a unit of length. 2 and 4 are **coefficients**[7] of the unit cm. "4 cm," like "4 apples," means four groups of 1 cm or (1 cm)(4) where one factor is concrete and the factor is abstract. We can also say, "4 times 1 cm equals 4 cm." Therefore, (4)(1 cm) = 4 cm, just as "4 times 1 apple equals 4 apples." Note also that 1 cm means 1 group of 1 cm. If we write the symbol cm, we mean 1 cm. The coefficient of cm is 1.

---

[6] *Congruent* comes from the Latin *congruere* meaning "to come together" or "to fall upon."

[7] In Algebra, a coefficient is a number or quantity placed in front of and multiplying another quantity, a quantity usually represented by letters or units in measurement. *Coefficient* is formed from two parts, the prefix and the root. The prefix *co* means "part, auxiliary, subsidiary, or complement." *Efficient* comes from the Latin *facere* meaning "to make." In algebra, a coefficient is "the first part that makes up." If we let the letter x represent apples, then $3x$ means "3 apples" and $3x$, in algebraic notation, is an **algebraic term**. A term is either (1) a single number (Lesson 2.4) or (2) algebraic where a variable (see note 8 for the definition of a variable) or numbers and variables are multiplied together.

If (2 cm)(4 cm) = 8 cm^2, then, if we let x, a **variable**[8] in Algebra, represent the the centimeter unit (Figure 7), what would you think is the product of $2x$ and $4x$, i.e., $(2x)(4x)$?

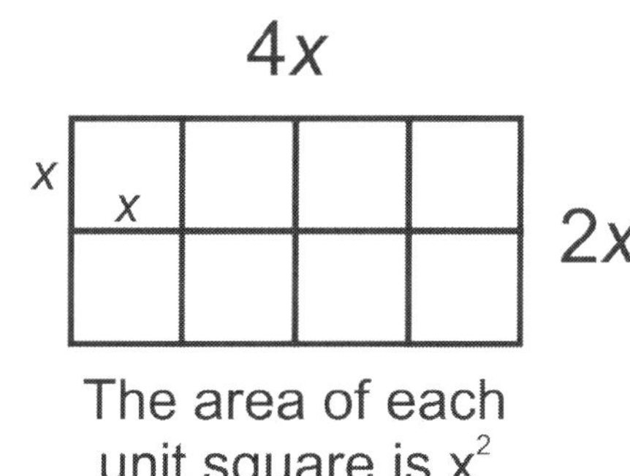

Figure 7. $4x$ by $2x$.

As with addition, we can rearrange, or regroup, the four factors $(2x)(4x)$, meaning "two multiplied by x multiplied by four multiplied by x."[9] This regrouping indicates that the associative law holds for multiplication. (See Lesson 5.15 where we will define this relationship.) We can regroup this way:

$(2x)(4x) = (2)(4)(x)(x)$

We obtain the product by multiplying the two coefficients (2 × 4) and then the two denominations ("x multiplied by x"). Since "x multiplied by x equals x^2," or $(x)(x) = x^2$, then:

$(2x)(4x) = 8x^2$

Congratulations, you are doing algebraic work!

AREA AND VOLUME FORMULAS

A **formula** is a law, rule, pattern, or principle. We frequently express formulas in symbols/variables. The formula for the area of a rectangle, in prose, is "the length multiplied by the width." If we let l = length, w = width, and A = area (three letters representing three measures), then:

A = lw (Remember, lw means "l multiplied by w.")

> [In the Greek mathematical world] there was no recognition of the enormous contribution that letters could make in increasing the effectiveness and generality of algebraic methodology.
>
> Morris Kline, *Mathematical Thought from Ancient to Modern Times* (1972), p. 144.

This formula, short and simple and expressed entirely by symbols, tells us everything that we need to know to calculate the area of a rectangle, i.e., find the product of its length and width. Note also the perichoretic nature of these dimensions. The area is contained in the rectangle's length and width, and the rectangle's length and width determine the area.

The formula for the volume of a rectangular solid, in prose, is "the length multiplied by the width multiplied by the height." If we let h = height and V = volume, then:

[8] A variable is a symbol, usually a letter, that stands for either an unknown number or for any number. Variable is Latin meaning "capable of varying in value." For the Italian mathematicians of the Renaissance, variables were originally named"things," or *cosa* in Italian. Cosa represented an "unknown quantity," and appears to have been in common use. *Cosa* was preserved for several centuries in the forms *coss* (algebra), *cossic* (algebraic), *cossist* (algebraist).

[9] Verbally, we say "two x times four x."

V = *lwh*

In a square, all sides are equal, or the length is equal to the width. If we let s = the measure of one of its sides, then the formula for the area of a square is:

A = *ss* (meaning "*s* multiplied by *s*"), or, making use of exponential form:

$A = s^2$

In a cube, all sides are equal. If we let s = the measure of one of its sides, then the formula for the volume of a cube is:

V = *sss* (meaning "*s* multiplied by *s* multiplied by *s*"), or, making use of exponential form:

$V = s^3$

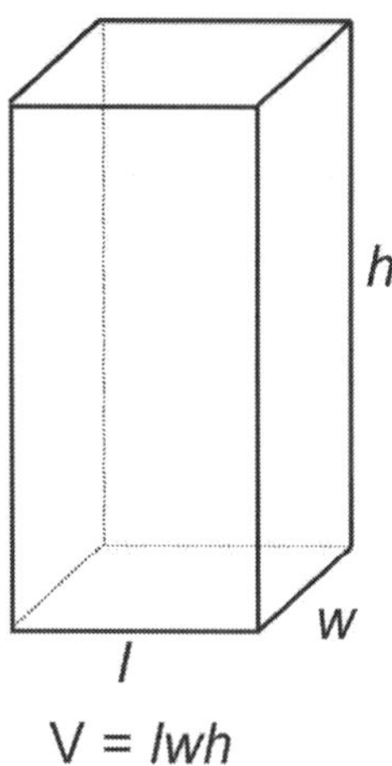

Figure 8.

EXERCISES

Define the following terms:

1. Exponentiation
2. Power
3. Base of the power
4. Exponent or degree of the power
5. Exponential form
6. Area
7. Unit of area
8. Unit square
9. Congruent figures
10. Solid
11. Volume
12. Unit of volume
13. Unit volume
14. Coefficient
15. Variable
16. Algebraic term
17. Formula
18. Googol (See homework exercise below.)
19. Googolplex See homework exercise below.)

> All counting consists in the repeated affirmation of unity. Only for the purpose of always knowing how often we have already affirmed unity do we mark it each time with another word: these are the numerals. Now repetition is only possible through succession. But succession, that is, being after one another, depends directly upon the intuition or perception of time. It is a conception which can only be understood by means of this; and thus counting also is only possible by means of time. This dependence of all counting upon time is also betrayed by the fact that in all languages multiplication is expressed by "time," ....
>
> Arthur Schopenhauer, *The World as Will and Idea* ([1877] 1909), p. 205.

Find (a) x in standard form and (b) rewrite x in exponential form of an integer in base 10:

20. $10^{15} = x$

Answers:

(a) 1,000,000,000,000,000

(b) 1×10^{15}

21. $2^4 = x$
22. $3^3 = x$
23. $3^2 = x$
24. $3^1 = x$
25. $3^0 = x$
26. $6 \times 10 = x$
27. $28 \times 10 = x$
28. $543 \times 100 = x$
29. $98 \times 1000 = x$
30. $2{,}013{,}000 \times 1{,}000{,}000 = x$
31. $9205 \times 10{,}000 = x$
32. $5 \times 10^0 = x$
33. $4{,}123{,}321 \times 10^1 = x$
34. $4{,}123{,}321 \times 16^0 = x$

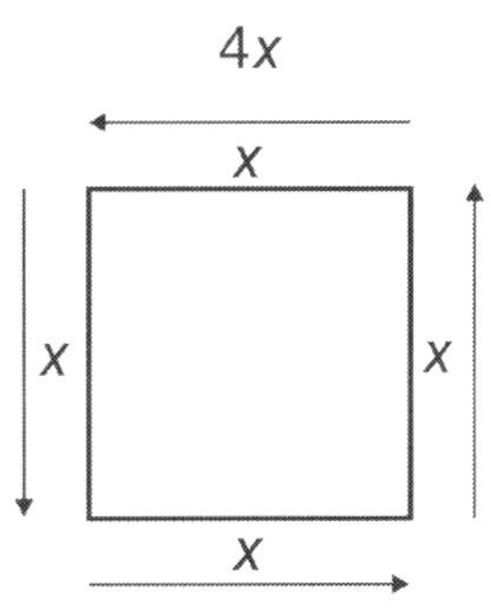

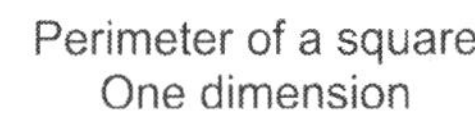

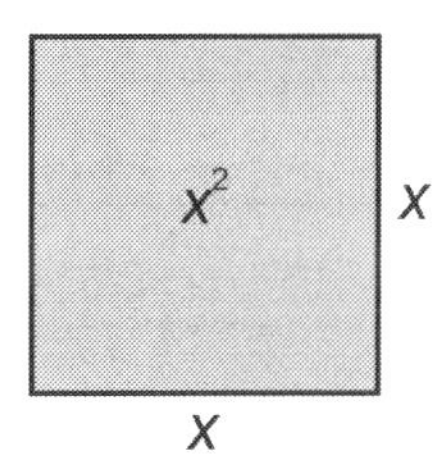

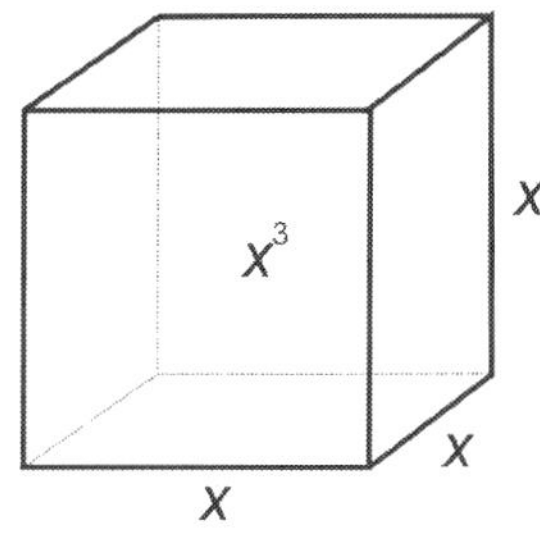

Figure 9.

Since exponentiation is a higher order operation than multiplication or addition/subtraction, we perform this operation first before multiplication and we perform the operation of multiplication before addition/subtraction. Evaluate the following numerical expressions: (Remember that parentheses change the order of operation, i.e., you do the operation within the parentheses first.)

35. $2^3 + 2 \times 2 + 8$
36. $2^2 + 2 \times 2^3 + 8$
37. $2^3 + (2 + 2)^2 + 8$
38. $(2^5 + 2) \times (2 + 8)$
39. $2 + 2^2 \times (2 + 8)$
40. $2^3 + 2^4 \times 2 + 8$
41. $2^5 + 2 \times (2 + 2)^3$

Comparing standard form to exponential form, calculate the savings in written symbols that the exponential form provides in Table 8. The answers to the first three rows are given. For the standard form of the last two rows, just write how many zeroes follow the digit 1.[10]

Table 8

Standard Form	Exponential Form	Savings in Symbols Written (including commas)	Savings Without Including Commas
100	10^2	$3 - 3 = 0$	$3 - 3 = 0$
1,000	10^3	$5 - 3 = 2$	$4 - 3 = 1$
10,000	10^4	$6 - 3 = 3$	$5 - 3 = 2$

[10] Astronomers estimate that the number of stars in the Milky Way galaxy is 10^{11}. Scientists estimate that the number of synaptic interconnections in a single human brain is 10^{15} and that the universe consists of about 10^{80} electrons (fundamental subatomic particles). 10^{100} has a special name attached to it – **googol**. A **googolplex** is 10^{googol}, or $10^{10^{100}}$.

Table 8			
Standard Form	**Exponential Form**	**Savings in Symbols Written (including commas)**	**Savings Without Including Commas**
100,000	10^5	42.	43.
1,000,000	10^6	44.	45.
10,000,000	10^7	46.	47.
100,000,000	10^8	48.	49.
1,000,000,000	10^9	50.	51.
	10^{12}	52.	53.
	10^{15}	54.	55.
	10^{18}	56.	57.
	10^{21}	58.	59.
	10^{24}	60.	61.
	10^{27}	62.	63.
	10^{30}	64.	65.
	10^{33}	66.	67.
	10^{80}	68.	69.
	10^{100}	70.	71.

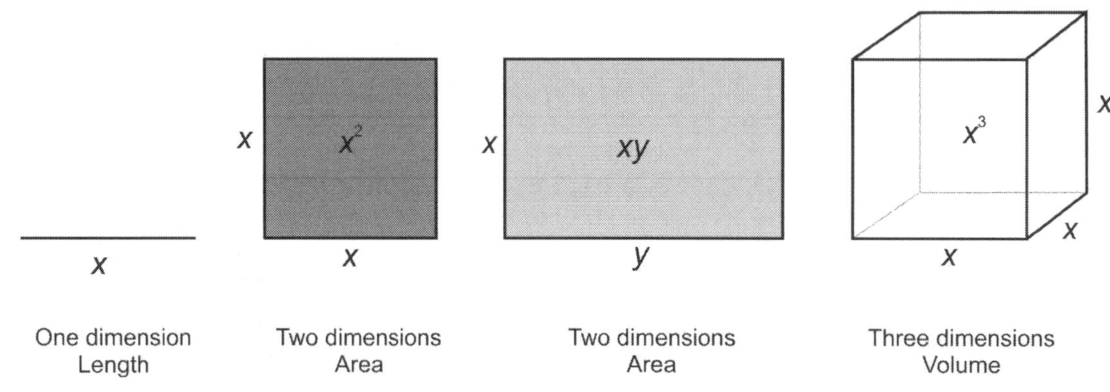

Figure 10.

4.3 The Roman Numeral System

Terms & Concepts Introduced
1. Hieroglyphics
2. Roman Numeral System
3. Varying number scale

The decimal scale is uniform (Lesson 1.3). There is a law of increase and decrease that is the same throughout the entire succession of numerals revealed by the notation. Consider 853. The succession of numerals from right to left:

- The numeral 3 represents 3 ones.
- The numeral 5 represents 5 tens.
- The numeral 8 represents 8 hundreds.

We see a consistent law of increase or multiplying by 10 through the succession of numerals: ones, tens, hundreds.

The succession of numerals from left to right:

- The numeral 8 represents 8 hundreds.
- The numeral 5 represents 5 tens.
- The numeral 3 represents 3 ones.

We see a consistent law of decrease or dividing by 10 (Step 5) through the succession of numerals: hundreds, tens, ones.

Varying Number Scale

> A uniform number system is a place value system while a varying number system is a fixed value system.

A **varying number scale** is where the law of increase and decrease is not the same throughout the entire sequence of units revealed by the notation. The ancient Romans used such a system, named a fixed-value system (Lesson 1.3) because the value of the numeral is always the same no matter where we place it in the notation. In other words, the Roman Numeral X is always 10, V is always 5, and so on.[1]

The ancient Romans used a system of numeration based upon the barred-gate system of counting. We are familiar with the **Roman Numeral System**, I, II, III, IV, V, VI, etc., because we still retain this system in dates on some books, movies, and buildings.

[1] The Roman Numeral X for 10 resembles the ends of a sawhorse and, in early 20th-century America, the ten-dollar bill was popularly called a "sawbuck" or a "saw" for short. The one-dollar bill was simply called a "buck." A twenty-dollar bill was called a "double saw" and a five-dollar bill a "half-saw," a fraction concept. The hundred-dollar bill (C = 100) was called a "C-bill," "C-note," or just a "C." Departing from this Roman practice, a thousand-dollar bill was known as a "grand" or a "G." These American idioms, buck, C-bill, C-note, G, grand, are still in use to name paper money in the United States. The thousand-dollar bill is no longer in circulation; it was discontinued in 1969.

In the finalized version of Roman notation, seven capital letters[2] are used to express numbers.

Table 1: Roman Numerals	
Numeral	Stands for:
I	1
V	5
X	10
L	50
C	100
D	500
M	1000

THE PRINCIPLES OF ROMAN NUMERALS

When used alone, each letter retains its given value. When we use the letters in groups, we invoke the following principles, all of which require a fundamental understanding of comparison, addition, subtraction, or multiplication:

Principle 1. (Repeated Addition or Multiplication Law). Repeating a letter repeats its value.
Examples:
XX = 20 (2 × 10)
CCC = 300 (3 × 100)

Principle 2. (Addition Law). When we place a letter after one of greater value (comparison), its value is added to that of the greater.
Examples:
VI = 6 (5 + 1)
XV = 15 (10 + 5)
LXX = 70 (50 + 20)
DC = 600 (500 + 100)

Principle 3. (Subtraction Law[3]). When we place a letter before one of greater value (comparison), its value is taken away from that of the greater (subtraction).
Examples:
IV = 4 (5 – 1)
IX = 9 (10 – 1)
XL = 40 (50 – 10)
XC = 90 (100 – 10)

Principle 4. (Addition and Subtraction Law). When we place a letter of any value between two letters, each of greater value, its value is taken from the sum of the other two.
Examples:
XIV = 14 (10 + 5 – 1)
LIX = 59 (50 + 10 – 1)

---

[2] Originally, these letters were not letters of the Latin alphabet.

[3] We use the same subtractive principle when we say "fifteen minutes to four o'clock" instead of "forty-five minutes past three o'clock." Subtraction was often used in ancient numeration systems to reduce the number of symbols required, and thus save space. For example, in the Roman numeral system, writing 4 as IV instead of IIII saves two symbols.

CXL = 140 (100 + 50 – 10)

Table 2 reveals more examples.

Table 2			
XX	20	MMM	3000
LX	60	CCIX	209
CC	200	MDXXXVIII	1538
CD	400	MDCCCLI	1851
DCC	700	MCMXXXIX	1939
MM	2000	MMI	2001

The fifth principle of Roman numerals invokes multiplication:

Principle 5. (Multiplication Extension Law). A bar or dash placed over a letter or group of letters increases its value by one thousand times. The Romans did not use this notation. It was introduced during medieval times to write very large numbers.

Examples:
$\overline{\text{X}}$ = 10,000 (10 × 1000)
$\overline{\text{XC}}$ = 90,000 (90 × 1000)
$\overline{\text{DL}}$ = 550,000 (550 × 1000)
$\overline{\text{XXV}}$ = 25,000 (25 × 1000)
$\overline{\text{CXX}}$ = 120,000 (120 × 1000)
$\overline{\text{CLXIV}}$ = 164,000 (164 × 1000)

In your notebook, find the decimal equivalent:

Problem 1. $\overline{\text{CLXIV}}\text{I}$
Problem 2. $\overline{\text{DL}}\text{CXL}$
Problem 3. $\overline{\text{M}}\text{DXCV}$

Here are the solutions:
Solution 1. $\overline{\text{CLXIV}}\text{I}$= 164,001
CLXIV = 164; bar means 164 × 1000 = 164,000
I = 1
164,000 + 1 = 164,001
Solution 2. $\overline{\text{DL}}\text{CXL}$ = 550,140
DL = 550; bar means 550 × 1000 = 550,000
CXL = 140
550,000 + 140 = 550,140
Solution 3. $\overline{\text{M}}\text{DXCV}$ = 1,000,595
M = 1000; bar means 1000 × 1000 = 1,000,000
DXCV = 595
1,000,000 + 595 = 1,000,595

Note that the Roman numeral system utilizes only three arithmetical operations: addition, subtraction, and multiplication. The operation of division is absent since division was a very difficult operation to perform in ancient times.

In Table 3, we see how to count with Roman numerals.

Table 3: Counting with Roman Numerals			
I	1	XI	11
II	2	XII	12
III	3	XIII	13
IV	4	XIV	14
V	5	XV	15
VI	6	XVI	16
VII	7	XVII	17
VIII	8	XVIII	18
IX	9	XIX	19
X	10	XX	20

EXERCISES

Define the following terms:

1. Varying number scale
2. Roman Numeral system
3. Hieroglyphics (See homework exercise below.)

Express decimal numerals as Roman Numerals:

4. The atomic number of nickel is 28.
5. The atomic number of tellurium is 52.
6. The atomic number of indium is 49.
7. Paul, the Apostle, was martyred in Rome in 64.
8. The American Interstate 96 is a freeway in the state of Michigan.
9. Psalm 111 begins with "Praise the Lord."
10. Express Bus Route 553 travels from downtown Minneapolis to Bloomington and back.
11. In 1699, *The Edinburgh Gazette* was first published.
12. The printout states that line 4409 in the code of the computer program is in error.
13. John Chrysostom, archbishop of Constantinople, was born in 349.
14. The first general purpose computer, the Mark I, started functioning in 1951.
15. The construction of the Golden Gate Bridge in San Francisco began in 1933.
16. There are 60,004 lines of code in the Accounts Payable system.
17. 35,014
18. 59,325
19. 89,042

Express Roman Numerals as decimal numerals in standard notation:

20. The movie *Ben Hur* was released in MCMLIX.
21. The movie *Sound of Music* was released in MCMLXV.
22. The move *Wizard of Oz* was released in MCMXXXIX.
23. Leonardo Da Vinci (1452-1519) lived LXVII years.
24. 985 – 851 = CXXXIV
25. In CCXLIX, the Roman Emperor Decius began persecuting the Christians and others refusing to participate in Emperor worship.
26. In MXIX, the Saracens attacked the French port of Narbonne.

27. Mt. Everest, located in Nepal, is the tallest mountain in the world at $\overline{\text{XXIX}}$XXIX feet.
28. In MCMLXXX, Julian Nott rode a hot air balloon from 0 to $\overline{\text{LV}}$CXXXIV feet in exactly I hour IX minutes and XLII seconds.
29. Li Xian, prince of the Tang Dynasty in China, was born in DCLIII.
30. $\overline{\text{C}}$I
31. $\overline{\text{V}}$DCCCXIII
32. $\overline{\text{DLXIV}}$
33. $\overline{\text{XXXI}}$DXX
34. $\overline{\text{LIX}}$CCCXLIV
35. $\overline{\text{XV}}$DCXLIX
36. $\overline{\text{M}}$MMMXCII
37. $\overline{\text{LXV}}$CCCLI
38. $\overline{\text{MDXXV}}$MCDLXIX

Another example of a varying number scale comes from ancient Egypt and it is one of the first recorded number systems in history. These people used **hieroglyphics**[1] as a means of writing. For numbers, they used pictures of objects or of animals. Some of the symbols used in these writings were a vertical staff, something that looked like a horseshoe, a crook, a man with hands outstretched, etc. They placed these symbols beside each other, right to left, higher to lower value. You simply add the numbers they represent to get the total. There is no subtractive or multiplicative principle in their number system. Table 4 reveals these symbols along with their decimal equivalent.

Table 4

Symbol	Represents, in the Decimal System:	Symbol	Represents, In the Decimal System:
(Rod)	1	(Finger)	10,000
(Cattle Hobble)	10	(Tadpole or frog)	100,000
(Coiled rope)	100	(Egyptian god with raised arm)	1,000,000
(Lotus flower)	1000		

Notice how each symbol represents a grouping of ten, showing a similarity to our decimal system. To represent the number four, the rod symbol is repeated four times:

[1] *Hieroglyphics* literally means "holy or sacred writings."

Table 5 shows you other hieroglyphical representations. Note that they can be arranged in any sequence. Wherever you see the symbol, it always retains its designated value.

Table 5

Decimal Number	Egyptian Representation
13	
22	
223	
999	
1243	
452,020	

Determine the savings in symbols written when you compare the decimal system with Egyptian hieroglyphics for:

39. 13
40. 22
41. 223
42. 999
43. 1243
44. 452,020
45. 98
46. 192
47. 1,810,405

Figure 1. Egyptian hieroglyphics. Source: Egypt Archive.

4.4 Napier's Bones and the Distributive Law

Terms & Concepts Introduced
1. Diagonal
2. Distributive Law of Multiplication over Addition
3. Distributive Law of Multiplication over Subtraction

In this and succeeding lessons, we are going to learn how to apply the basic multiplication facts, along with a principle that catches the pulse of what multiplication is all about, to both streamline and expedite the calculation of products.

Let's explore some history first. The Scottish mathematician John Napier (1550-1617) was an unusual man. For example, Napier once suspected one of his servants of theft. He brought all his servants together and announced that he had a black rooster that would identify the guilty party. He ordered his servants into a dark room and asked each to pat the rooster on its back. Not known to the servants, Napier had coated the bird's back with a layer of chimney soot. Upon leaving the room, each servant had to give a "show of hands." The guilty one, fearing to touch the rooster, came out "clean-handed" thus sealing his culpability. Yes, as Napier predicted, his rooster did identify the thief.

Figure 1. John Napier. Source: Public Domain.

He also created logarithms,[1] a tool for quickly multiplying and dividing using addition and subtraction (see Step 12). Until the invention of electronic calculators (ca. 1970s), every math student learned how to use logarithms to multiply and divide. This method was so important for the mathematical sciences, i.e., Astronomy, Physics, and Chemistry, that Florian Cajori (1859-1930), observed, in the late 19th century:

> "The miraculous powers of modern calculation are due to three inventions: the Arabic Notation, Decimal Fractions, and Logarithms."[2]

Logarithmic calculation methods, particularly with its multipage tables, tables that adorn the back of old mathematics textbooks, have virtually disappeared. Calculating with logarithms is no longer invoked, but logarithm principles serve as a unifying connection in a multitude of mathematical and scientific contexts. In the words of mathematics professor Eli Maor (1937-):

> "But if logarithms have lost their role as the centerpiece of computational mathematics, the logarithmic function remains central to almost every branch of mathematics, pure or applied. It shows up in a host of applications, ranging from physics and chemistry to biology, psychology, art, and music."[3]

Among other inventions, Napier developed a mechanical calculating device for multiplication, Napier's bones.[4]

---

[1] Logarithm literally means the "word concerning number" (*logos* + *arithmos*) and is generally translated to mean "ratio number" or "number of the speech or expression." In Lesson 12.1, we shall see how logarithms mutually indwell exponentiation, i.e., the rope and post illustration in Lesson 4.2 will appear again.

[2] Florian Cajori, *A History of Mathematics* (London: The Macmillan Company, [1893] 1909), p. 187.

[3] Eli Maor, *e: The Story of a Number* (Princeton: Princeton University Press, 1994), p. 16. We will explore functions later (Step 10). Functions are the staple of algebraic study.

[4] This method of multiplication was widely used for over 300 years, well into the 20th century.

THE CONSTRUCTION AND USE OF NAPIER'S BONES

First, we construct this board:

Figure 2. Board.

Next, we build a set of rods or strips, the "bones." Each strip reflects the basic facts of the multiplication table from 0×2 to 0×9 (second column) to 9×2 to 9×9 (last column).

Set of Rods

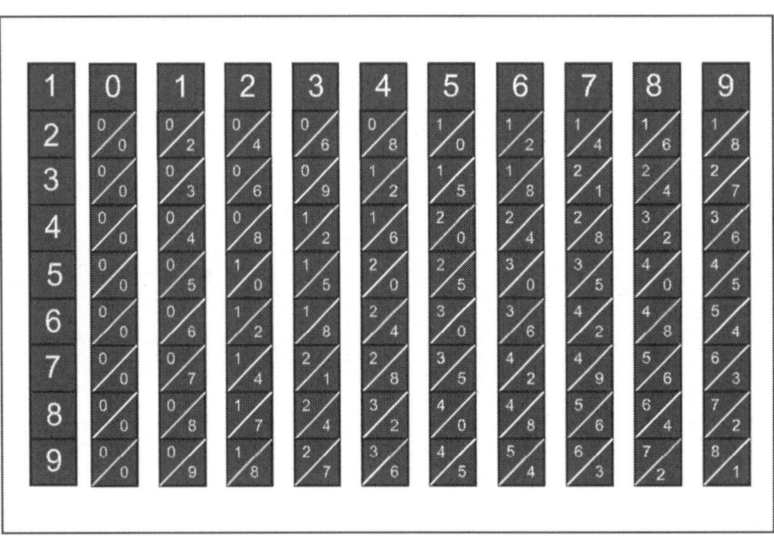

Figure 3. Napier's Nine Rods, representing the times tables for one-digit factors from 0 to 9.

Note that all the products are written in two digits. To find the product of 5 and 3, find the two numbers at the intersection of row five, representing the factor 5, and column five, representing the factor 3. Do you see 15, one ten and five ones? Do you also see $2 \times 3 = 06$ (zero tens and six ones)? Note the 3×2

– 06. The Naperian slants are named, in geometry, **diagonals**.[5] Note that the diagonal divides each square into two triangles,[6] the upper left triangle representing the tens positon and the lower right triangle representing the ones position.

How do we find the product of 15 and 5? First, we lay out the board (Figure 5). Second, we position the one rod and the five rod, representing one ten and five ones (15), side by side on the board. Third, we read the product from the four numbers positioned to the right of 5, the second factor in the multiplication problem, on the board (Figure 4):

Figure 4.

We see three slanted columns from left to right. In the first column, we see 0, in the second 5 and 2 (adding these two numbers gives us 7), and in the third 5.

The 0 represents the hundreds position, the 5 + 2 = 7 the tens position and the 5 the ones position. Our product:

$15 \times 5 = 75$

Why? Let's set up the problem vertically:

$$\begin{array}{r} 15 \\ \times\ 5 \\ \hline \end{array}$$

Napier's bones does this: $15 \times 5 = (10 + 5) \times 5$

As in the case of the paradigm we established with complementary addition, we are going to work from left to right, digit by digit, place value by place value. We have two products to find (in order, 1 × 5 and 5 × 5) pictured by the 1-rod matched to 5 and the 5-rod matched to 5.

Multiplying 15 × 5

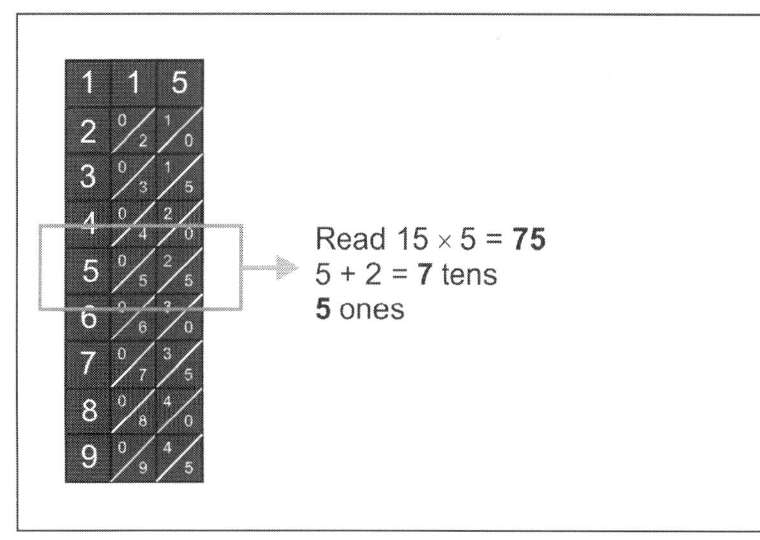

Figure 5.

Step 1. Find the product of 1 × 5. 1 × 5 = 5, and we see 05 in Napier's bones. The 0 means 0 in the hundreds position, and the 5 means 5 in the tens position. We are multiplying the tens position by the ones position, i.e., we are finding the product of 10 × 5 = 50. Make sure that you see the place value 1 represents.

[5] In general, a diagonal is a line segment connecting any two vertices of a polygon that are not adjacent. Remember, each vertex and its two adjoining line segments constitutes an angle. Diagonal is transliterated directly from the Greek *diagonalis* and means "from angle to angle."

[6] Because of the square arrangement, the two triangles are right triangles.

Step 2. Find the product of 5×5.

We are multiplying the ones position by the ones position. $5 \times 5 = 25$ (2 tens and 5 ones) and we see 25 in Napier's bones, 2 in the tens position and 5 in the ones position.

Step 3. We add by place value. Reading the slanted columns from left to right, 0 hundreds, $5 + 2 = 7$ tens, 5 ones. Answer: 75. (Note: it is not necessary to write the leading 0 in the product.)

Let's dig a little deeper into what is happening. Let's consider a plot of farmland divided by a road (Figure 6). On the left side of the road, we have trees arranged ten columns by five rows. On the right side of the road, the trees are arranged five columns by five rows.

How many trees on the left of the road?
$10 \times 5 = 50$

How many trees on the right of the road?
$5 \times 5 = 25$

How many trees altogether?
$50 + 25 = 75$

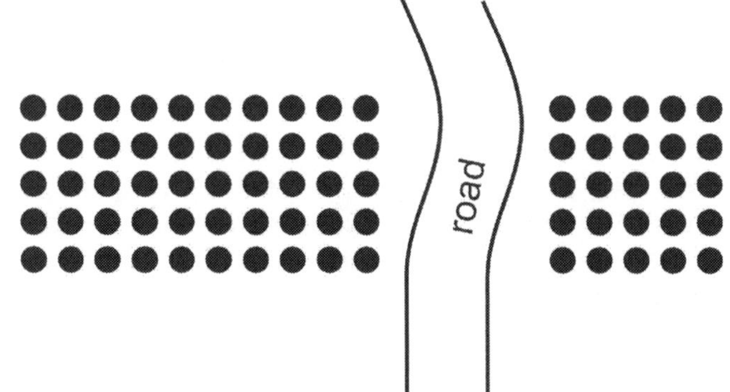

Figure 6.

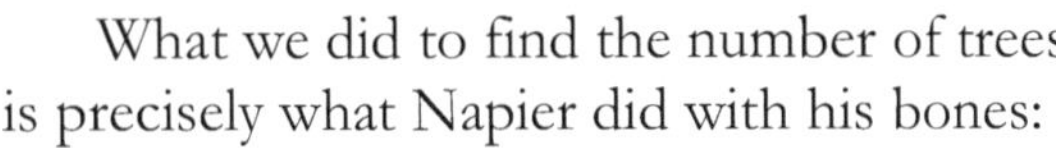

What we did to find the number of trees is precisely what Napier did with his bones:

1. Find two products.
2. Add those products.

Here is what we did, expressed in symbols and written horizontally:

$15 \times 5 = (10 + 5) \times 5 = (10 \times 5) + (5 \times 5) = 50 + 25 = 75$

As we have noted (Lesson 3.1), 15×5, $(10 + 5) \times 5$, $(10 \times 5) + (5 \times 5)$, and $50 + 25$ are numeric expressions. Let's walk through what's happening step by step.

Step 1. $15 = 10 + 5 \Rightarrow$

$15 \times 5 = (10 + 5) \times 5$

Step 2. Note the parentheses. In the past, parentheses told us to "do what is inside first," but doing this would send us back to where we started. Note that we are multiplying 5 by the sum of two numbers. We are going to distribute 5 to each of these numbers, i.e., multiply 10 by 5 and then 5 by 5, or:

$10 \times 5 = 50$

$5 \times 5 = 25$

Step 3. We add these products to get our final answer:

$$\begin{array}{r} 50 \\ +\ 25 \\ \hline 75 \end{array}$$

We can remove the parentheses from the numerical expression $(10 \times 5) + (5 \times 5)$, i.e., $10 \times 5 + 5 \times 5$, and still get the same result.

Multiplication is a higher-order operation; it is counting forward faster by repetition. In a numerical expression like $10 \times 5 + 5 \times 5$, since multiplication is a higher-order operation than addition, we multiply first, and then add second. Why?

We turn $10 \times 5 + 5 \times 5$ into counting by first rewriting multiplication as multiple additions:

$10 \times 5 + 5 \times 5 = (10 + 10 + 10 + 10 + 10) + (5 + 5 + 5 + 5 + 5)$

(Note: We are adding 5 groups of 10 to 5 groups of 5.)

Next, we turn these additions into counting exercises; i.e., we could rewrite 10 and 5 as:

$10 = 1 + 1 + 1 + 1 + 1 + 1 + 1 + 1 + 1 + 1$
$5 = 1 + 1 + 1 + 1 + 1$

Substituting the ten addends, the five 10s and the five 5s, with counting by ones, we produce a string of counting by ones consisting of 75 addends. (Note: Before we can count by ones, we must rewrite the multiplication *first*.[7])

By performing the operations in this manner, we have introduced another very important law by which we can turn, or distribute, a multiplication problem into an addition problem. Multiplication and addition are connected by this law, the **Distributive Law of Multiplication over Addition**.

Remember our work from Lesson 4.1?

If we multiply by 6, we multiply the other factor by 5 and then add that factor to the result.

Example 1. $6 \times 7 = 7 \times 6 = 7 \times (5 + 1) = 7 \times 5 + 7 \times 1 = 35 + 7 = 42$

We note that $6 = 5 + 1$ and then we distributed 7 over $5 + 1$, or:

$7 \times 5 + 7 \times 1 = 35 + 7 = 42$

Example 2. $3 \times 6 = 3 \times (5 + 1) =$

$3 \times 5 + 3 \times 1 = 15 + 3 = 18$

We note that $6 = 5 + 1$ and then we distributed 3 over $5 + 1$, or:

$3 \times 5 + 3 \times 1 = 15 + 3 = 18$

THE DISTRIBUTIVE LAW

In general, when we multiply the sum of two numbers b and c by a given number a, we can distribute the product by adding the products of a by b and a by c. In symbols:

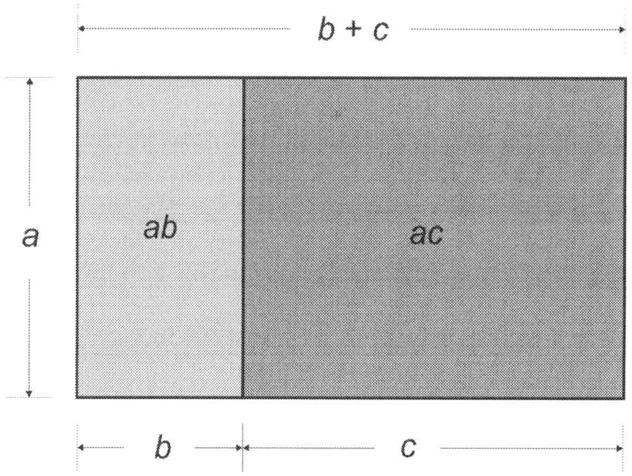

$a \times (b + c) = ab + ac$, or (dropping the multiplication sign):

$$a(b + c) = ab + ac$$

(Note: We say "*a* times the quantity[8] *b* plus *c* equals *ab* plus *ac*.")

Figure 7.

Make sure you learn to understand this symbolic representation. Figure 7 shows what is happening in the context of geometric areas.

[7] In subsequent lessons, we will expand upon the meaning of the order of operations for numeric expressions.

[8] In mathematics, saying "quantity" is intended to alert the listener to the use of parentheses.

Consider:

$4(5 + 3) = (4 \times 5) + (4 \times 3) = 20 + 12 = 32$

Notice, especially, that $4 \times (5 + 3) = 4(5 + 3)$. The parentheses around the "5 + 3" indicates that we are multiplying that sum by 4. We drop the × in this situation.

To cement this law into our minds, consider this illustration:

4(🕯 + ✏) = 4 × 🕯 + 4 × ✏

On the left side, we have first grouped a candle and a pencil in one set and then repeated it four times. On the right side, we took four candles and four pencils and then joined the two groups. Both processes generate the same result.

We can use this law to make some multiplication problems easier.

Consider 13 × 8. Since 13 is 3 more than 10, we can set 13 = 10 + 3. Our problem becomes: 10 + 3)8. By the Commutative Law of Multiplication (Lesson 5.15), we note:[9] (10 + 3)8 = 8(10 + 3). We distribute the 8 over the (10 + 3):

$(10 + 3)8 = 8(10 + 3) = (8 \times 10) + (8 \times 3)$

We get:
$8 \times 10 = 80$
$8 \times 3 = 24$
Adding, we get 104:

```
  80
+ 24
----
 104
```

We can also note that 8 is two less than 10, so we can see 8 = 10 – 2. Our problem becomes 13(10 – 2). We can distribute the 13 over the (10 – 2) or:

$13(10 - 2) = (13 \times 10) - (13 \times 2)$

We get:
$13 \times 10 = 130$
$13 \times 2 = 26$

Subtracting, we get:

```
  130
–  26   The difference is 104
-----
  114
```

The above reasoning is an example of the **Distributive Law of Multiplication over Subtraction**. In general, when you multiply the difference of two numbers *b* and *c* by a given number *a*, you can distribute the product by subtracting the product of *a* by *c* from *a* by *b*. In symbols:

$$a(b - c) = ab - ac$$

[9] Since $ab = ba$, we can let $a = 10 + 3$ and $b = 8$. Therefore, by substitution, $(10 + 3)8 = 8(10 + 3)$.

(Note: We say "*a* times the quantity *b* minus *c* equals *ab* minus *ac*.")

For example:

$9(7 - 2) = (9 \times 7) - (9 \times 2) = 63 - 18 = 45.$

The Distributive Law is a very powerful tool in mathematics. You can use it in many creative ways, not only as a means to solve arithmetic problems. You will encounter it time and time again as you learn Algebra, Geometry, Trigonometry, Calculus, etc. Make sure you understand it.

Table 1

Problem	Numeric Expression	Algebraic Expression	Solution
82×11			
	$82(10 + 1)$	$a(b + c)$	
	$82 \times 10 + 82 \times 1$	$ab + ac$	
			$820 + 82 = 902$
15×99			
	$15(100 - 1)$	$a(b - c)$	
	$15 \times 100 - 15 \times 1$	$ab - ac$	
			$1500 - 15 = 1485$

EXERCISES

1. Define diagonal.

In words and symbols, explain the following:

2. The Distributive Law of Multiplication over Addition
3. The Distributive Law of Multiplication over Subtraction

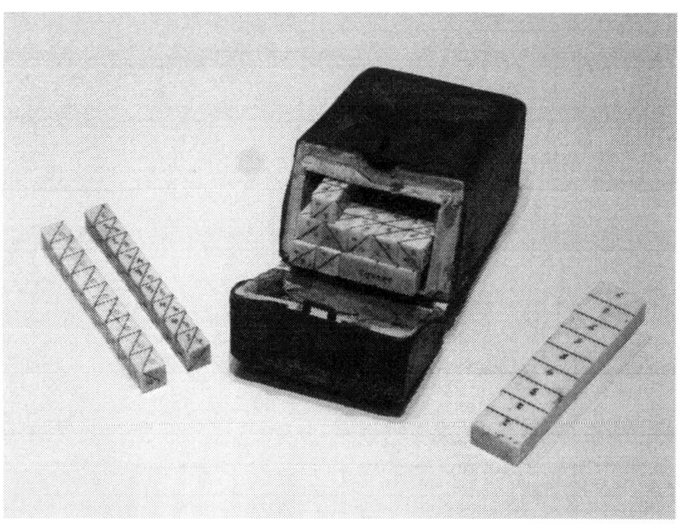

Figure 8. An ivory set of Napier's Bones (ca. 1650). Source: Wikimedia Commons.

Verbally answer the following: (You should be able to see the answer immediately.)

4. 8×4
5. 7×7

6. 6×9
7. 0×5
8. 10×8
9. 10×0
10. 9×5
11. 11×10
12. 10×12
13. 6×7
14. 8×6
15. 5×9
16. 8×8
17. 9×3
18. 4×6
19. 3×8
20. 5×5
21. 9×8
22. 7×5
23. 10×5
24. 4×7
25. 8×7
26. 9×9
27. 10×10

Find the product in two ways: (1) Use one form of the Distributive Law on the first factor and (2) use the other form on the second factor. Which way is easier?

28. 7×11

    Answer.
    $7(10 + 1) = 70 + 7 = 77$
    $11(10 - 3) = 110 - 33 = 77$
    The first way is easier.
29. 12×6
30. 11×8
31. 9×12
32. 9×11
33. 31×9
34. 111×99

Use the Distributive Law of Multiplication over Addition, i.e., partition or "break up" the 11 factor to 10 + 1, to find the following products: (Refer to Lesson 4.11 where we will see this procedure again.)

35. 12×11
36. 13×11
37. 14×11
38. 15×11

39. 16 × 11
40. 17 × 11
41. 18 × 11
42. 19 × 11
43. Look carefully at your answers to the above multiplication problems. Next, look at the digits in the multiplicand. Do you see a pattern between the product and these digits? What is it?

Answer the following questions: (Use the Distributive Law to help you.)

44. At nine cents each, what is the cost of seven widgets?

    Answer. Since one widget costs nine cents,[10] 7 widgets cost 7 × 9 cents = 63 cents.
45. What is the cost of four books at $8 a book?
46. What will seven yards of material cost at 11 cents a yard?
47. 12 inches equal one foot. How many inches in nine feet?
48. At $9 per ticket, what will 13 tickets cost?
49. What is the cost of 29 pounds of fish at (a) $10 per pound? (b) $9 per pound? (c) $11 per pound?
50. Seven days make a week. How many days in 52 weeks?
51. 100 years make a century. How many years in 19 centuries?
52. 100 centimeters make a meter. How many centimeters in 1215 meters?
53. 1000 meters make a kilometer.[11] How many meters in 262 kilometers?
54. If the average attendance at a baseball game is 42,500, how many people will attend 111 games?

Key words indicating multiplication:

- Cubed
- Double
- Factor
- Multiple of
- Multiply
- Quadruple
- Squared
- Times
- Triple
- Twice
- Total

[10] One widget costing nine cents can be rewritten as "nine cents per widget," a phrase that represents a rate (Lesson 7.11).

[11] The prefix *kilo* comes from the Greek and means "1000." One kilometer means 1000 times one meter, or one thousand meters.

4.5 Multiplication by One Digit

In this lesson, we will apply the Distributive Law of Multiplication over Addition to multiplication problems where one factor is one digit in length. The other factor can be any number of digits. We will also continue to invoke our left-to-right paradigm when doing these types of problems.

First, make sure that you have mastered Table 1. You need to see these products need instantly in your mind.

Table 1								
×	**2**	**3**	**4**	**5**	**6**	**7**	**8**	**9**
2	4	6	8	10	12	14	16	18
3		9	12	15	18	21	24	27
4			16	20	24	28	32	36
5				25	30	35	40	45
6					36	42	48	54
7						49	56	63
8							64	72
9								81

Also, remember that any number times one is that number, any number times zero is zero, and the Commutative Law of Multiplication holds, e.g., $3 \times 4 = 4 \times 3$.

Five Problems

Here are a few types of problems:

Problem 1. 15×9 (two digits by one digit)
Problem 2. 72×6 (two digits by one digit)
Problem 3. 325×8 (three digits by one digit)
Problem 4. 1543×7 (four digits by one digit)
Problem 5. $42{,}188 \times 6$ (five digits by one digit)

We note the following:

Problem 1. $15 \times 9 = 9 \times 15 = 9(10 + 5)$
15 is one ten and five ones, $1 \times 10^1 + 5 \times 10^0$, or, since $10^1 = 10$ and $10^0 = 1$:
$15 = 10 + 5$

Problem 2. $72 \times 6 = 6 \times 72 = 6(70 + 2)$
72 is seven tens and two ones, $7 \times 10^1 + 2 \times 10^0$, or:
$72 = 70 + 2$

Problem 3. $325 \times 8 = 8 \times 325 = 8(300 + 20 + 5)$
325 is three hundreds, two tens, and five ones, $3 \times 10^2 + 2 \times 10^1 + 5 \times 10^0$, or:
$325 = 300 + 20 + 5$

Problem 4. $1543 \times 7 = 7 \times 1543 = 7(1000 + 500 + 40 + 3)$
1543 is one thousand, five hundreds, four tens, and three ones, $1 \times 10^3 + 5 \times 10^2 + 4 \times 10^1 + 3 \times 10^0$, or: $1543 = 1000 + 500 + 40 + 3$

Problem 5. $42{,}188 \times 6 = 6 \times 42{,}188 = 6(40{,}000 + 2000 + 100 + 80 + 8)$
42,188 is four ten thousands, two thousands, one hundred, eight tens, and eight ones, i.e.,

$4 \times 10^4 + 2 \times 10^3 + 1 \times 10^2 + 8 \times 10^1 + 8 \times 10^0$, or:
42,188 = 40,000 + 2000 + 100 + 80 + 8

Now, we apply the Distributive Law:

Problem 1. $9(10 + 5) = 9 \times 10 + 9 \times 5 = 90 + 45 = 135$

Problem 2. $6(70 + 2) = 6 \times 70 + 6 \times 2 = 420 + 12 = 432$

Problem 3. $8(300 + 20 + 5) = 8 \times 300 + 8 \times 20 + 8 \times 5 = 2400 + 160 + 40 = 2600$
Note:
$8 \times 300 = 8 \times 3 \times 100 = 24 \times 100 = 2400$
$8 \times 20 = 8 \times 2 \times 10 = 16 \times 10 = 160$

Problem 4. $7(1000 + 500 + 40 + 3) = 7 \times 1000 + 7 \times 500 + 7 \times 40 + 7 \times 3 = 7000 + 3500 + 280 + 21 = 10{,}801$
Note:
$7 \times 500 = 7 \times 5 \times 100 = 35 \times 100 = 3500$
$7 \times 40 = 7 \times 4 \times 10 = 28 \times 10 = 280$

Problem 5. $6(40{,}000 + 2000 + 100 + 80 + 8) =$
$6 \times 40{,}000 + 6 \times 2000 + 6 \times 100 + 6 \times 80 + 6 \times 8 = 240{,}000 + 12{,}000 + 600 + 480 + 48 = 253{,}128$
Note:
$6 \times 40{,}000 = 6 \times 4 \times 10{,}000 = 24 \times 10{,}000 = 240{,}000$
$6 \times 2000 = 6 \times 2 \times 1000 = 12 \times 1000 = 12{,}000$
$6 \times 80 = 6 \times 8 \times 10 = 48 \times 10 = 480$

As we distribute the single-digit factor to the numbers within the parentheses, the subproducts that we need to add decrease in value. Why? We are multiplying from left to right, from higher order to lower order. By doing this, as we sum from left to right, the cumulative sum is getting more and more precise. For example, consider 240,000 + 12,000 + 600 + 480 + 48:

1. The first addend, 240,000, is our first estimate.
2. Adding 12,000 to 240,000, we get 252,000, our second and better estimate.
3. Adding 600 to 252,000, we get 252,600, our third and even better estimate.
4. Adding 480 to 252,600, we get 253,080, our fourth and better yet estimate.
5. Finally, adding 48 to 253,080, we get 253,128, the exact product.

There is a streamlined way to calculate these products. Instead of representing the work horizontally, we represent it vertically.

Problem 1. Find the product of 15 and 9.
We write the problem vertically:

$$\begin{array}{r} 15 \\ \times 9 \\ \hline \end{array}$$

We first distribute the 9 to the 1 in the tens position. Since $9 \times 1 = 9$, we write the product as follows:

$$\begin{array}{r} 15 \\ \times 9 \\ \hline 9 \end{array}$$ Our first estimate is 90, or 9 tens.

Second, we distribute 9 to the 5 in the ones position. Since 9 × 5 = 45, we note that 45 = 40 + 5 (4 tens and 5 ones). We write:

$$\begin{array}{r} 15 \\ \times 9 \\ \hline \overset{4}{9}5 \end{array}$$ Product: 135

We write the 4 tens as a smaller 4 above[1] the 9 tens and the 5 ones directly to the right of 9. Summing, we get our exact product: 135 (4 tens and 9 tens are 13 tens or 1 hundred and 3 tens.)

Problem 2. Find the product of 6 and 72.

The answer determined horizontally: 6 × 72 = 6(70 + 2) = 6 × 70 + 6 × 2 = 420 + 12 = 432
Note, especially, the appearance of 420 and 12 as our work unfolds. We write the problem vertically:

$$\begin{array}{r} 72 \\ \times\ 6 \\ \hline \end{array}$$

First, we first distribute the 6 to the 7 in the tens position. Since 6 × 7 = 42, we write the product as follows:

$$\begin{array}{r} 72 \\ \times\ 6 \\ \hline 42 \end{array}$$ Our first estimate is 420, or 42 tens, or 4 hundreds and 2 tens.

Second, we distribute 6 to the 2 in the ones position. Since 6 × 2 = 12, we note that 12 = 10 + 2 (1 ten and 2 ones). We write:

$$\begin{array}{r} 72 \\ \times 6 \\ \hline 4\overset{1}{2}2 \end{array}$$ Product: 432

We write the 1 ten as a smaller 1 above the 2 tens and the 2 ones directly to the right of the 2 tens.

Summing, we get our exact product: 432 (1 ten and 2 tens are 3 tens.)

Notice the beautiful dance as the product unfolds in our work, from left to right, higher order place value to lower order place value. To generate the final product all we must know is the product of a single digit number by a single digit number, i.e., our multiplication table.

[1] This method is like the record 10 method we used in complementary addition (Step 2).

Problem 3. Let's try 325 × 8 with our streamlined method. We first recall that:

$325 \times 8 = 325 \times 8 = 8(300 + 20 + 5) = 8 \times 300 + 8 \times 20 + 8 \times 5 = 2400 + 160 + 40 = 2600$

Note, especially, the appearance of 2400, 160, and 12 as our work unfolds. We write the problem vertically:

$$\begin{array}{r} 325 \\ \times \quad 8 \\ \hline \end{array}$$

We first distribute the 8 to the 3 in the hundreds position. Since 8 × 3 = 24, we write the product as follows:

$$\begin{array}{r} 325 \\ \times \quad 8 \\ \hline 24 \end{array}$$

Our first estimate is 2400, or 24 hundreds, or 2 thousands and 4 hundreds. Do you see it?

Second, we distribute 8 to the 2 in the tens position. Since 8 × 2 = 16, a double-digit answer, we write the 1 above the 4 and the 6 immediately to the right of the 4:

$$\begin{array}{r} 325 \\ \times \quad 8 \\ \hline 2\overset{1}{4}6 \end{array}$$

Our second estimate is 2560. Do you see it that we added 160 to 2400?

Third, we distribute 8 to the 5 in the ones position. Since 8 × 5 = 40, a double-digit answer, we write the 4 above the 6 and the 0 immediately to the right of the 6:

$$\begin{array}{r} 325 \\ \times \quad 8 \\ \hline 2\overset{1}{4}\overset{4}{6}0 \end{array}$$

Product: 2600

We add the digits from left to right and our exact product is 2600.

Problem 4. Let's find 1543 × 7.

We note:

$7 \times 1543 = 7(1000 + 500 + 40 + 3) = 7000 + 3500 + 280 + 21 = 10{,}801$

Watch, again, the appearance of 7000, 3500, 280, and 12 as our work unfolds. We write the problem vertically:

$$\begin{array}{r} 1543 \\ \times \quad 7 \\ \hline \end{array}$$

We first distribute the 7 to 1 in the thousands position. Since 7 × 1 = 7, we write the product as follows:

$$\begin{array}{r} 1543 \\ \times \quad 7 \\ \hline 7 \end{array}$$

Our first estimate is 7000.

Second, we distribute the 7 to 5 in the hundreds position. Since 7 × 5 = 35, a double-digit answer, we write the 3 above the 7 and the 5 immediately to the right of the 7:

1543

× 7 Our second estimate is 10,500. Do you see it that we added 3500 to 7000?

$\overset{3}{7}5$

Third, we distribute the 7 to 4 in the tens position. Since 7 × 4 = 28, a double-digit answer, we write the 2 above the 5 and the 8 immediately to the right of the 5:

1543

× 7 Our third estimate is 10,780. Do you see it that we added 280 to 10,500?

$\overset{3}{7}\overset{2}{5}8$

Fourth, we distribute the 7 to 3 in the ones position. Since 7 × 3 = 21, a double-digit answer, we write the 2 above the 8 and the 1 immediately to the right of the 8:

1543

× 7 Product: 10,801

$\overset{3}{7}\overset{2}{5}\overset{2}{8}1$

We add the digits from left to right and our exact product is 10,801.

Problem 5. Let's do one more: 42,188 × 6.

We note:

6 × 40,000 + 6 × 2000 + 6 × 100 + 6 × 80 + 6 × 8 = 240,000 + 12,000 + 600 + 480 + 48 = 253,128

Again, 240,000, 12,000, 600, 480, and 48 will appear as our work unfolds. We write the problem vertically:

42188

× 6

We first distribute the 6 to the 4 in the ten thousands position. Since 6 × 4 = 24, we write the product as follows:

42188

× 6 Our first estimate is 240,000.

24

Second, we distribute 6 to the 2 in the thousands position. Since 6 × 2 = 12, a double-digit answer, we write the 1 above the 4 and the 2 immediately to the right of the 4:

```
  42188
×     6     Our second estimate is 252,000. Do you see it that we added 12,000 to 240,000?
 ------
  1
 242
```

Third, we distribute 6 to the 1 in the hundreds position. Since 6 × 1 = 6, a single-digit answer, we just write 6 immediately to the right of the 2:

```
  42188
×     6     Our third estimate is 252,600. Do you see it that we added 600 to 252,000?
 ------
  1
 2426
```

Fourth, we distribute the 6 to 8 in the tens position. Since 6 × 8 = 48, a double-digit answer, we write the 4 above the 6 and the 8 immediately to the right of the 6:

```
  42188
×     6     Our fourth estimate is 253,080. Do you see it that we added 480 to 252,600?
 ------
  1  4
 24268
```

Fifth, we distribute 6 to the 8 in the ones position. Since 6 × 8 = 48, a double-digit answer, we write the 4 above the 8 and the 8 immediately to the right of the 8:

```
  42188
×     6     Product: 253,128
 ------
  1  44
 242688
```

We add the digits from left to right and our exact product is 253,128. Do you see it?

THE DANCE OF MULTIPLICATION

There is a pulse to this procedure. With this method, we are in sync with the dance of the product as it unfolds from left to right, higher-order place value to lower-order place value.

The Distributive Law of Multiplication over Addition governs any multiplication problem, as we shall see in the lessons to come. This law mutually indwells the algorithm of multiplication, a beautiful example of perichoresis in action.

THE DISTRIBUTIVE LAW FOR ONE-DIGIT FACTORS

In general, we can represent multiplying a single digit a by multiple digits (b, c, d, etc.) symbolically as follows:

$$a(b + c) = ab + ac$$

$$a(b + c + d) = ab + ac + ad$$

$$a(b + c + d + e) = ab + ac + ad + ae$$

$$a(b + c + d + e + f) = ab + ac + ad + ae + af$$

etc.

Even this symbolic description has a pulsating rhythm to it, the heartbeat of multiplication over addition.

CHECKING MULTIPLICATION ACCURACY BY CASTING OUT NINES

Before we end this lesson, let's check the accuracy of our multiplication work the same way we checked the accuracy of addition, by casting out nines. Let's see how it works.

Problem 1. 15 × 9
- Product: 135
- Check:
  - Sum the digits in 15: 1 + 5 = 6
  - Cast out the nine is the other factor: 0
  - Multiply: 6 × 0 = 0
  - Sum the digits in the product: 1 + 3 + 5 = 9
  - Cast out that nine: 0
  - 0 = 0. Check.

Problem 2. 72 × 6
- Product: 432
- Check:
  - Sum the digits in 72: 7 + 2 = 9
  - Casting out that nine: 0
  - Second factor: 6
  - Multiply: 0 × 6 = 0
  - Sum the digits in the product: 4 + 3 + 2 = 9
  - Cast out that nine: 0
  - 0 = 0. Check.

Problem 3. 325 × 8
- Product: 2600
- Check:
  - Sum the digits in 325: 3 + 2 + 5 = 10 and 1 + 0 = 1
  - Second factor: 8
  - Multiply: 1 × 8 = 8
  - Sum the digits in the product: 2 + 6 = 8
  - 8 = 8. Check.

Problem 4. 1543 × 7
- Product: 10,801
- Check:
  - Sum the digits in 1543: (cast out 5 + 4) 1 + 3 = 4
  - Second factor: 7
  - Multiply: 4 × 7 = 28
  - Sum these digits: 2 + 8 = 10 and 1 + 0 = 1
  - Sum the digits in the product: (cast out 8 + 1) 1
  - 1 = 1. Check.

Problem 5. 42,188 × 6
- Product: 253,128
- Check:

Sum the digits in 42,188: (cast out 1 + 8) 4 + 2 + 8 = 14 and 1 + 4 = 5
Second factor: 6
Multiply: 5 × 6 = 30
Sum these digits: 3 + 0 = 3
Sum the digits in the product: (cast out 5 + 3 + 1) 2 + 2 + 8 = 12 and 1 + 2 = 3
3 = 3. Check.

EXERCISES

Find the following products using the left-to-right method taught in this lesson and check your work by casting out nines:

1. 53 × 7
2. 78 × 8
3. 247 × 4
4. 1111 × 9
5. 2478 × 5
6. 5109 × 4
7. 212,678 × 3
8. 3885 × 6
9. 8129 × 2
10. 412 × 90 (Hint: 90 = 9 × 10)
11. 20,005 × 80 (Hint: 80 = 8 × 10)
12. 8218 × 300 (Hint: 300 = 3 × 100)
13. 4631 × 40
14. 1295 × 8
15. 53,189 × 5000
16. If 62 men can build a wall in seven days, in how many days can one man build it? (Hint: We assume that it will take one man 62 times as many days as it will take 62 men to build the wall.)
17. If seven men can do a similar piece of work in 25 days, how many days will it take one man to do the same work?
18. How many people will consume as much water in one day as 49 people will consume in eight days?
19. If eight sacks of flour last 23 persons for six months, then how long will the same quantity of flour last for one person?

Answer these questions:

20. Explain why casting out nines does not show that 826 × 9 = 7344 is wrong.
21. The following three tables contain all the possible combinations, a total of 84, of single-digit factors. You should be able to see immediately the products. If not, keep working on these until you do.

Table 2 (24 products)							
7 ×4	4 ×7	1 ×2	6 ×4	8 ×5	4 ×3	3 ×6	8 ×2
2 ×5	9 ×3	1 ×4	2 ×7	5 ×3	9 ×5	1 ×8	5 ×8
6 ×6	4 ×1	6 ×7	3 ×2	8 ×9	9 ×7	5 ×5	7 ×2

Table 3 (24 products)							
8 ×7	7 ×3	4 ×9	3 ×3	6 ×1	4 ×6	3 ×4	1 ×1
6 ×5	5 ×9	3 ×5	2 ×8	2 ×9	7 ×8	1 ×6	8 ×4
8 ×1	6 ×9	5 ×2	4 ×5	3 ×8	5 ×7	2 ×2	2 ×1

Table 4 (36 products)								
1 ×9	9 ×2	3 ×9	5 ×6	1 ×7	3 ×1	4 ×2	9 ×8	8 ×6
9 ×9	6 ×8	3 ×7	6 ×3	5 ×4	4 ×4	8 ×3	1 ×3	4 ×8
7 ×6	2 ×3	7 ×5	6 ×2	9 ×1	5 ×1	7 ×1	2 ×4	9 ×4
8 ×8	9 ×6	2 ×6	1 ×5	7 ×7	7 ×8	9 ×8	7 ×9	4 ×9

22. Select 50 pairs of one-digit factors from the above tables and mentally sum the product. For example, sum the products of 8 × 8 and 7 × 4 in your head. Do the same with 49 other pairs.
23. Select another 50 pairs, all different from the previous exercise, of one-digit factors from the above tables and mentally sum the product.

Consider the multiplication problem 29 × 82. Let 29 = 20 + 9 and 82 = 80 + 2. Answer this question:

24. How would you use the Distributive Law of Multiplication over Addition to find 29 × 82? (Hint: Let $a = 20 + 9$ and find $a(80 + 2)$. After you distribute the a to 80 and 2, substitute a with 20 + 9 and distribute again.)

The Interesting 1089

	Digits of Product Reversed
1089 × 1 = 1089	9801
1089 × 2 = 2178	8712
1089 × 3 = 3267	7623
1089 × 4 = 4356	6534
1089 × 5 = 5445	5445
1089 × 6 = 6534	4356
1089 × 7 = 7623	3267
1089 × 8 = 8712	2178
1089 × 9 = 9801	1089

What patterns do you see?

4.6 Multiplication: Two-Digit Factors

Terms, Symbols & Concepts Introduced
1. ≈
2. Concave polygon
3. Convex polygon
4. Decagon
5. Dodecagon
6. Equiangular triangle
7. Equilateral triangle
8. Heptagon
9. Nonagon
10. Polygon
11. Regular polygon
12. Sides of a polygon
13. Vertices of a polygon

In past lessons, especially in our discussion of perimeter and area, we have encountered some well-known geometric shapes: triangle, square, rectangle, quadrilateral, hexagon, and pentagon. Let's use these shapes to develop a general definition.

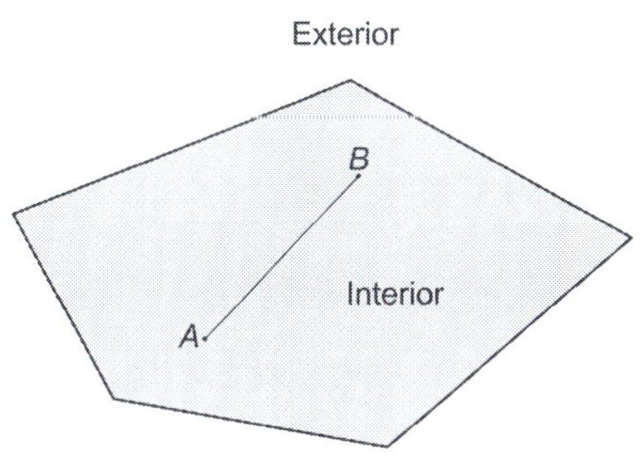

Figure 1. Convex polygon.

We define a **polygon**, derived from a Greek word meaning "many angled," as the joining of several line segments end to end to completely enclose an area. Note that we have already encountered examples of polygons, e.g., quadrilaterals like the square and rectangle, the triangle, the pentagon, the hexagon, and the octagon.

Most useful polygons are **convex**[1] polygons. In a convex polygon, the line segment connecting any two points inside the polygon will always remains entirely inside the polygon.

A polygon that is not convex is **concave**,[2] i.e., it is indented, bent inward. In a concave polygon, you can draw a line segment with end points inside the polygon that it will not always remain entirely inside. The line segments that make up a polygon are its **sides**, and their endpoints are its **vertices** (Lesson 2.7).

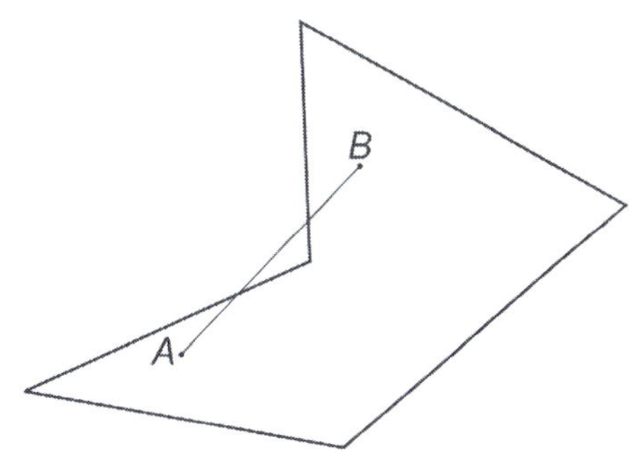

Figure 2. Concave polygon.

We have already explored the regular hexagon, e.g., honeycomb cells. In geometry, regular indicates equality. Therefore, a **regular polygon** is a convex polygon in which all sides are the same length, and all interior angles are of the same measure.

Here are some of the more popular regular polygons, named mostly with Greek prefixes:

- **Equilateral** (meaning "equal sides") or **equiangular** ("equal angles") triangle (the regular 3-gon[3]). Also note that an equilateral triangle is equiangular. In Lesson 13.5, we will prove this.
- Square (the regular 4-gon).
- Pentagon (regular 5-gon). Many flowers are in the shape of a regular 5-gon.
- Hexagon (regular 6-gon). The shape of honeycomb cells and some flowers (Figure 3).
- **Heptagon** (regular 7-gon).

Figure 3. Hexagonal flower structure. Source: James D. Nickel.

---

[1] *Convex* is derived from the Latin *convexus,* meaning "rounded."
[2] *Concave* is derived from the Latin *concavus,* meaning "bent."
[3] The suffix *gon* is from the Greek *gonia*, meaning "corner."

- Octagon (regular 8-gon). See Lesson 2.7.
- **Nonagon** (regular 9-gon).[4]
- **Decagon** (regular 10-gon).
- **Dodecagon** (regular 12-gon).

Can you see in your mind's eye that we can consider a circle as a regular "infinity-gon," a regular polygon with an infinite number of sides, albeit of very, very short or of infinitesimal length?[5]

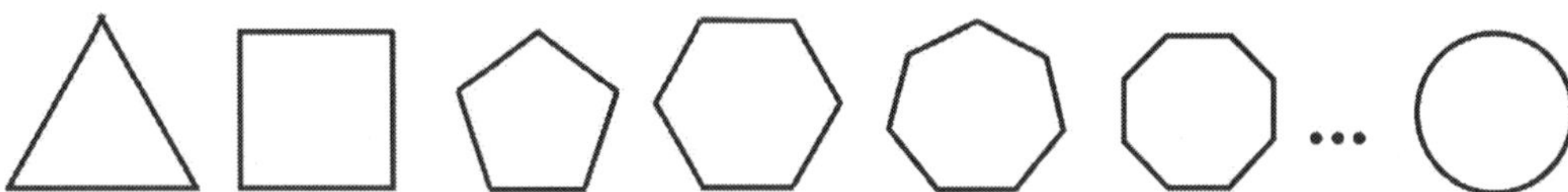

Figure 4. The infinity-gon.

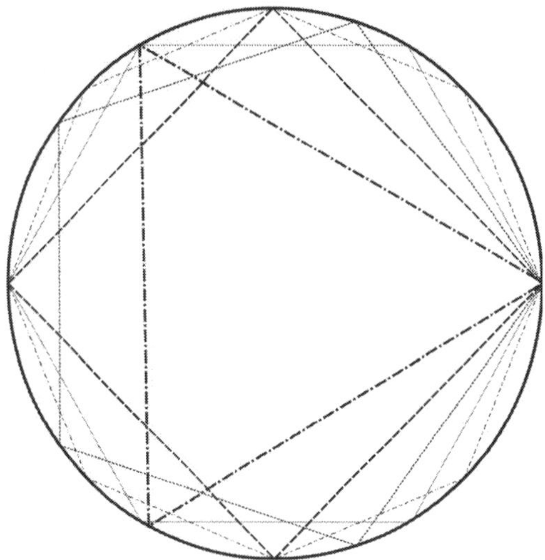

Figure 5. Regular polygons inscribed in a circle.

Let's say that we know the length of the side of a regular 24-gon; it is 49 cm. What is its perimeter? In other words, what is 49 × 24?

NAPIER'S BONES

Let's use Napier's bones as our first method (Figure 6). Note the two sections enclosed by rectangles. We are going to use the bones to determine:

$49(20 + 4) = 49 \times 20 + 49 \times 4$

We can see in Figure 6 that $49 \times 2 = 98$. We acutally want 49×20. Since $20 = 2 \times 10$, we must multiply the place values in the bones by 10. $8 + 1 = 9$ and 9 is in the tens place. Multiplying the place value by

[4] The prefix in *nonagon* is Latin (*nonus* meaning "ninth"), not Greek. *Enneagon* (*ennea* means "nine") is the Greek equivalent and is less commonly used than nonagon. By the 16th century, the name used for a 9-gon in France was *nonogone*. By the 17th century, *nonagon* was in use in England.

[5] You are now thinking conceptually of the results of an infinite process. You are on your way to understanding Calculus, the branch of study in mathematics that builds upon Arithmetic, Algebra, Geometry, and Trigonometry.

10, 9 is now in the hundreds place. Likewise, 0 is in the tens place. Or, we simply append 0 to 98 and get this:

49 × 20 = 980 (9 hundreds plus 8 tens)

Next, we see this:

49 × 4 = 196 (1 hundred plus 9 tens plus 6 ones)

Summing 980 and 196, we get 1176. The perimeter of the regular 24-gon is 1176 cm. Note that we are multiplying a concrete number, 49 cm, by an abstract number, 24. The product is concrete.

In our last lesson, one homework question asked us to develop the Distributive Law to find the product of a two-digit number by a two-digit number. We have the same situation here. We have already partitioned 24 into 20 + 4; i.e., 24 = 20 + 4. We can also partition 49 into 40 + 9; i.e., 49 = 40 + 9. Therefore, we have this:

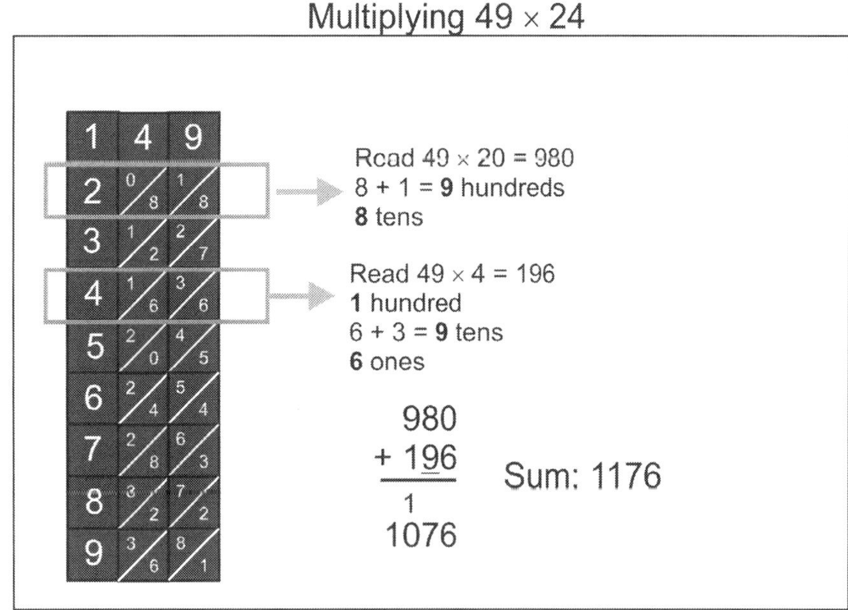

Figure 6.

49 × 24 =
(40 + 9)(20 + 4)

Let's apply the Distributive Law of Multiplication over Addition. We know that $a(b + c) = ab + ac$. We let $a = 40 + 9$, $b = 20$, and $c = 4$. First, we distribute 40 + 9 to 20. We get:

(40 + 9)20

Next, we distribute 40 + 9 to 4. We get:

(40 + 9)4

We have done this:

49 × 24 = (40 + 9)(20 + 4) = (40 + 9)20 + (40 + 9)4

Now, we distribute the 20 to the 40 + 9 and the 4 to the 40 + 9. We get:

(40 + 9)20 = (40)(20) + (9)(20)
(40 + 9)4 = (40)(4) + (9)(4)

Multiplying, we get:

(40)(20) + (9)(20) = 800 + 180
(40)(4) + (9)(4) = 160 + 36

Now, we add the four addends:

800 + 180 + 160 + 36 = 1176

Do you see the 980 and the 196 from Napier's bones? 800 + 180 = 980 and 160 + 36 = 196. As we have already noted, Napier's bones mutually indwells the Distributive Law of Multiplication over Addition.

DISTRIBUTIVE LAW: TWO-DIGITS BY TWO DIGITS

If we note that 49 × 24 = (40 + 9)(20 + 4), we can apply the Distribute Law involving four terms: 40, 9, 20, and 4. In general, let a, b, c, and d be those four terms. We now have:

$(a + b)(c + d)$

Distributing $a + b$ to c, we get:

$(a + b)c$

Distributing $a + b$ to d, we get:

$(a + b)d$

We now have this:

$(a + b)(c + d) = (a + b)c + (a + b)d$

Distributing c to $a + b$, we get:

$ac + bc$

Distributing d to $a + b$, we get:

$ad + bd$

We now have this:

$(a + b)(c + d) = (a + b)c + (a + b)d = ac + bc + ad + bd$

Notice that we have four subproducts, ac, bc, ad, and bd that form addends of a sum that equals the product of $a + b$ and $c + d$.

Congratulations on your algebraic work!

How can we put this extension of the Distributive Law to work when multiplying a two-digit factor by a two-digit factor? Let's work through another example. We want to find the area of a rectangular garden that measures 57 feet by 23 feet (57 × 23). Our product will be the area, and it is measured in square feet or ft^2.

57 × 23 = (50 + 7)(20 + 3) = (50 + 7)20 + (50 + 7)3 = (50)(20) + (7)(20) + (50)(3) + (7)(3)

We have four subproducts to sum:

(50)(20) + (7)(20) + (50)(3) + (7)(3) = 1000 + 140 + 150 + 21

Adding, we get:

```
  1000
   150
   140      57 × 23 = 1311 (The area of the garden is 1311 ft².)
+   21
  ----
   1
  1211
```

Notice the place-value designation of these four addends:

1000	=	10	hundreds	
150	=	15	tens	We note:
140	=	14	fens	15 tens + 14 tens = 29 tens
21	-	21	ones	

SPEED PARADIGM FOR MULTIPLYING A TWO-DIGIT FACTOR BY A TWO-DIGIT FACTOR

Investigate Figure 7. It illustrates the streamlined method we will use to multiply such factors. The four circles on the right form a square that represent the four digits in a two-digit by two-digit multiplication problem. We illustrate the four subproducts to the right of the original square.

For the first subproduct (the first double arrow in Figure 7), we multiply the tens position by the tens position (5 in 57 and 2 in 23) to get the number of hundreds in our product. There is only one subproduct like this.

For the second and third subproducts (the crossed double arrows), we multiply the tens position by the ones position and the ones position by the tens position to get the number of tens in our product. There are two subproducts like this.

For the fourth subproduct (the last double arrow), we multiply the ones position by the ones position to get the number of ones in our product. There is only one subproduct like this.

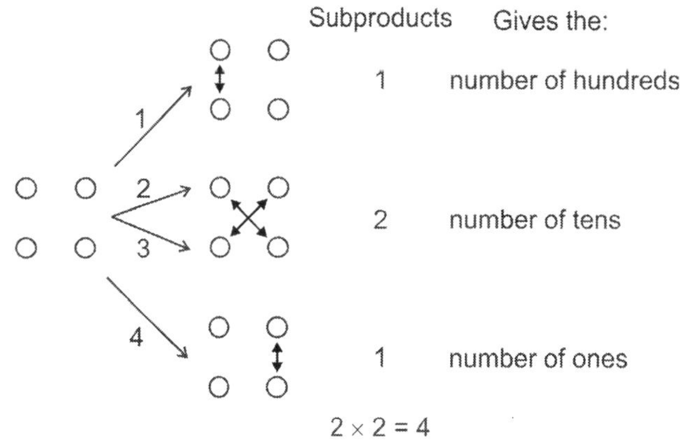

Figure 7. Two-Digit by Two-Digit Paradigm.

Adding the number of products calculated, we get the four subproducts that we must sum (see the figure):

$1 + 2 + 1 = 4$

Multiplying the number of digits by the number of digits, 2 × 2, gives us the number of subproducts, 4, to sum (see the figure):

$2 \times 2 = 4$

(We note: $1 + 2 + 1 = 2 \times 2$)

Let's see how the dance of this streamlined method unfolds the product, from left to right, from hundreds to tens to ones. We set up our problem vertically and then notice how 1000, 290, and 21 appear, step by step (Figure 8):

$$\begin{array}{r} 57 \\ \times 23 \\ \hline \end{array}$$

First, we multiply the tens position of the top factor by the tens position of the bottom factor, or 5 × 2 = 10 (10 hundreds). We write this subproduct like this:

$$\begin{array}{r} 57 \\ \times 23 \\ \hline 10 \end{array}$$

The 10 stands for 10 hundreds, or 1000, our first estimate.

In symbols, ≈ means approximately equal.

We can say that 1000 is approximately equal to the product of 57 and 23.

Second and third, we multiply the tens position of the top factor by the ones position of the bottom factor, or 5 × 3 = 15 (15 tens). Next, we multiply the ones position of the top factor by the tens position of the bottom factor, or 7 × 2 = 14 (14 tens). Adding tens (do this sum in your head: 15 + 14), we get 29 tens or 2 hundreds and 9 tens. We write this subproduct like this, with the 2 over the 0 in the hundreds position and 9 to the right of 0 in the tens position:

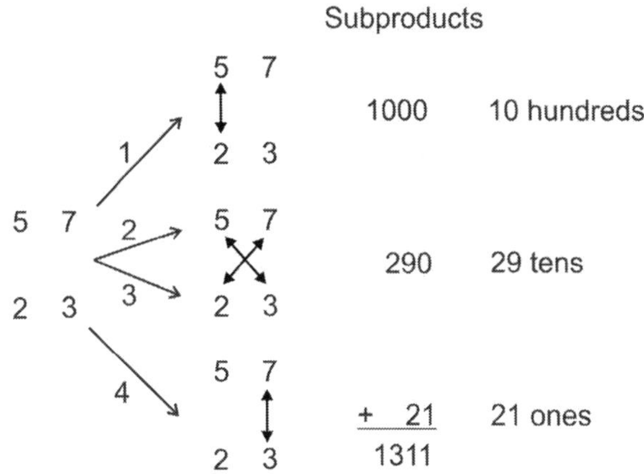

Figure 8. 57 × 23 = 1311

$$\begin{array}{r} 57 \\ \times 23 \\ \hline 1\overset{2}{0}9 \end{array}$$

The 29 stands for 29 tens (290), and our second estimate is 1290.

Fourth, we multiply the ones position of the top factor by the ones position of the bottom factor, or 7 × 3 = 21 (21 tens). We can rewrite 21 tens as 2 tens and 1 one. We write this subproduct like this, with the 2 over the 9 in the tens position and 1 to the right of 9 in the ones position:

$$\begin{array}{r} 57 \\ \times 23 \\ \hline 1\overset{2}{0}\overset{2}{9}1 \end{array}$$

The 21 stands for 21 ones and our exact product, read left to right, is 1311.

Review this process, from start to finish.
Check by casting out nines:

- Digit sum of first factor: 5 + 7 = 12 and 1 + 2 = 3
- Digit sum of second factor: 2 + 3 = 5
- Multiply the results: 3 × 5 = 15 and 1 + 5 = 6
- Sum the digits of the product: 1 + 3 + 1 + 1 = 6
- 6 = 6. Check.

Next, multiply 49 by 24 in your notebook and compare your results with those shown below.

$$\begin{array}{r} 49 \\ \times\ 24 \\ \hline \end{array}$$

$$\begin{array}{r} 49 \\ \times\ 24 \\ \hline 8 \end{array} \quad \text{(8 hundreds)}$$

$$\begin{array}{r} 49 \\ \times\ 24 \\ \hline \overset{3}{8}4 \end{array} \quad \text{(34 tens: 16 + 18)}$$

$$\begin{array}{r} 49 \\ \times\ 24 \\ \hline \overset{3}{8}\overset{3}{4}6 \end{array} \quad \text{(36 ones)} \qquad \text{Product: 1176}$$

$$\begin{array}{r} 800 \\ 340 \\ +\ 36 \\ \hline 1176 \end{array}$$

Remember that we perform the second and third subproducts together. Make sure that you add them in your head (16 + 18 = 34). If you applied yourself to the homework in the last lesson, you should be able to do these mental sums with ease. Also, make sure you understand the place value that each digit holds, as the product unfolds from left to right.

Check by casting out nines:

Digit sum of first factor: 4 + 9 = 13 and 1 + 3 = 4
Digit sum of second factor: 2 + 4 = 6
Multiply the results: 4 × 6 = 24 and 2 + 4 = 6
Sum the digits of the product: 6 (cast out 1 + 1 + 7)
6 = 6. Check.

Let's try some more problems:

Problem 1. 23 × 35
Problem 2. 36 × 74
Problem 3. 45 × 45
Problem 4. 22 × 93
Problem 5. 83 × 19
Problem 6. 77 × 27
Problem 7. 99 × 99
Problem 8. 11 × 11

Do them all in your notebook, check by casting out nines, and then compare your answers with the demonstrations below. Make sure to sum the second and third subproducts in your head. Here are the results:

Problem		Product	Problem		Product
1.	23 × 35 1 1 695 Check: 5 × 8 = 40 and 4 + 0 = 4	805 Check: 8 + 5 = 13 and 1 + 3 = 4	2.	36 × 74 5 2 2144 Check: 0 × 2 = 0	2664 Check: 18 = 1 + 8 = 9 = 0 (cast out the 9)
3.	45 × 45 4 2 1605 Check: 0 × 0 = 0	2025 Check: 2 + 2 + 5 = 9 = 0 (cast out the 9)	4.	22 × 93 2 1846 Check: 4 × 3 = 12 and 1 + 2 = 3	2046 Check: 2 + 4 + 6 = 12 and 1 + 2 = 3
5.	83 × 19 7 2 857 Check: 2 × 1 = 2	1577 Check: 1 + 5 + 7 + 7 = 20 And 2 + 0 = 2	6.	77 × 27 6 4 1439 Check: 5 × 0 = 0	2079 Check: 0 (cast out 2 + 7 and 9)
7.	99 × 99 1 6 8 8121 See note.[1] Check: 0 × 0 = 0	9801 Check: 0 (cast out 9 and 8 + 1)	8.	11 × 11 121 Check: 2 × 2 = 4	121 Check: 1 + 2 + 1 = 4

What do we do when we multiply a number by a multiple of 10, i.e., 10, 20, 30, 40, 50, 60, 70, 80, or 90? We already know how to multiply by 10. Just append zero to the other factor, e.g., 86 × 10 = 860. With the other multiples of 10, we note:

20 = 2 × 10
30 = 3 × 10
40 = 4 × 10
etc.

We just multiply by the single digit and then append 0 to the result. For example, 45 × 80 is:

[1] For the second and third subproduct, 81 + 81 = 162. 162 is 162 tens, or 1620. We note the place value designation: 162 = 1 thousand, 6 hundreds, 2 tens. Therefore, we write the smaller 1 above the 8 (thousands place), the 6 above the 1 (hundreds place), and the 2 in tens place to the right of the 1.

```
  45
× 8     Product: 360 × 10 = 3600
 4
320
```

We could also do this:

```
  45
× 80    Product: 3600
 4
3200
```

The first subproduct: 4 × 8 = 32, or 32 hundreds = 3200
The second the third subproducts: 4 × 0 = 0 tens and 8 × 5 = 40 tens. Sum: 40 tens or 400
The fourth subproduct: 5 × 0 = 0, or 0 ones

Anytime zero appears in the digits, the computation of the subproducts is much easier because zero times any number is zero.

We could also set up 36 × 7 like this:

```
  36
× 07    Product: 252
 2 4
 012
```

The first subproduct: 3 × 0 = 0, or 0 hundreds
The second and third subproducts: 3 × 7 = 21 tens and 0 × 6 = 0 tens. Sum: 21 tens or 210
The fourth subproduct: 6 × 7 = 42, or 42 ones, or 4 tens and 2 ones

Casting out nines check:
- Digit sum of first factor: 0 (cast out 3 + 6)
- Digit sum of second factor: 7
- Multiply the results: 0 × 7 = 0
- Sum the digits of the product: 2 + 5 + 2 = 0 (cast out the 9)
- 0 = 0. Check.

This method of multiplication is simple to master. Always keep clear in your mind what is happening with place values. You want to be able to calculate with understanding, not blindly. Finally, I hope that you see that computing multiplication problems like this is fun. Hold on to your seat, more such fun is coming!

> Any fool can know. The point is to understand.
>
> Attributed to Albert Einstein (1879-1955).

EXERCISES

Define the following terms:

1. Polygon
2. Convex polygon
3. Concave polygon
4. Sides of a polygon
5. Vertices of a polygon
6. Regular polygon
7. Equilateral triangle
8. Equiangular triangle
9. Heptagon
10. Nonagon
11. Decagon
12. Dodecagon
13. $\approx$

Multiply and check by casting out nines:

14. 42 times 97
15. 56 times 29
16. 72 times 18
17. 80 times 99
18. 92 times $36
19. 69 times $77
20. 56 (a) by 40, (b) by 41, (c) by 42, and (d) by 43
21. 83 (a) by 70, (b) by 69, (c) by 68, and (d) by 67
22. $45 (a) by 7, (b) by 43, (c) by 64, and (d) by 99
23. 83 feet by (a) 4, (b) by 81, (c) by 38, and (d) by 72
24. 69^2
25. 79^2
26. 89^2

Answer the following questions and check multiplication by casting out nines:

27. A side of a regular nonagon measures 89 mm. What is its perimeter?
28. What is the cost of 85 widgets at $58 per widget?
29. What is the cost of 75 pounds of nails at 13 cents a pound?
30. Saving $74 a month, how much is saved in 12 months?
31. Saving $27 a month, how much is saved in 24 months?
32. Saving $69 a month, how much is saved in 36 months?

Complete column three in the following table: (The first three rows are completed for you.)

Period	Name	Exponential Form
1st	Ones	10^0
2nd	Thousand	10^3
3rd	Million	10^6
4th	Billion	33.
5th	Trillion	34.
6th	Quadrillion	35.
7th	Quintillion	36.
8th	Sextillion	37.
9th	Septillion	38.
10th	Octillion	39.
11th	Nonillion	40.
12th	Decillion	41.
13th	Undecillion	42.
14th	Duodecillion	43.
15th	Tredecillion	44.
16th	Quattuordecillion	45.
17th	Quindecillion	46.
18th	Sexdecillion	47.
19th	Septendecillion	48.
20th	Octodecillion	49.
21st	Novendecillion	50.
22nd	Vigintillion	51.

Recall the meaning of these prefixes.

Prefix	Meaning
bi-	two
tri-	three
quad-	four
quint-	five
sex-	six
sep-	seven
oct-	eight
non-	nine
dec-	ten

Determine the meaning of these prefixes:

Prefix	Meaning
undec-	52.
duodec-	53.
tredec-	54.

Prefix	Meaning
quattuordec-	55.
quindec-	56.
sexdec-	57.
septendec-	58.
octodec-	59.
novendec-	60.
vigint-	61.

62. Find the pattern between the exponent (the power of 10) and the number that acts as a prefix to "illion." For example, the exponent of 10^{12} is 12.

Refer to the three tables in the homework for Lesson 4.5 (Table 1, Table 2, and Table 3). Do the following exercises: (Double-check your mental work with pencil and paper.)

63. Select 25 triplets of one-digit factors from these tables and mentally sum the three products. For example, sum the products of 3 × 2, 1 × 6, and 9 × 9 in your head. Do the same with 24 other pairs.
64. Select another 25 triplets of one-digit factors from these tables and mentally sum the three products.

The Wonders of 11

11×11 = 121
11×11×11 = 12321
11×11×11×11 = 1234321
11×11×11×11×11 = 123454321
11×11×11×11×11×11 = 12345654321
11×11×11×11×11×11×11 = 1234567654321
11×11×11×11×11×11×11×11 = 123456787654321
11×11×11×11×11×11×11×11×11 = 12345678987654321
Wow!

4.7 Multiplication: Three-Digit Factors

Let's dig deeper. What if we wanted to multiply a three-digit number by a three-digit number, e.g., 627 × 453? We write 627 = 600 + 20 + 7 and 453 = 400 + 50 + 3. Therefore:

$627 \times 453 = (600 + 20 + 7)(400 + 50 + 3)$

Let's work through the general case first:

$(a + b + c)(d + e + f)$

Let $x = d + e + f$. We now have:

$(a + b + c)(x)$

We distribute x to a, b, and c as follows:

$(a + b + c)(x) = ax + bx + cx$

Now, we substitute x with $d + e + f$. We get:

$ax + bx + cx = a(d + e + f) + b(d + e + f) + c(d + e + f)$

We now distribute a, b, and c, respectively, to $(d + e + f)$. We get:

$$a(d + e + f) + b(d + e + f) + c(d + e + f) =$$
$$ad + ae + af +$$
$$bd + be + bf +$$
$$cd + ce + cf$$

Therefore, $(a + b + c)(d + e + f) =$

$$ad + ae + af +$$
$$bd + be + bf +$$
$$cd + ce + cf$$

How many subproducts must we calculate?

$3 \times 3 = 9$

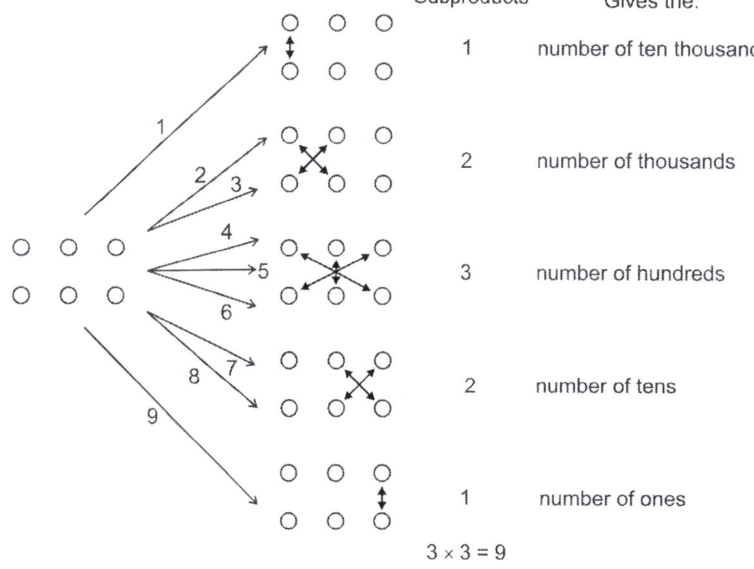

Figure 1. Three-Digit by Three-Digit Paradigm.

Speed Paradigm for Multiplying a Three-Digit Factor by a Three-Digit Factor

Inspect Figure 1 and reflect on the beauty of its symmetry, i.e., the pulse of its pattern. Do you see the nine subproducts that we generate from the three-by-three array?

The first subproduct (the first double arrow in the figure), hundreds by hundreds (the place value for 6 in 627 and 4 in 453 in our example), generates the number of ten thousands. There is only one subproduct like this.

The second and third subproducts (the first crossed double arrows), hundreds by tens and tens by hundreds, generate the number of thousands. There are two subproducts like this.

The fourth, fifth, and sixth subproducts (the three double arrows), hundreds by ones, ones by hundreds, and tens by tens, generate the number of hundreds. There are three subproducts like this.

The seventh and eighth subproducts (the second crossed double arrows), tens by ones and ones by tens, generate the number of tens. There are two subproducts like this.

The ninth subproduct (the last double arrow), ones by ones, generates the number of ones. There is only one subproduct like this.

Adding the number of products calculated, we get nine subproducts that we must sum:

$1 + 2 + 3 + 2 + 1 = 9$

Multiplying the number of digits by the number of digits, 3×3, gives us the number of subproducts, nine, to sum:

$3 \times 3 = 9$

(We note: $1 + 2 + 3 + 2 + 1 = 3 \times 3$)

Let's inspect these nine subproducts as we work through our example: 627×453. Since $627 = 600 + 20 + 7$ and $453 = 400 + 50 + 3$, we write:

$(600 + 20 + 7)(400 + 50 + 3)$

Applying the Distributive Law, we get:

$600(400 + 50 + 3) + 20(400 + 50 + 3) + 7(400 + 50 + 3)$

Applying the Distributive Law a second time, we get:

$600 \times 400 + 600 \times 50 + 600 \times 3 +$
$20 \times 400 + 20 \times 50 + 20 \times 3 +$
$7 \times 400 + 7 \times 50 + 7 \times 3$

It is a matter of calculating the nine subproducts and then adding the results. Do that in your notebook now. Notice especially how the place values unfold, along with the symmetry of the addends (1, 2, 3, 2, 1):

Step 1. $6 \times 4 = 24$
24 ten thousands = 240,000 (one addend)

Step 2. $6 \times 5 + 2 \times 4 = 30 + 8 = 38$
38 thousands = 30,000 + 8000 = 38,000 (two addends)

Step 3. $6 \times 3 + 2 \times 5 + 7 \times 4 = 18 + 10 + 28 = 56$
56 hundreds = 1800 + 1000 + 2800 = 5600 (three addends)

Step 4. $2 \times 3 + 7 \times 5 = 6 + 35 = 41$
41 tens = 60 + 350 = 410 (two addends)

Step 5. $7 \times 3 = 21$
21 ones = 21 (one addend)

Now, let's apply our new algorithm to the work. Note how the numbers 24, 38, 56, 41, and 21 unfold.

$$\begin{array}{r} 627 \\ \times\ 453 \\ \hline \end{array}$$

Step 1 (the first double arrow: one addend). 6 × 4, or 6 hundreds by 4 hundreds = 24 ten thousands. We write:

627
× 453 Our first estimate: 240,000
24

Step 2 (the first crossed double arrows: two addends). 6 × 5 = 30 (hundreds by tens), 2 × 4 = 8 (tens by hundreds). Do this sum in your head: 30 + 8 = 38, or 38 thousand. We write:

627
× 453 Our second estimate,
3
248
reading from left to right:
278,000

Step 3 (the three double arrows: three addends). 6 × 3 = 18 (hundreds by ones), 7 × 4 = 28 (ones by hundreds), 2 × 5 = 10 (tens by tens). Do this sum in your head: 18 + 28 + 10 = 56, or 56 hundreds. We write:

627
× 453 Our third estimate,
3 5
2486
reading from left to right:
283,600

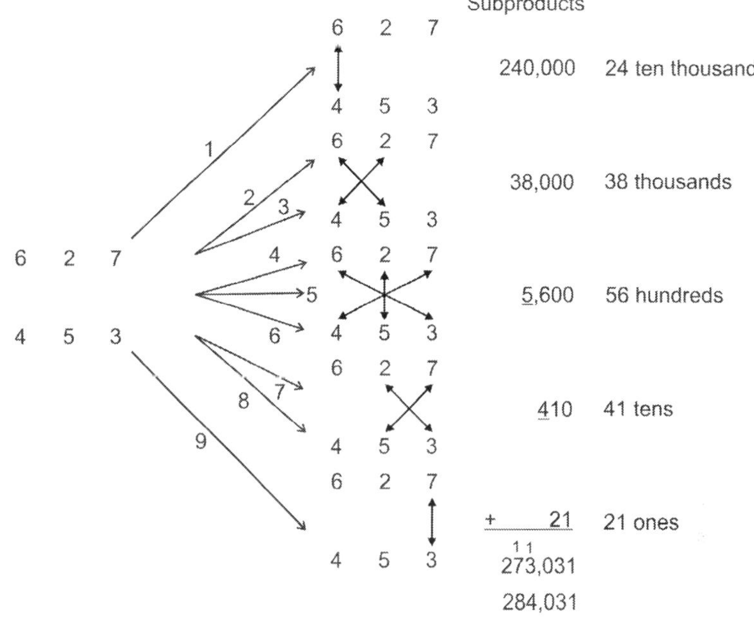

Figure 2. 627 × 453 = 284,031

Step 4 (the second crossed double arrows: two addends). 2 × 3 = 6 (tens by ones), 5 × 7 = 35 (ones by tens). Do this sum in your head: 6 + 35 = 41, or 41 tens. We write:

627
× 453 Our fourth estimate, reading from left to right: 284,010
3 5 4
24861

Step 5 (the last double arrow: one addend). 7 × 3 = 21, or 21 ones. We write:

627
× 453 Our exact product, reading from left to right: 284,031
3 5 4 2
248611

Review these steps to catch their rhythm. Observe how the estimates get better and better, from (1) ten thousands to (2) thousands to (3) hundreds to (4) ten to (5) ones. What symmetry … what beauty!

Let's check our product by casting out nines.

Multiply the factors after adding the digits and casting out nines:

6 (cast out 2 + 7) × 3 (cast out 4 + 5) = 18 and 1 + 8 = 9 or 0 (cast out the 9)

Sum the digits in the product: 2 + 8 + 4 + 3 + 1 = 0 (cast out 8 + 1 and 2 + 4 + 3)

0 = 0. Check.

In your notebook, find these products and check by casting out nines:

Problem 1. 326 × 457

Problem 2. 123 × 987

Problem 3. 908 × 326

Problem 4. 998 × 6, or 998 × 006

Problem 5. 304 × 907

Problem 6. 111 × 111

Problem 7. 222 × 222

Problem 8. 333 × 333

As you do your work, note the unfolding of your answer, from left to right, from ten thousands to ones. Make sure you add the triplet of the fourth, fifth, and sixth products in your head. Here is the final work (casting out nines is left for you to do):

Problem		Product	Problem		Product
1.	326 × 457 2 5 4 4 123542	148,982	2.	123 × 987 2 5 3 2 96081	121,401
3.	908 × 326 1 7 1 4 278868	296,008	4.	998 × 006 5 4 005448	5988
5.	304 × 907 5 2 270708	275,728	6.	111 × 111 12321	12,321
7.	222 × 222 1 48284	49,284	8.	333 × 333 1 2 1 98789	110,889

Multiplying a four-by-four array is next.

EXERCISES

Answer these problems and check your work by casting out nines:

1. What is the cost of 36 sofas at $485 each?
2. What is the cost of 182 liters of diesel fuel at 77 cents a liter? (Note: 4 liters is approximately equal to 1 gallon.)
3. What is the cost of constructing a fiber optic cable 129 miles long at $963 a mile?

4. If a manufacturing plant produces 899 yards of canvas a day, then how many yards can it make in 260 days?
5. What is the value of 256 boxes, each containing 64 books worth \$15 a book?
6. It requires 108 tons of iron rail for 1 mile of railroad track. How many tons will be required for 5 miles, and what will be its cost at \$395 a ton?[1]
7. What is 999^2?
8. Two jets start from the same location and fly in opposite directions, one at the rate of 563 miles an hour and the other at a rate of 456 miles an hour. How far apart will they be at the end of (a) 3 hours? (b) 6 hours?
9. If a worker in an assembly line makes 113 widgets in an hour and 98 widgets per hour are inspected, how many widgets will not be inspected in 112 hours?
10. What is the difference between the cost of 254 yards of cloth at \$95 a yard and 254 yards of cloth at \$125 a yard?
11. If Henry earns \$81 an hour and James \$68 an hour, how much will both earn in 72 hours?
12. Compute $12 \times 89 \times 887$.
13. Find x: $1 + 2 + 3 + 4 + \ldots + 96 + 97 + 98 + 99 + 100 = x$. The famous mathematician Carl Friedrich Gauss (1777-1855) was asked to do this as a grammar school student and within seconds he had the answer. (Hint: Note that $1 + 100$ (the first and the last number in the sum) $= 101$. Then, note that $2 + 99$ (the second and second to last number in the sum) $= 101$. Continuing, $3 + 98 = 101$, $4 + 97 = 101$, etc. The last sum is $50 + 51 = 101$. After 50 such additions, the sums taken from the beginning and the end finally meet in the middle. Therefore, we have 50 sums all totaling 101. What is the sum? Compute the product of 50 and 101 two different ways, both making use of the Distributive Law.)

In cellular telephony, geographical areas are divided into many clusters. Each cluster consists of seven cells, each one roughly hexagonal in shape (Figure 3). Each cell is served by a base station, a low-powered radio transmitter or tower. Each tower is allocated about 47 two-watt radio channels. Answer this question:

14. How many channels does a seven-cell cluster serve?

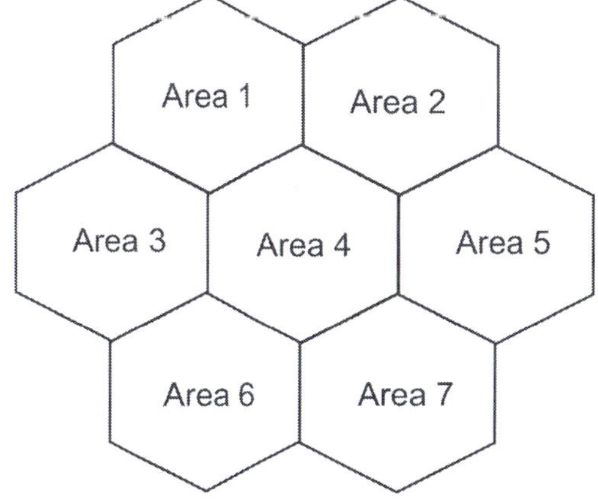

Figure 3.

A typical cell has a radius of 6 to 12 miles (10 to 20 km, see Step 7 for the conversion factor). When a cell phone user leaves the area of one cell, the user is switched to one of the channels in the adjacent cell. Cellular telephones get their name from the same arrangement of cells that bees use in building honeycombs.

Refer to the three tables in the homework for Lesson 4.5 (Table 1, Table 2, and Table 3). Do the following exercises: (Double-check your mental work with pencil and paper.)

15. Select 15 quadruplets of one-digit factors from these tables and mentally sum the four products. For example, sum the products of 7×6, 5×6, 8×7, and 9×9 in your head. Do the same with 14 other pairs.
16. Select another 15 quadruplets of one-digit factors from these tables and mentally sum the four products.

[1] 1 ton = 2000 pounds (Step 7).

Figure 3. Source: United States Postal Service.

A postage stamp, issued in 1981, that honors the poet Edna St. Vincent Millay (1892-1950). In 1923, she published the following poem about the work of the Greek geometer Euclid, and won the Pulitzer Prize for Poetry:

Euclid alone has looked on Beauty bare.
Let all who prate of Beauty hold their peace,
And lay them prone upon the earth and cease
To ponder on themselves, the while they stare
At nothing, intricately drawn nowhere
In shapes of shifting lineage; let geese
Gabble and hiss, but heroes seek release
From dusty bondage into luminous air.
O blinding hour, O holy, terrible day,
When first the shaft into his vision shone
Of light anatomized! Euclid alone
Has looked on Beauty bare. Fortunate they
Who, though once only and then but far away,
Have heard her massive sandal set on stone.

4.8 Multiplication: Four-Digit Factors

In this lesson, we will learn how to multiply a four-digit number by a four-digit number, e.g., 4523 × 5297. We write 4523 = 4000 + 500 + 20 + 3 and 5297 = 5000 + 200 + 90 + 7. Therefore:

$4523 \times 5297 = (4000 + 500 + 20 + 3)(5000 + 200 + 90 + 7)$

As we did in the last lesson, let's work through the general case first:

$(a + b + c + d)(e + f + g + h)$

Let $x = e + f + g + h$. We now have:

$(a + b + c + d)(x)$

We distribute x to a, b, c, and d as follows:

$(a + b + c + d)(x) = ax + bx + cx + dx$

Now, we substitute x with $e + f + g + h$. We get.

$ax + bx + cx + dx = a(e + f + g + h) + b(e + f + g + h) + c(e + f + g + h) + d(e + f + g + h)$

We now distribute a, b, c and d, respectively, to $(e + f + g + h)$. We get:

$$\begin{aligned} a(e + f + g + h) + b(e + f + g + h) + c(e + f + g + h) + d(e + f + g + h) = \\ ae + af + ag + ah + \\ be + bf + bg + bh + \\ ce + cf + cg + ch + \\ de + df + dg + dh \end{aligned}$$

Therefore, $(a + b + c + d)(e + f + g + h) =$

$$\begin{aligned} ae + af + ag + ah + \\ be + bf + bg + bh + \\ ce + cf + cg + ch + \\ de + df + dg + dh \end{aligned}$$

How many subproducts must you calculate?

4 × 4 = 16
(We note:
1 + 2 + 3 + 4 + 3 + 2 + 1 =
4 × 4)

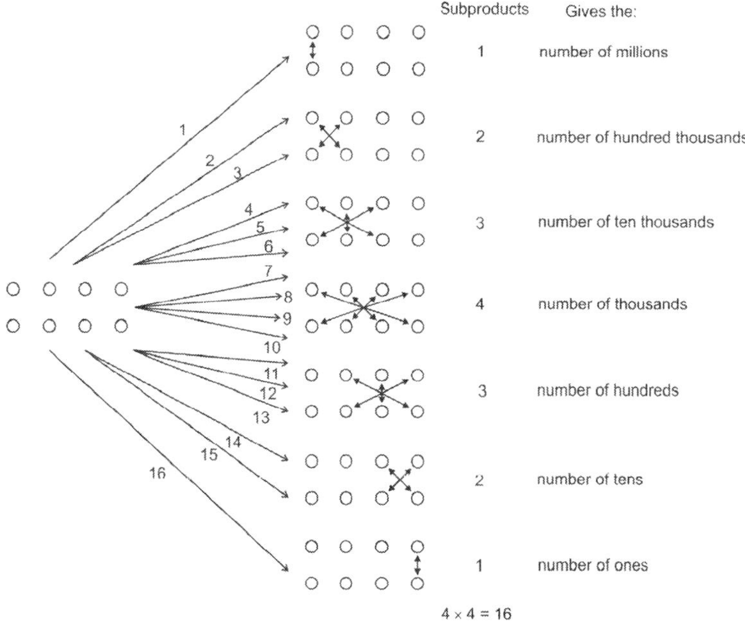

Figure 1. Four-Digit by Four-Digit Paradigm.

Speed Paradigm for Multiplying a Four-Digit Factor by a Four-Digit Factor

Inspect Figure 1 and reflect on the beauty of this four-by-four symmetry, i.e., the beautiful pulse of its pattern. Do you see the sixteen subproducts generated from the four-by-four array along with the symmetry of the number of addends resulting from each subproduct (1, 2, 3, 4, 3, 2, 1)?

Since $4523 \times 5297 = (4000 + 500 + 20 + 3)(5000 + 200 + 90 + 7)$, we let $x = 5000 + 200 + 90 + 7$. We write:

$$(4000 + 500 + 20 + 3)x$$

Distributing x, we get:

$$(4000 + 500 + 20 + 3)x = 4000x + 500x + 20x + 3x$$

Replacing x with $5000 + 200 + 90 + 7$, we get:

$$\begin{aligned}&4000x + 500x + 20x + 3x = \\ &\quad 4000(5000 + 200 + 90 + 7) + \\ &\qquad 500(5000 + 200 + 90 + 7) + \\ &\qquad\quad 20(5000 + 200 + 90 + 7) + \\ &\qquad\qquad 3(5000 + 200 + 90 + 7)\end{aligned}$$

Distributing 4000, 500, 20, and 3, respectively, we get:

$$\begin{aligned}&4000(5000 + 200 + 90 + 7) + \\ &\quad 500(5000 + 200 + 90 + 7) + \\ &\qquad 20(5000 + 200 + 90 + 7) + \\ &\qquad\quad 3(5000 + 200 + 90 + 7) = \\ &4000 \times 5000 + 4000 \times 200 + 4000 \times 90 + 4000 \times 7 + \\ &\quad 500 \times 5000 + 500 \times 200 + 500 \times 90 + 500 \times 7 + \\ &\qquad 20 \times 5000 + 20 \times 200 + 20 \times 90 + 20 \times 7 + \\ &\qquad\quad 3 \times 5000 + 3 \times 200 + 3 \times 90 + 3 \times 7\end{aligned}$$

Multiplying the subproducts, we get:

$$\begin{aligned}&4000 \times 5000 + 4000 \times 200 + 4000 \times 90 + 4000 \times 7 + \\ &\quad 500 \times 5000 + 500 \times 200 + 500 \times 90 + 500 \times 7 + \\ &\qquad 20 \times 5000 + 20 \times 200 + 20 \times 90 + 20 \times 7 + \\ &\qquad\quad 3 \times 5000 + 3 \times 200 + 3 \times 90 + 3 \times 7 = \\ &20{,}000{,}000 + 800{,}000 + 360{,}000 + 28{,}000 + \\ &\quad 2{,}500{,}000 + 100{,}000 + 45{,}000 + 3500 + \\ &\qquad 100{,}000 + 4000 + 1800 + 140 + \\ &\qquad\quad 15{,}000 + 600 + 270 + 21\end{aligned}$$

The sum of these sixteen addends is 23,958,331. Confirm this sum by doing the addition in your notebook. Determining this product requires a lot of written work. The streamlined method will shorten the writing but require more mental effort as a trade-off.

Figure 2 unfolds the method. Note the pulse of descending order of place values:

Step 1. $4 \times 5 = 20$
20 millions = 20,000,000 (one addend)

Step 2. $4 \times 2 + 5 \times 5 = 8 + 25 = 33$
33 hundred thousands = 800,000 + 2,500,000 = 3,300,000 (two addends)

Step 3. $4 \times 9 + 5 \times 2 + 5 \times 2 = 36 + 10 + 10 = 56$
56 ten thousands = 360,000 + 100,000 + 100,000 (three addends)

Step 4. 4 × 7 + 5 × 3 + 5 × 9 + 2 × 2 = 28 + 15 + 45 + 4 = 92
92 thousands = 28,000 + 45,000 + 4000 + 15,000 (four addends)

Step 5. 5 × 7 + 2 × 3 + 2 × 9 = 35 + 6 + 18 = 59
59 hundreds = 3500 + 1800 + 600 (three addends)

Step 6. 2 × 7 + 9 × 3 = 14 + 27 = 41
41 tens = 140 + 270 (two addends)

Step 7. 3 × 7 = 21
21 ones (one addend)

Let's apply it to our problem. We set up the factors vertically:

$$\begin{array}{r} 4523 \\ \times 5297 \\ \hline \end{array}$$

Watch how the numbers 20, 33, 56, 92, 59, 41, and 21 are generated, from left to right. The method will take seven steps:

Step 1 (the first double arrow: one addend). 4 × 5, or 4 thousands by 4 thousands = 20 millions. We write:

$$\begin{array}{r} 4523 \\ \times 5297 \\ \hline 20 \end{array}$$

Our first estimate: 20 million, or 20,000,000

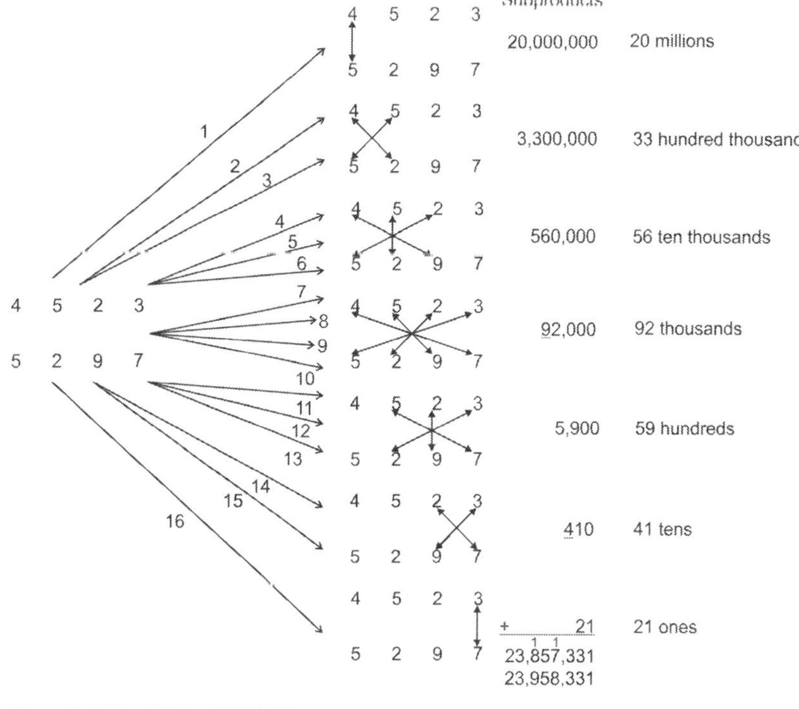

Figure 2. 4523 × 5297 = 23,958,331

Step 2 (the first crossed double arrows: two addends). 4 × 2 = 8 (thousands by hundreds), 5 × 5 = 25 (hundreds by thousands). Do this sum in your head: 8 + 25 = 33, or 33 hundred thousand. We write:

$$\begin{array}{r} 4523 \\ \times 5297 \\ \hline 2\overset{3}{0}3 \end{array}$$

Our second estimate, reading from left to right: 23,300,000

Step 3 (the first three double arrows: three addends). 4 × 9 = 36 (thousands by tens), 2 × 5 = 10 (tens by thousands), 5 × 2 = 10 (hundreds by hundreds). Do this sum in your head: 36 + 10 + 10 = 56, or 56 ten thousands. We write:

$$\begin{array}{r} 4523 \\ \times 5297 \\ \hline 2\overset{3}{0}\overset{5}{3}6 \end{array}$$

Our third estimate, reading from left to right: 23,860,000

Step 4 (the quadruple double arrows: four addends). 4 × 7 = 28 (thousands by ones), 3 × 5 = 15 (ones by thousands), 5 × 9 = 45 (hundreds by tens), 2 × 2 = 4 (tens by hundreds). Do this sum in your head: 28 + 15 + 45 + 4 = 92, or 92 thousands. We write:

4523
×5297 Our fourth estimate, reading from left to right: 23,952,000
3 5 9
20362

Step 5 (the last three double arrows: three addends). 5 × 7 = 35 (hundreds by ones), 3 × 2 = 6 (ones by hundreds), 2 × 9 = 18 (tens by tens). Do this sum in your head: 35 + 6 + 18 = 59, or 59 hundreds. We write:

4523
× 5297 Our fifth estimate, reading from left to right: 23,957,900
3 5 9 5
203629

Step 6 (the last crossed double arrows: two addends). 2 × 7 = 14 (tens by ones), 3 × 9 = 27 (ones by tens). Do this sum in your head: 14 + 27 = 41, or 41 tens. We write:

4523
×5297 Our sixth estimate, reading from left to right: 23,958,310
3 5 9 5 4
2036291

Step 7 (the last double arrow: one addend). 3 × 7 = 21, or 21 ones. We write:

4523
×5297 Our precise product, reading from left to right: 23,958,331
3 5 9 5 4 2
20362911

Review these steps carefully to catch their movement. Again, observe how the estimates get better and better, from (1) millions to (2) hundred thousands to (3) ten thousands to (4) thousands to (5) hundreds to (6) tens to (7) ones. Again, we see the beauty of symmetry.

Let's check our work by casting out nines:

Multiply the factors after adding the digits and casting out nines: 5 × 5 = 25 and 2 + 5 = 7
Sum the digits in the product: 2 + 5 = 7 (cast out 9, 8 + 1, and 3 + 3 + 3)
7 = 7. Check.

You must work hard on doing the subproducts in your head. In your notebook, (1) continue to develop this skill by finding these products and (2) check your answer by casting out nines:

Problem 1. 1111 × 1111
Problem 2. 2222 × 2222
Problem 3. 3333 × 3333
Problem 4. 7643 × 2185
Problem 5. 7156 × 238, or 7156 × 0238

Problem 6. 2709 × 75, or 2709 × 0075

As you do your work, note the unfolding of your answer, from left to right, from millions to ones. Make sure you add the multiple addends in your head. Take your time. Here is the final work (casting out nines is left for you to do):

Problem		Product	Problem		Product
1.	1111 ×1111 1234321	1,234,321	2.	2222 ×2222 1 1 1 4826284	4,937,284
3.	3333 ×3333 1 2 3 2 1 9876789	11,108,889	4.	7643 × 2185 1 7 9 6 4 1 14903545	16,699,955
5.	7156 × 0238 2 6 3 5 4 01439588	1,703,128	6.	2709 ×0075 5 3 6 4 00149535	203,175

In our next lesson, we will use our multiplication skills to help us decipher Babylonian cuneiform numbers.

EXERCISES

Calculate the following and check your answers by casting out nines:

1. Multiply 6734 by (a) 5298, (b) 76, (c) 208, and (d) 8.
2. What is the value of 3016 square feet of marble slab at $164 a square foot?
3. If 1208 pounds of cotton are raised from an acre of land, how many pounds will 259 acres produce? (Note: 1 acre = 43,560 square feet.)
4. What is the cost of 725 butter dishes at $11.75 each? (Hint: Change $11.15 to 1105 cents). To answer in dollars, write the decimal point two places from the right.
5. What is the value of 78 mauls[1] at $42.25 a maul?
6. How many square inches in 6240 square feet if one foot equals 12 inches?
7. How many cubic inches (volume) in 124 cubic feet?
8. In cubic feet, what is the volume of a cord[2] of wood (Figure 3)?
9. How many cubic feet are in 16 cords of wood?
10. How much does 16 cords of wood cost at $16.25 a cord?
11. Compute 9999^2.

Figure 3. A cord of wood. Source: Public Domain.

[1] A maul is a heavy hammer used to drive stakes or split wood.

[2] A cord of wood is a stack of split wood four feet wide, four feet high, and eight feet long. In early American history, the average colonial house used 15 to 20 cords of firewood a year, enough to empty local supplies quickly in most areas of the east coast.

Refer to the three tables in the homework for Lesson 4.5 (Table 1, Table 2, and Table 3). Do the following problems: (Make sure to double-check your mental work with pencil and paper.)

12. Select ten quintuplets of one-digit factors from the above tables and mentally sum the five products. For example, sum the products of 5 × 4, 2 × 9, 6 × 6, 7 × 9, and 4 × 8 in your head. Do the same with nine other pairs.
13. Select another ten quintuplets of one-digit factors from the above tables and mentally sum the five products.
14. Select five sextuplets of one-digit factors from the above tables and mentally sum the six products. For example, sum the products of 1 × 8, 8 × 9, 3 × 4, 5 × 7, 2 × 3, and 8 × 9 in your head. Do the same with four other pairs.
15. Select another five sextuplets of one-digit factors from the above tables and mentally sum the six products.

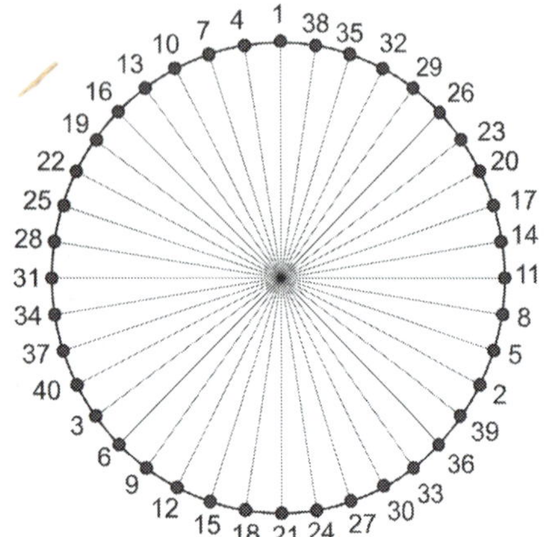

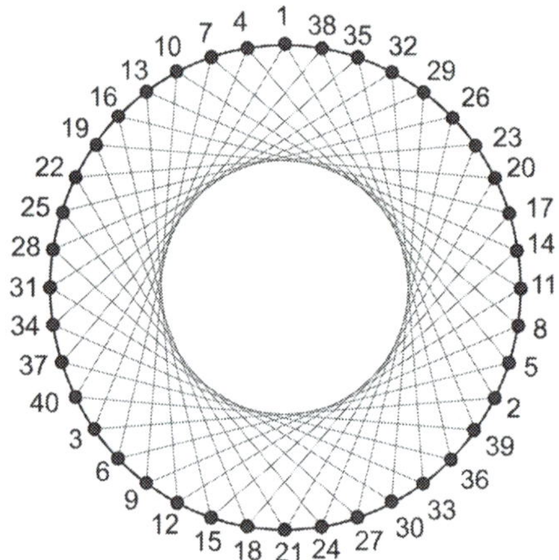

Figure 4. In Lesson 3.7, we saw how the astronomer/mathematician Johannes Kepler (1571-1630) developed a geometric method to plot the places in the heavens when Jupiter and Saturn passed near each other in conjunction. Kepler looked at the diagram and decided to draw line segments from conjunction 1 to conjunction 2, conjunction 2 to conjunction 3, etc., noting how an inner circle came into view. He discovered that if the inner circle approximated the orbit of the inner planet, Jupiter, and the outer circle, twice the circumference of the inner circle, approximated the orbit of Saturn, an orbit nearly twice the circumference of the orbit of Jupiter. Reflecting his distinctive spirit, Kepler expressed ravishing delight in the geometric revelation of this connection.

4.9 The Babylonian Number System

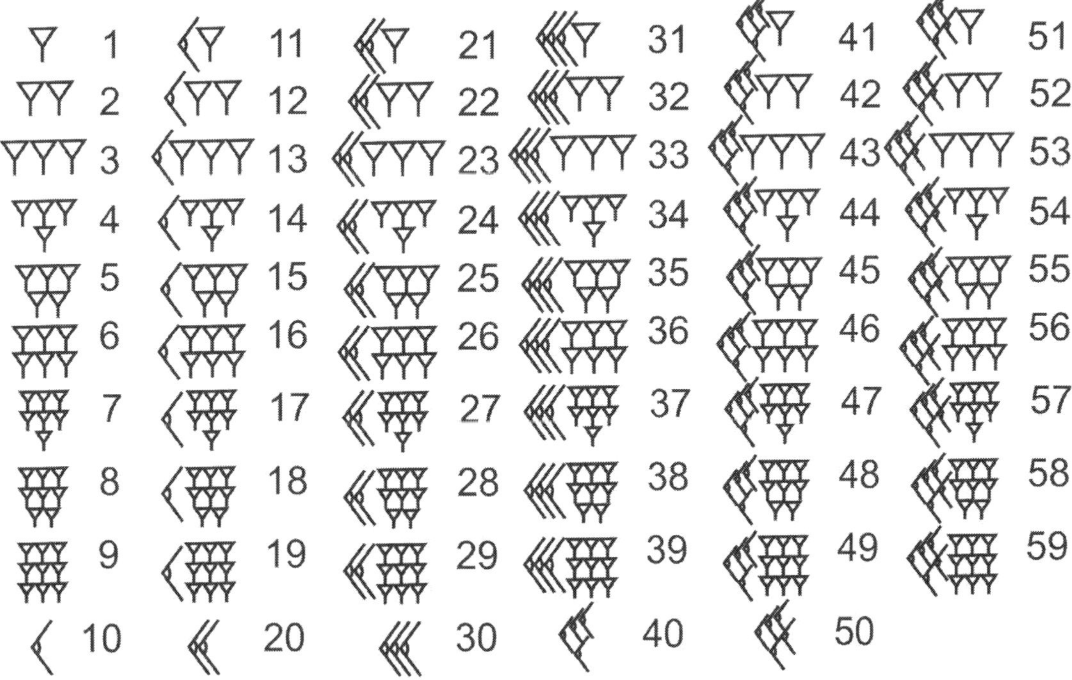

Figure 1. Sexagesimal digits.

Terms & Concepts Introduced
1. Calculi
2. Cuneiform
3. Sexagesimal system
4. Stylus

As we have noted, our decimal system employs a uniform number scale based upon groupings by 10, or it is a base 10 system. In this number system, each place has a value based on powers of 10.

The Babylonians (ca. 1800 BC) employed a uniform number scale based upon groupings of 60, or base 60, in Latin, **sexagesimal**. They also used blanks to denote that a position has no value (e.g., zero 60s), but using an empty space is suspect to misinterpretation. In our time, we still retain vestiges of this system. We see it in circle measurement; i.e., there are 360° in one circle. We also see it when we tell time; i.e., one hour consists of 60 minutes, and one minute consists of 60 seconds.

The Sumerians, one of the earliest civilizations, created different clay tokens, named **calculi**, to represent different groups of objects. By 3200 BC, these people recorded a quantity of goods by sealing a collection of these tokens in a clay jar. Before placing them inside the jar, they made an impression of each on the outside of the jar. These token markings later came to represent numbers themselves,[1] eliminating the need for the jars, so the Babylonians enhanced the use of these symbols, named **cuneiform**, by writing them on flat clay tablets with a blunt reed named a **stylus**.[2]

These markings led to one of the first numeral systems and to what many historians believe to be the first form of written language.

[1] Notice the transition in thinking from (1) a real world object like sheep or money to (2) clay impressions representing these objects to (3) the abstract idea of number.

[2] The impressions left by the stylus looked like wedges and *cuneiform* is Latin meaning "wedge shaped."

Two symbols were used to denote numbers (Figure 1):

1. A vertical wedge represented the number one; it looks somewhat like a nail.
2. A horizontal wedge represented the number ten; it looks very much like a chevron, a badge consisting of stripes meeting at an angle, worn on the sleeve by non-commissioned officers, police officers, etc., as an indication of rank, service, or the like.

The table shows you how they wrote the numbers from 1 to 59. Each of the compound symbols is one numeral, the Babylonian numeral.

If we let $a_1a_2 \ldots a_n$ be Babylonian numerals, we can generalize its representation in the base 60 positional system:

$$a_1a_2a_3 \ldots a_n = a_1 60^{n-1} + a_2 60^{n-2} + a_3 60^{n-3} + \ldots + a_{n-3} 60^3 + a_{n-2} 60^2 + a_{n-1} 60 + a_n$$

Compare this representation with a numeral $a_1a_2 \ldots a_n$ in base 10 (Lesson 4.2):

$$a_1a_2a_3 \ldots a_n = a_1 10^{n-1} + a_2 10^{n-2} + a_3 10^{n-3} + \ldots + a_{n-3} 10^3 + a_{n-2} 10^2 + a_{n-1} 10 + a_n$$

For a four-digit Babylonian numeral, we have:

$$a_1a_2a_3a_4 = a_1 60^3 + a_2 60^2 + a_3 60 + a_4$$

Notice, in the first row,  stands for $10 + 1 = 11$. If a number was greater than 59, it was positioned by place value.

What does  stand for?

$1 \times 60^1 + 10 = 60 + 10 = 70$

EXERCISES

Define the following terms:

1. Calculi
2. Sexagesimal system
3. Cuneiform
4. Stylus

Do the following problems:

5. Compute (a) 60^3 and (b) 60^4.
6. Complete the following table (For the Translation column, we separate place values with a comma.):

Cuneiform	Translation	Positional Expansion	Decimal Number
	1, 10	$1 \times 60^1 + 10$	70
	1, 20		
	1, 30		
	1, 40		

Cuneiform	Translation	Positional Expansion	Decimal Number
	1, 50		
See note.[3]	0, 2 or 2, 0		
(2 symbols)			
(2 symbols)			
(2 symbols)			

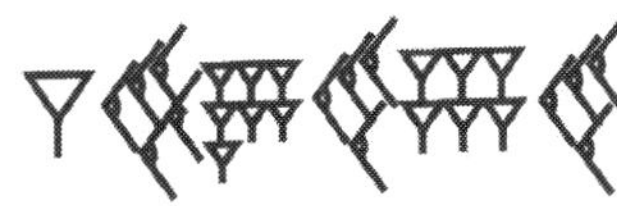

7. The above Babylonian number means 1, 57, 46, 40. Write it as a decimal number. (Hint: Compute $1 \times 60^3 + 57 \times 60^2 + 46 \times 60 + 40$)

When translated, a Babylonian numeral means: (Write each as a decimal number.)

8. 6, 28
9. 23, 13
10. 1, 45
11. 42, 6, 58
12. 3, 18, 21, 55
13. 1, 1, 1, 1, 1

> 111, in base 2, means:
> $1 \times 2^2 + 1 \times 2^1 + 1 \times 2^0 = 1 \times 4 + 1 \times 2 + 1 \times 1 = 4 + 2 + 1 = 9$ (base 10)
> This is how we convert a number in any base to base 10, or to a decimal number.

Answer the following questions for the bases indicated:

14. What is the decimal number for 110011 (Base 2)?
15. What is the decimal number for 324 (Base 5)?
16. What is the decimal number for 100436 (Base 8)?

---

[3] Whether means 2 or 120 must be determined by the Babylonian context. So, give two possible answers.

Convert 91 to Base 2:
To start, find the highest power of 2 in 91:
$2^6 = 64$
What is left over?
$91 - 64 = 27$
Find the highest power of 2 in 27:
$2^4 = 16$
What is left over?
$27 - 16 = 11$
Find the highest power of 2 in 11:
$2^3 = 8$
What is left over?
$11 - 8 = 3$
Find the highest power of 2 in 3:
$2^1 = 2$
What is left over?
$3 - 2 = 1$
Finally, find the highest power of 2 in 1:
$2^0 = 1$
What is left over?
$1 - 1 = 0$
Since $0 < 2^0$, we are finished. We can add powers of 2 to get 91:
$91 = 64 + 16 + 8 + 2 + 1 = 2^6 + 2^4 + 2^3 + 2^1 + 2^0$
Noting that there are zero 5th and 2nd powers of 2 in this sum, we write:
$1011011 = 1 \times 2^6 + 0 \times 2^5 + 1 \times 2^4 + 1 \times 2^3 + 0 \times 2^2 + 1 \times 2^1 + 1 \times 2^0$
[Note: What we are doing is division (Step 5, the next step in our dance).]

Convert the following numbers, written in Base 10, to the base indicated:

17. Write the decimal number 88 in Base 2.
18. Write the decimal number 56 in Base 2.
19. Write the decimal number 112 in Base 2.
20. Write the decimal 565 in Base 2.
21. Write the decimal number 56 in Base 5.
22. Write the decimal number 112 in Base 5.
23. Write the decimal number 112 in Base 8.
24. Write the decimal 565 in Base 5.
25. Write the decimal 565 in Base 8.

A day in the life of a Babylonian schoolboy, 4000 years ago.
I recited my tablet, ate my lunch, prepared my (new) tablet, wrote it, finished it; then my model tablets were brought to me; and in the afternoon my exercise tablets were brought to me. When school was dismissed, I went home, entered the house, and found my father sitting there. I explained my exercise-tablets to my father, recited my tablet to him, and he was delighted.
Cited in John Tabak, *Numbers: Computers, Philosophers & the Search for Meaning* (2004), p. 13.

4.10 Multiplication: Five-Digit Factors

In this lesson, we will explore the rhythm of multiplying a five-digit number by a five-digit number, e.g., 12,345 × 54,321. We write 12,345 = 10,000 + 2000 + 300 + 40 + 5 and 54,321= 50,000 + 4000 + 300 + 20 + 1. Therefore:

12,345 × 54,321 = (10,000 + 2000 + 300 + 40 + 5)(50,000 + 4000 + 300 + 20 + 1)

The general case follows the same procedure as in previous lessons. How many subproducts must we calculate?

5 × 5 = 25

(We note: 1 + 2 + 3 + 4 + 5 + 4 + 3 + 2 + 1 = 5 × 5)

Speed Paradigm for Multiplying a Five-Digit Factor by a Five-Digit Factor

Again, inspect Figure 1 and reflect on the beauty of this five-by-five symmetry, i.e., the marvelous tempo of its pattern. Do you see the twenty-five subproducts that we generate from the five-by-five array? Note the pattern of the descending order of place values, along with the symmetry of the number of subproducts (1, 2, 3, 4, 5, 4, 3, 2, 1).

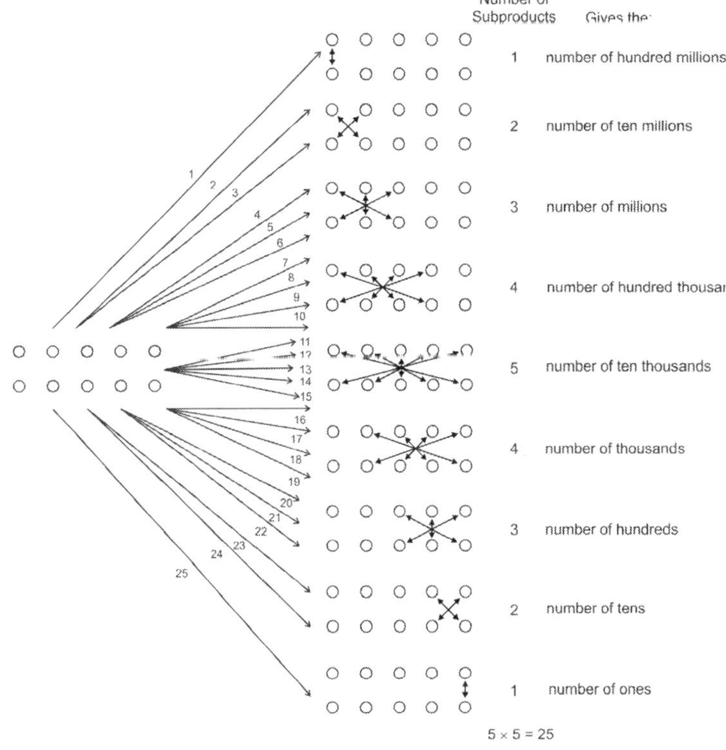

Figure 1. Five-Digit by Five-Digit Paradigm.

Let's apply this paradigm for 12,345 × 54,321. Figure 2 on the next page unfolds the method:

Step 1. 1 × 5 = 5
5 hundred millions = 500,000,000 (one addend)

Step 2. 1 × 4 + 2 × 5 = 4 + 10 = 14
14 ten millions = 40,000,000 + 100,000,000 = 140,000,000 (two addends)

Step 3. 1 × 3 + 2 × 4 + 3 × 5 = 3 + 8 + 15 = 26
26 millions = 3,000,000 + 8,000,000 + 15,000,000 = 26,000,000 (three addends)

Step 4. 1 × 2 + 2 × 3 + 3 × 4 + 4 × 5 = 2 + 6 + 12 + 20 = 40
40 hundred thousands = 200,000 + 600,000 + 1,200,000 + 2,000,000 = 4,000,000 (four addends)

Step 5. 1 × 1 + 2 × 2 + 3 × 3 + 4 × 4 + 5 × 5 = 1 + 4 + 9 + 16 + 25 = 55
55 ten thousands = 10,000 + 40,000 + 90,000 + 160,000 + 250,000 = 550,000 (five addends)

Step 6. 2 × 1 + 3 × 2 + 4 × 3 + 5 × 4 = 2 + 6 + 12 + 20 = 40
40 thousands = 2000 + 6000 + 12,000 + 20,000 = 40,000 (four addends)

Step 7. $3 \times 1 + 4 \times 2 + 5 \times 3 = 3 + 8 + 15 = 26$
26 hundreds = 300 + 800 + 1500 = 2600 (three addends)

Step 8. $4 \times 1 + 5 \times 2 = 4 + 10 = 14$
14 tens = 40 + 100 = 140 (two addends)
Step 9. 5 ones = 5 (one addend)

Let's now work the problem in streamlined fashion. We set up the factors vertically:

```
  12345
× 54321
```

Step 1. We engage the first double arrow, one addend: hundred millions. 1×5 (ten thousands by ten thousands), or 5 hundred millions. We write:

```
  12345
× 54321     Our first estimate: 5
  5
```

hundred million, or 500,000,000

Step 2. We engage the first crossed double arrows, two addends: ten millions. $1 \times 4 = 4$ (ten thousands by thousands), $2 \times 5 = 10$ (hundreds by ten thousands). Do this sum in your head: $4 + 10 = 14$, or 14 ten millions. We write:

```
  12345
× 54321     Our second esti-
  1
  54
```

mate, reading from left to right: 640,000,000

Step 3. We engage the first three double arrows, three addends: millions. $1 \times 3 = 3$ (ten thousands by hundreds), $3 \times 5 = 15$ (hundreds by ten thousands), $2 \times 4 = 8$ (thousands by thousands). Do this sum in your head: $3 + 15 + 8 = 26$, or 26 millions. We write:

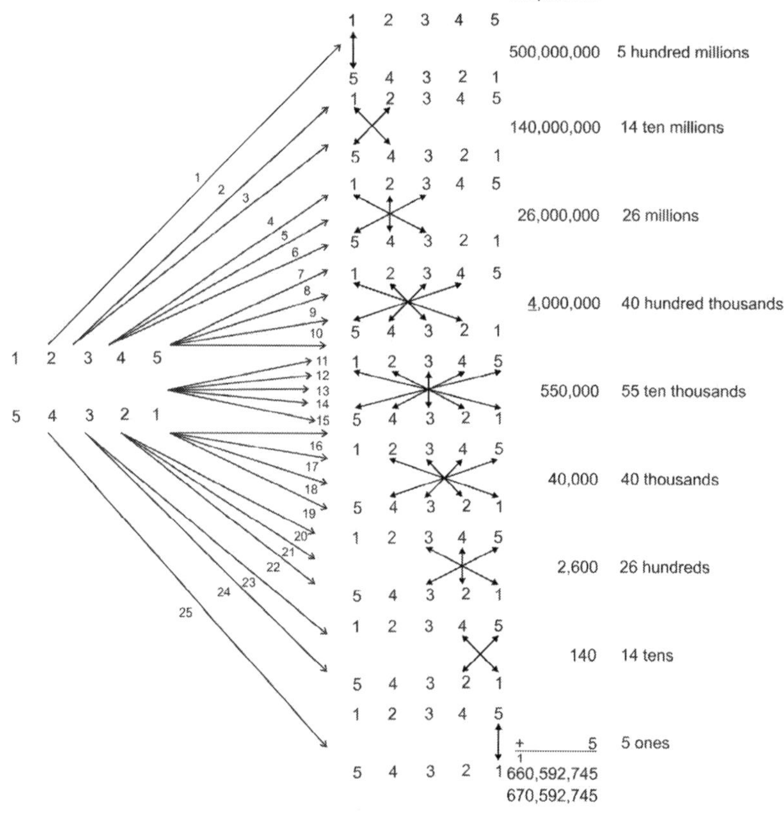

Figure 2. 12,345 × 54,321 = 670,592,745

```
  12345
× 54321     Our third estimate, reading from left to right: 666,000,000
  1 2
  546
```

Step 4. We engage the first quadruple double arrows, four addends: hundred thousands. 1 × 2 = 2 (ten thousands by tens), 4 × 5 = 20 (tens by ten thousands), 2 × 3 = 6 (thousands by hundreds), 3 × 4 = 12 (hundreds by thousands). Do this sum in your head: 2 + 20 + 6 + 12 = 40, or 40 hundred thousands. We write:

```
 12345
×54321    Our fourth estimate, reading from left to right: 670,000,000
 1 2 4
5460
```

Step 5. We engage the quintuple double arrows, five addends: ten thousands. 1 × 1 = 1 (ten thousands by ones), 5 × 5 = 25 (ones by ten thousands), 2 × 2 = 4 (thousands by tens), 4 × 4 = 16 (tens by thousands), and 3 × 3 = 9 (hundreds by hundreds). Do this sum in your head: 1 + 25 + 4 + 16 + 9 = 55, or 55 ten thousands. We write:

```
 12345
×54321    Our fifth estimate, reading from left to right: 670,550,000
 1 2 4 5
54605
```

Step 6. We engage the last quadruple double arrows, four addends: thousands. 2 × 1 = 2 (thousands by ones), 5 × 4 = 20 (ones by thousands), 3 × 2 = 6 (hundreds by tens), 4 × 3 = 12 (tens by hundreds). Do this sum in your head: 2 + 20 + 6 + 12 = 40, or 40 thousands. We write:

```
 12345
×54321    Our sixth estimate, reading from left to right: 670,590,000
 1 2 4 5 4
546050
```

Step 7. We engage the last three double arrows, three addends: hundreds. 3 × 1 = 3 (hundreds by ones), 5 × 3 = 15 (ones by hundreds), 4 × 2 = 8 (tens by tens). Do this sum in your head: 3 + 15 + 8 = 26, or 26 hundreds. We write:

```
 12345
×54321    Our seventh estimate, reading from left to right: 670,592,600
 1 2 4 5 4 2
5460506
```

Step 8. We engage the last crossed double arrows, two addends: tens. 4 × 1 = 4 (tens by ones), 5 × 2 = 10 (ones by tens). Do this sum in your head: 4 + 10 = 14, or 14 tens. We write:

```
 12345
×54321    Our eighth estimate, reading from left to right: 670,592,740
 1 2 4 5 4 2 1
54605064
```

Step 9. We engage the last double arrow, one addend: ones. 5 × 1 = 5 (ones by ones), or 5 ones. We write:

```
  12345
× 54321
1 2 4 5 4 2 1
546050645
```

Our precise product, reading from left to right: 670,592,745

Check:

First factor: 1 + 2 + 3 (cast out 4 + 5) = 6
Second factor: 3 + 2 + 1 (cast out 5 + 4) = 6
Multiply: 6 × 6 = 36 and 3 + 6 = 9 or 0 (cast out the 9)
Sum of the digits in the product: 6 + 5 + 7 (cast out 7 + 2, 5 + 4, and 9) = 18 and 1 + 8 = 9 or 0
0 = 0. Check.

Do you see an extra symmetry in these nine steps? Look at the sums: 5, 14, 26, 40, 55, 40, 26, 14, 5. Does this have something to do with 12,345 × 54,321? Do you see that the symmetry of the subproducts reflects the symmetry of the factors? Isn't this perichoretic pattern wonderful? Multiplication, seen rightly, is indeed beautiful.

In your notebook, multiply 65,119 by 21,005. All your work should unfold on one line and, when completed, should look like:

```
  65119
× 21005
 1   4 3     4
1267394555
1,367,824,595
```

Check:

First factor: 6 + 5 + 1 + 1 (cast out the 9) = 13 and 1 + 3 = 4
Second factor: 2 + 1 + 5 = 8
Multiply: 4 × 8 = 32 and 3 + 2 = 5
Sum of the digits in the product: 7 + 2 + 5 (cast out 3 + 6, 1 + 8, 4 + 5, and 9) = 14 and 1 + 4 = 5
5 = 5. Check.

SIX-DIGIT FACTOR BY A SIX-DIGIT FACTOR

Can you develop a picture for a six-digit by six-digit problem? Go ahead and do it in your notebook. How many steps? How many subproducts? Use these patterns to help you find the answers:

Table 1

Factors	Steps		Pattern	Number of Subproducts				More Patterns
2 by 2	3	=	2 + 2 – 1	4	=	2 × 2	=	1 + 2 + 1
3 by 3	5	=	3 + 3 – 1	9	=	3 × 3	=	1 + 2 + 3 + 2 + 1
4 by 4	7	=	4 + 4 – 1	16	=	4 × 4	=	1 + 2 + 3 + 4 + 3 + 2 + 1
5 by 5	9	=	5 + 5 – 1	25	=	5 × 5	=	1 + 2 + 3 + 4 + 5 + 4 + 3 + 2 + 1
6 by 6	?	=	?	?	=	?	=	?

In your notebook, calculate 123456 × 654321. Here is the final work:

```
  123456
× 654321        Product: 80,779,853,376

1 3 5 7 9 7 5 3 1
67200100276
```

Check:

First factor: 1 + 2 = 3

Second factor: 1 + 2 = 3

Multiply: 3 × 3 = 9 or 0 (cast out the 9)

Sum of the digits in the product: 8 + 7 + 7 + 8 + 5 + 3 + 7 (cast out 9 and 3 + 6) = 45 and 4 + 5 = 9 or 0

0 = 0. Check.

The joy of this method is in seeing the answers unfold on one line, instead of the traditional methods that require multiple lines. And, did you notice the symmetry in the subproducts? 6, 17, 32, 50, 70, 91, 70, 50, 32, 17, 6.

Multiplication, like all arithmetic operations, requires careful attention to detail, and to do problems like this takes mental focus. This type of mental work is good for you. You won't regret having learned this method; it will pay off in rich dividends for future work in mathematics.

In subsequent lessons, we shall learn some shortcut methods for multiplication but the streamlined algorithm we have learned in this lesson and past lessons (Lesson 4.5, 4.6, 4.7, and 4.8) will work for every situation you encounter.

EXERCISES

Do the following problems:

1. (a) How many steps will it take and (b) how many subproducts will you need to calculate to determine the product of a six-digit number by a six-digit number?
2. (a) How many steps will it take and (b) how many subproducts will you need to calculate to determine the product of a seven-digit number by a seven-digit number?
3. Compute $99{,}999^2$ (a) using our speed paradigm (check by casting out nines) and (b) the Distributive Law: 99,999(100,000 – 1). Which method is easier and faster?
4. Describe the patterns revealed in this table:

Table 2

Multiplication Problem	Product
9^2	81
99^2	9801
999^2	998,001
9999^2	99,980,001
$99{,}999^2$	9,999,800,001
$999{,}999^2$	999,998,000,001

5. Find $9{,}999{,}999^2$ based upon the patterns revealed in Table 2.

Create a multiplication problem, find the product according to the following criteria, and check by casting out nines:

6. One-digit factor by five-digit factor
7. One-digit factor by six-digit factor
8. One-digit factor by a seven-digit factor
9. Two-digit factor by five-digit factor
10. Two-digit factor by six-digit factor
11. Three-digit factor by five-digit factor
12. Three-digit factor by six-digit factor
13. Four-digit factor by five-digit factor
14. Four-digit factor by six-digit factor
15. Five-digit factor by five-digit factor
16. Five-digit factor by six-digit factor
17. Six-digit factor by six-digit factor

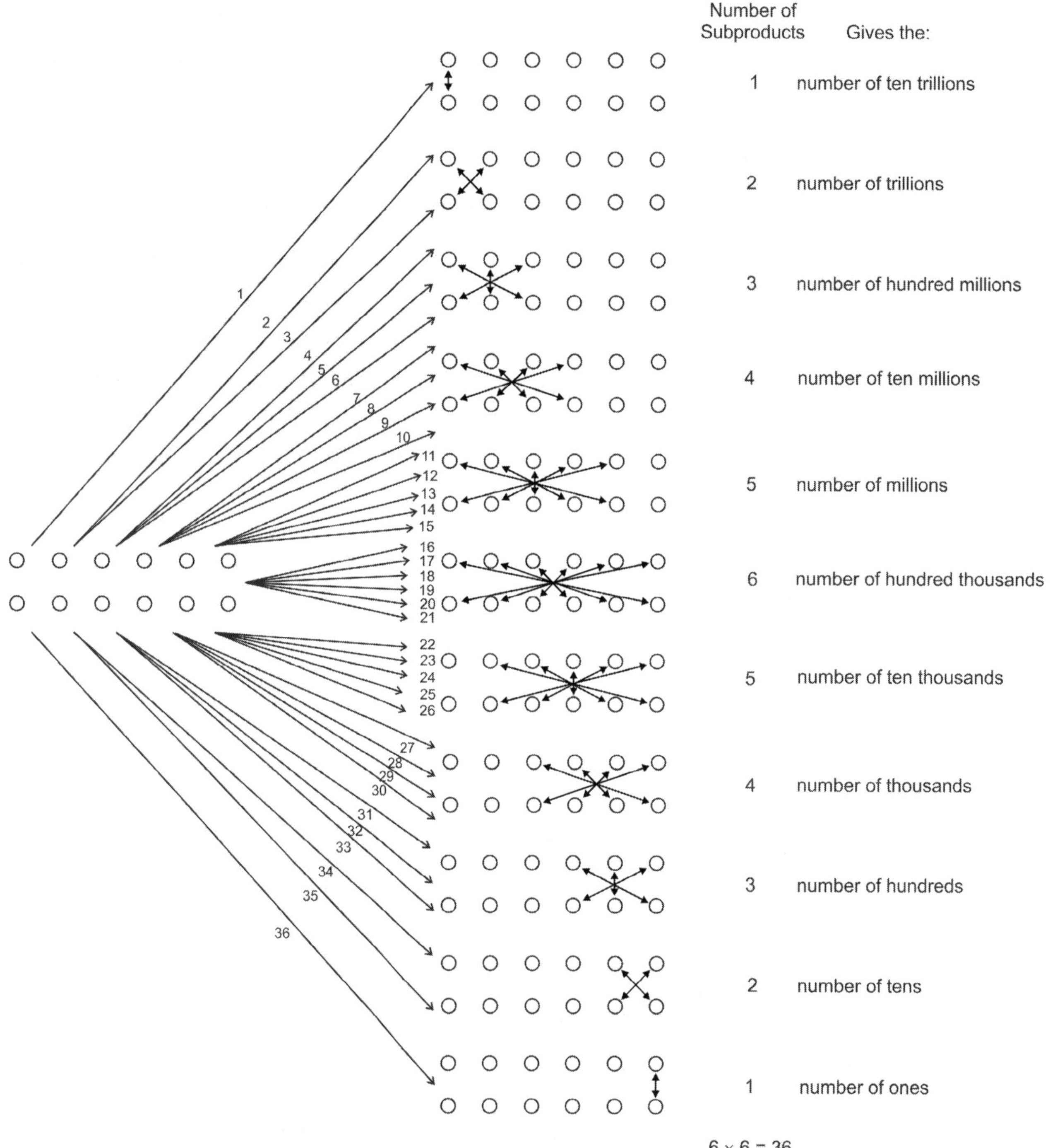

Figure 3. Six-Digit by Six-Digit Paradigm.

4.11 Four Multiplication Shortcuts

We have seen that the standard paradigm for multiplication is already inherently fast. In this lesson, we will explore four shortcut multiplication methods that make calculations even quicker:

Concepts Introduced
1. Factor Shortcut Law
2. Multiplying by 11 Law
3. Squaring Numbers That End with 5 Law

1. Factor Shortcut Law
2. Extension of the Distributive Law
3. Multiplying by 11 Law
4. Squaring Numbers That End with 5 Law

We will investigate two more shortcuts in a subsequent lesson (Lesson 4.13).

Learn when to use these shortcuts instead of the standard paradigm. Before you do any multiplication, think first!

Factor Shortcut Law

We have already noted some quicker ways to multiply (Lesson 4.2). For example, when either factor has appended zeroes, indicating powers of 10, do the following:

Step 1. Remove the zeroes and find the product of the significant figures.[1]
Step 2. Append as many zeroes to the product as you removed in Step 1.

Here are two examples:

Example 1. Multiply 875 by 20,000
 Step 1. 875 × 2 = 1750
 Step 2. Answer: 17,500,000

We simplified our work by factoring 20,000 where one of the factors is a power of 10, i.e., 20,000 = 2 × 10,000. We multiply 875 by 2 (1750) and then we multiply 1750 by 10,000 (17,500,000).

Example 2. Multiply 24,500 by 600
 Step 1.

```
   245
 ×   6   Product: 1470
 -----
   2 3
  1240
```

Check:
Multiply factors: 2 × 6 = 12 and 1 + 2 = 3
Sum digits of product: 1 + 4 + 7 = 12 and 1 + 2 = 3
3 = 3. Check.

 Step 2. Answer: 14,700,000

We factored both 24,500 and 600 into factors where one factor is a power of 10:

24,500 = 245 × 100 and 600 = 6 × 100

We can restate the original problem like this:

24,500 × 600 = 245 × 100 × 6 × 100

[1] In this context, a significant figure is any non-zero digit. We will talk more about significant figures in Lesson 7.4.

We can regroup as follows:[2]

$245 \times 100 \times 6 \times 100 = 245 \times 6 \times 100 \times 100 = 245 \times 6 \times 10{,}000 = 1470 \times 10{,}000 = 14{,}700{,}000$

INSPECTING 60

Let's now consider multiplying a number by 60, the base of the Babylonian number system. What factors of 60 should we consider? It should be obvious that 1 and 60 are factors of 60 ($1 \times 60 = 60$), but these factors will not help us streamline a multiplication problem. We know that $2 \times 30 = 60$. We can write this relationship as follows:

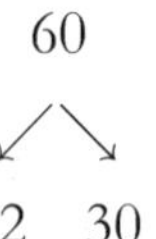

We know that $2 \times 15 = 30$. We write:

60
↙↘
2 30
↙↘
2 15

Finally, $3 \times 5 = 15$. We write:

60
↙↘
2 30
↙↘
2 15
↙↘
3 5

$60 = 2 \times 2 \times 3 \times 5$

There is more than one way to get $2 \times 2 \times 3 \times 5$. We know that $6 \times 10 = 60$. We then write:

60
↙↘
6 10
↙↘ ↙↘
2 3 2 5

$60 = 2 \times 2 \times 3 \times 5$

The numbers 2, 3, and 5 are special numbers and we will learn more about them in Step 6. For now, note that the only factors of 2, 3, and 5 are:

$1 \times 2 = 2$

[2] This regrouping demonstrates the Associative Law of Multiplication (Lesson 5.15).

1 × 3 = 3
1 × 5 = 5

Consider all the possible multiplicative combinations of 2, 2, 3, and 5. Here are two:

2 × 2 = 4
3 × 5 = 15

Therefore, 4 × 15 = 60 (4 and 15 are factors of 60). Try more (2 × 3, 2 × 5, etc.). One of the possible reasons why the Babylonians chose 60 as their number base is because 60 has so many factors.

MORE IMPORTANT FACTORS

There is a difference between the factors of a number and its addends: We multiply factors and add addends. Because of this, a factor is always an addend but an addend is not always a factor. For example, the factors of 36 are 2 and 18; 3 and 12; 4 and 9; and 6 and 6. Some of the addends of 36 are 2 and 34, 10 and 26, 18 and 18, 6 and 30, etc.

Note the following:

- Multiplying by 81 is the same as multiplying by 9 × 9.
- Multiplying by 64 is the same as multiplying by 8 × 8.
- Multiplying by 48 is the same as multiplying by 6 × 8.
- Multiplying by 45 is the same as multiplying by 9 × 5.
- Multiplying by 27 is the same as multiplying by 9 × 3.
- Multiplying by 15 is the same as multiplying by 3 × 5.
- Multiplying by 36 is the same as multiplying by 6 × 6.
- Multiplying by 36 is the same as multiplying by 12 × 3.
- Multiplying by 75 is the same as multiplying by 15 × 5.
- Multiplying by 150 is the same as multiplying by 3 × 5 × 10.
- Multiplying by 250 is the same as multiplying by 5 × 5 × 10.

The Factor Shortcut Law states that the product of any number of factors will be the same in whatever order we multiply them. To use this law, follow these steps:

Step 1. Partition one of the two factors of a multiplication problem into two or more of its factors.
Step 2. Multiply the other factor by one of the factors found in Step 1, the resulting product by another factor, and so forth until you have used all the factors in Step 1. The last product is the product of the original problem.

For example, we can multiply 478 by 45 by multiplying 478 by the factors of 45. Let's see how this works. We note:

45 = 9 × 5

First, let's multiply 478 by 9:

```
 478
×  9     Product: 4302
────
  6 7
3632
```

Second, we multiply 4302 by 5:

```
  4302
×    5     Product: 21,510
 -----
  1 1
20500
```

To verify, we multiply 478 by 45 using our standard paradigm:

```
  478
×  45     Product: 21,510
 -----
 4 6 4
16870
```

Check:

Multiply factors: 1 × 0 = 0
Sum digits of product: 2 + 1 + 5 + 1 = 9 or 0
0 = 0. Check.

Use this shortcut to find these products:

Problem 1. 2115 × 250
Problem 2. 121,181 × 36

First, determine which factors you want to use. For Problem 1, 250 = 5 × 5 × 10 but, for Problem 2, 36 has several possibilities: 36 = 4 × 9 or 36 = 6 × 6. You are free to use any set. Here are the answers:

Problem 1. 2115 × 250 = 528,750
2115 × 5 = 10,575, 10,575 × 5 = 52,875, and 52,875 × 10 = 528,750
Check:
Multiply factors: 0 × 7 = 0
Sum digits of product: 5 + 8 + 5 = 18 and 1 + 8 = 9 or 0
0 = 0. Check.
Problem 2. 121,181 × 36 = 4,362,516
121,181 × 4 = 484,724 and 484,724 × 9 = 4,362,516
121,181 × 6 = 727,086 and 727,086 × 6 = 4,362,516
Check:
Multiply factors: 5 × 0 = 0
Sum digits of product: 0
0 = 0. Check.

Again, you must learn when to apply this law and when to perform other methods. Your skill in choosing shortcuts is refined by practice and plenty of it.

EXTENSION OF THE DISTRIBUTIVE LAW

To review, we can judiciously use the Distributive Law. Use it when one factor is close to multiples of 10 (e.g., 19, 21, 39, 41, 99, 101, 999, 1001, etc.). Why? It is very easy to multiply any number by 10 or by a multiple of 10.

Example 1. 75 × 19 =
75(20 – 1) = 75 × 20 – 75

$75 \times 20 = 75 \times 2 \times 10 = 150 \times 10 = 1500$
$1500 - 75 = 1425$

Example 2. $75 \times 21 =$
$75(20 + 1) = 75 \times 20 + 75$
$75 \times 20 = 1500$
$1500 + 75 = 1575$

Example 3. $326 \times 99 =$
$326(100 - 1) = 326 \times 100 - 326 =$
$32{,}600 - 326 = 32{,}274$

Example 4. $326 \times 101 =$
$326(100 + 1) = 326 \times 100 + 326 =$
$32{,}600 + 326 = 32{,}926$

Example 5. $999{,}999^2 =$
$999{,}999(1{,}000{,}000 - 1) = 999{,}999{,}000{,}000 - 999{,}999 =$

$$\begin{array}{r} 999999000000 \\ -\quad 999999 \\ \hline 999999111111 \end{array}$$

Difference: 999,998,000,001

In your notebook, verify these products with our standard paradigm and check by casting out nines.

MULTIPLYING BY 11 LAW

In previous work, we encountered an interesting property of 11 when multiplying it by a two-digit number.

$11 \times 11 = 121$
$11 \times 12 = 132$
$11 \times 13 = 143$
$11 \times 14 = 154$
$11 \times 15 = 165$
$11 \times 16 = 176$
$11 \times 17 = 187$
$11 \times 18 = 198$
$11 \times 19 = 209$
$11 \times 20 = 220$
$11 \times 36 = 396$
$11 \times 81 = 891$
$11 \times 55 = 605$

If one factor is 11, to calculate the product all we need to do is add the other factor's first and last digits and put this sum between them.

What is happening in 11×19? First, let's use our standard paradigm:

$$\begin{array}{r} 19 \\ \times 11 \\ \hline \overset{1}{1}09 \end{array}$$

Product: 209

First, we write 1 and 9 with a space between them. Adding the first digit to the last digit, we get $1 + 9 = 10$. We write 0 between 1 and 9 and record one hundred or add 1 to the hundreds position to get 209.

Step 1. 1 9

Step 2. $\overset{1}{1}09 = 209$

Likewise, for 11 × 55, we see this:

$$\begin{array}{r} 55 \\ \times 11 \\ \hline \overset{1}{5}05 \end{array}$$

Product: 605

We write 5 and 5 with a space between them. Adding the first digit to the last digit, we get 5 + 5 = 10. Again, we write 0 between 5 and 5 and record one hundred or add 1 to the hundreds position to get 605.

Step 1. 5 5

Step 2. $\overset{1}{5}05 = 605$

Let's play with the dance that is 11. Mathematicians discover many laws by experimenting, or by playing with numbers. Let's consider multiplying a three-digit number by 11. What pattern do you see?

Consider 123 × 11. Let's multiply using our standard paradigm:

$$\begin{array}{r} 123 \\ \times\ 11 \\ \hline 1353 \end{array}$$

See something? Our product begins with 1 and ends with 3. Adding the hundreds position to the tens position, we get 1 + 2 = 3. Next, adding the tens position to the ones position, we get 2 + 3 = 5. Can you state the rule?

Use your rule to find the following products. Cover the solutions, do them in your notebook using our standard paradigm, and check:

Problem 1. 712 × 11

Problem 2. 523 × 11

Problem 3. 398 × 11

Solutions:

Problem 1. 712 × 11 = 7832
- First digit (thousands): 7
- Last digit (ones): 2
- We write: 7 2
- Hundreds digit: 7 + 1 = 8
- We write: 78 2
- Tens digit: 1 + 2 = 3
- We write: 7832
- In your notebook, check by casting out nines.

Problem 2. 523 × 11 = 5753
- First digit (thousands): 5
- Last digit (ones): 3
- We write: 5 3
- Hundreds digit: 5 + 2 = 7
- We write: 57 3

Tens digit: 2 + 3 = 5
We write: 5753
In your notebook, check by casting out nines.

Problem 3. 398 × 11 = 4378
First digit (thousands): 3
Last digit (ones): 8
We write: 3   8
Hundreds digit: 3 + 9 = 2, record one thousand
We write: $\overset{1}{3}2\;8$
Tens digit: 8 + 9 = 7, record one hundred
We write: $\overset{1}{3}\overset{1}{2}78 = 4378$
In your notebook, check by casting out nines.

Create several more examples multiplying three-digit factors by 11 so that you get used to the process.

Now, solve this: 67,928 × 11. Cover the solution, do it in your notebook, and check your answer using the standard paradigm.

Solution:
67,928 × 11 = 747,208
Step 1. 6   8
Step 2. 6 + 7 = 13: $\overset{1}{6}3\quad 8$
Step 3. 7 + 9 = 16: $\overset{1}{6}\overset{1}{3}6\;\; 8$
Step 4. 9 + 2 = 11: $\overset{1}{6}\overset{1}{3}\overset{1}{6}1\; 8$
Step 5. 2 + 8 = 10: $\overset{1}{6}\overset{1}{3}\overset{1}{6}\overset{1}{1}08 = 747,208$

In your notebook, check by casting out nines.

Let's explore one more revelation of multiplication by 11. First, let's generate a table of powers of 11:

Table 1		
Exponential Form	**Meaning**	**Decimal Form**
11^0	1	1
11^1	11	11
11^2	11 × 11	121
11^3	11 × 11 × 11 = 121 × 11	1331
11^4	11 × 11 × 11 × 11 = 1331 × 11	14641
11^5	11 × 11 × 11 × 11 × 11 = 14641 × 11	?

Investigate the third column, the decimal form without commas. What patterns are unfolding? Besides invoking the rule for multiplication by 11, there is another way to determine 11^5 based up the row by row triangular pattern. Do you see how to get 2 in row 3 from 1 + 1 in row 2? Do you see how to get 3 in row 4 from 1 + 2 or 2 + 1 in row 3?

Write your observations as an answer to one of the exercises below (Question 1).

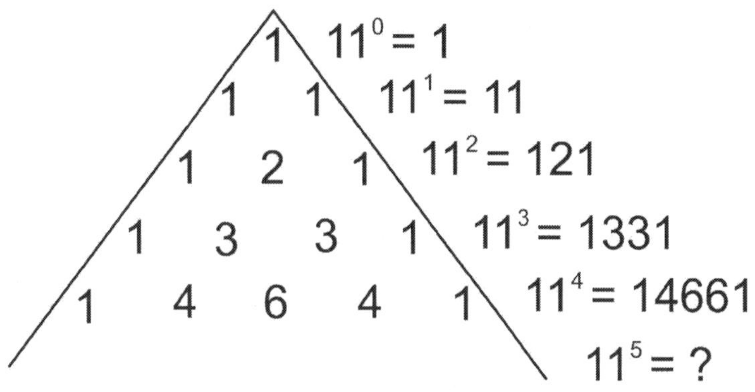

Figure 1. Powers of 11.

SQUARING NUMBERS THAT END WITH 5 LAW

Here is another fascinating number dance. Let's consider two-digit numbers that end in 5, and multiply each by itself. We want to find the second power, or the square, of 15, 25, 35, 45, etc. In your notebook, check each product by casting out nines.

We first calculate 15 × 15:

  15
×15 Product: 225
1 2
105

Next, 25 × 25:

  25
×25 Product: 625
2 2
405

Next, 35 × 35:

  35
×35 Product: 1225
3 2
905

Finally, 45 × 45:

  45
×45 Product: 2025
4 2
1605

Let's put our results in a table:

Table 2	
Factors	**Product**
15 × 15	225
25 × 25	625
35 × 35	1225
45 × 45	2025

Take some time to inspect these results. Do you see the dance? First, note that the last two digits are always 25. By applying our standard paradigm, you should see that this will always be the case. How can one determine the digits in front of the 25?

Table 3				
Factors	**Product**	**Digits in Front of 5 in the Factors**	**Add 1 to the First Digit and Multiply the Digits**	**Digit(s) in Front of 25 in the Product**
15 × 15	225	1 and 1	2 × 1	2
25 × 25	625	2 and 2	3 × 2	6
35 × 35	1225	3 and 3	4 × 3	12
45 × 45	2025	4 and 4	5 × 4	20

From the rule revealed in column four of Table 3, we know that 55 × 55 = 3025 since 6 × 5 = 30. Also:

65 × 65 = 4225 (since 7 × 6 = 42)
75 × 75 = 5625 (since 8 × 7 = 56)
85 × 85 = 7225 (since 9 × 8 = 72)
95 × 95 = 9025 (since 10 × 9 = 90)
105 × 105 = 11,025 (since 11 × 10 = 110)

In your notebook, check these problems by casting out nines.

Like a domino falling on another and triggering another to fall,[1] the pattern continues *ad infinitum*.

In conclusion, we have explored four laws:

1. Factor Shortcut Law
2. Extension of the Distributive Law
3. Multiplying by 11 Law
4. Squaring Numbers That End with 5 Law

Part of the fun of doing arithmetic is finding situations where we can apply these speedy techniques.

[1] There is a method of proof in mathematics named Mathematical Induction (Step 13). The domino pattern shown here is, in an elementary way, an example of this procedure. A proof is a way of reasoning demonstrating that certain mathematical patterns are universally true for all situations and all times. The methods of mathematical proof are one way in which the Triune God reveals to man truth about the way things are in the created reality–things visible and invisible, concrete and abstract.

EXERCISES

Write a short essay:

1. Explain the two patterns revealed in the third column of the Power of 11 Table, Table 1, in this lesson.

Multiply by the Factor Shortcut Law and check your final answer by casting out nines:

2. $46.22 by 48 (Hint: 48 = 8 × 6)
3. 21,112 by 72 (Hint: 72 = 6 × 4 × 3)
4. 92,003 by 84 (Hint: 84 = 7 × 4 × 3)
5. $213.45 by 108 (You discover the factors.)
6. 520,019 by 120 (You discover the factors.)

Answer the following using the Factor Shortcut Law and check your answer by casting out nines:

7. What will 63 acres of land cost at $35,300 an acre?
8. At 36 cents a pound, what will be the cost of 48 sacks of lemonade powder, each containing 32 pounds?

Find the products and check your answer by casting out nines:

9. $8.17 by 100
10. $80.60 by 60
11. $3600 by 700
12. 250 times 8000
13. 800 times 52,000
14. 9,000,000 times 8070

Answer the following questions and check your answers by casting out nines:

15. There are 640 acres in 1 square mile. How many acres in:
    (a) 250 square miles?
    (b) 350 square miles?
    (c) 1100 square miles?
16. The salary of Mr. Jones is $62,000 a year. How much does he receive in nine years?
17. How many millimeters are in 800 meters?
18. How many meters are in 256 kilometers?
19. How many centimeters are in 516 meters?
20. How many centimeters are in 120 kilometers?
21. What is the volume in cubic centimeters of a cube whose side is 600 cm?
22. What is the volume in cubic inches of a rectangular solid whose length is 20 inches, width is 30 inches, and height is 50 inches?

Find the following products and check by casting out nines:

23. 22 × 11
24. 35 × 11
25. 45 × 11
26. 99 × 11
27. 123 × 11
28. 788 × 11
29. 2345 × 11
30. 85,517 × 11

31. 590,155 × 11
32. 67,189,329,113 × 11

The Wonders of Patterns
12345679 × 9 = 111111111
12345679 × 18 = 222222222
12345679 × 27 = 333333333
12345679 × 36 = 444444444
12345679 × 45 = 555555555
12345679 × 54 = 666666666
12345679 × 63 = 777777777
12345679 × 72 = 888888888
12345679 × 81 = 999999999

Use the Distributive Law to find the following products:
33. 85 × 31
34. 85 × 29
35. 365 × 102
36. 365 × 98
37. 698 × 1002
38. 698 × 998
39. 21,237 × 999,999
40. 650 × 100,000,002

Find the followings squares according to the rule in the text and check by casting out nines:
41. 115^2
42. 125^2
43. 245^2
44. 365^2

45. Based on Table 1 in the text, if $11^4 = 14{,}641$, explain the two shortcut ways to find 11^5.

More Patterns
7 × 9 = 63
77 × 99 = 7623
777 × 999 = 776,223
7777 × 9999 = 77,762,223
77,777 × 99,999 = 7,777,622,223
777,777 × 999,999 = 777,776,222,223
7,777,777 × 9,999,999 = 77,777,762,222,223
77,777,777 × 99,999,999 = 7,777,777,622,222,223
777,777,777 × 999,999,999 = 777,777,776,222,222,223
7,777,777,777 × 9,999,999,999 = 77,777,777,762,222,222,223

4.12 Multiplication of Integers

In this lesson, we will first review the powers of 11 and then determine the rules for multiplication of integers. With these rules, we can explore more multiplication shortcuts (Lesson 4.1).

Terms & Concepts Introduced
1. Multiplication of Integers
2. Pascal's Triangle

Pascal's Triangle

First, let's review the table of the powers of 11.

Table 1

Exponential Form	Meaning	Decimal Form
11^0	1	1
11^1	11	11
11^2	11 × 11	121
11^3	11 × 11× 11 = 121 × 11	1331
11^4	11 × 11 × 11 × 11 = 1331 × 11	14641

The digit arrangement in the third column of Table 1 forms **Pascal's Triangle,**[1] where each number not equal to 1 within the triangle is found by adding the pairs of numbers directly above it at the left and right. Note that 2 = 1 + 1, 3 = 1 + 2 = 2 + 1, 4 = 1 + 3 = 3 + 1, 6 = 3 + 3, etc. Do you see 11^0, 11^1, 11^2, 11^3, and 11^4 in rows 1 to 5?

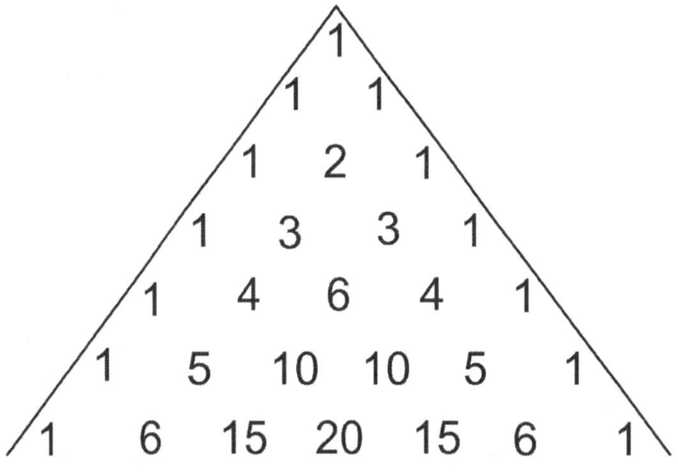

Figure 1. Pascal's Triangle.

What about 11^5? Since 11^4 = 14,641, then 11^5 = 14,641 × 11. Multiplying with our standard paradigm, we get:

```
  14641
× 00011    Product: 161,051
-------
   1 1
 150051
```

[1] This numerical triangle is named in honor of the French mathematician and philosopher Blaise Pascal (1623-1662).

Do you see row 6 in your work, the 1, 5, 10, 10, 5, and 1?

15(10)(10)51
   1   1

161051

Let's repeat this procedure for $11^6 = 11^5 \times 11$. We can multiply 161,051 by 11 this way: 161,050(10 + 1) = 1,610,500 + 161,050:

$$\begin{array}{r} 1610500 \\ +\ 161050 \\ \hline 1771561 \end{array}$$

Multiplying by our standard paradigm, we get:

$$\begin{array}{r} 161051 \\ \times\ 000011 \\ \hline 177161 \end{array}$$

Do you see 1, 6, 15, 20, 15, 6, and 1 in this work? Add the 1 in the first 15 to the first 6 to get 7. Add the 2 in the 20 to 5 in the first 15 to get 7. Add the 1 in the second 15 to 0 in the 20 to get 1. Or:

16(15)(20)(15)61
   7   7   1

1771561

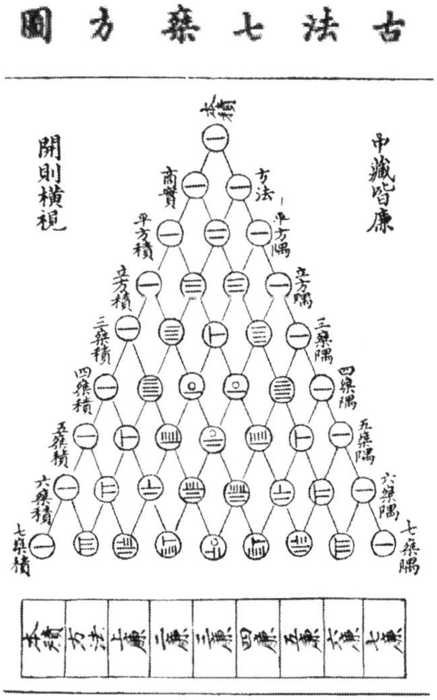

Figure 2. Pascal's Triangle generated by the Chinese Mathematician Yang Hui (1303). Source: Public Domain.

Pascal's Triangle is embodied in multiple mathematical applications illustrating the principle of unity in diversity – applications ranging from an algebraic proof by Isaac Newton to the counting principles of probability, a branch of mathematics named combinatorics.[2]

> The eternal silence of these infinite spaces frightens me (p. 61).
>
> Two extremes: to exclude reason, to admit reason only (p. 74).
>
> The last proceeding of reason is to recognize that there is an infinity of things which are beyond it (p. 77).
>
> Blaise Pascal, *Pensées*, trans. W. F. Trotter (1958).

MULTIPLYING INTEGERS: FOUR CASES

In Lesson 3.3, we determined the rules for addition and subtracting integers. We shall now discover the rules for multiplying integers.

We have four possibilities to consider:

Case 1. Multiplying a positive integer by a positive integer

Case 2. Multiplying a positive integer by a negative integer

Case 3. Multiplying a negative integer by a positive integer

Case 4. Multiplying a negative integer by a negative integer

[2] Specifically, the groundwork of probability theory (Step 9) is Isaac Newton's Binomial Theorem (Lesson 4.13) along with the theory of combinations and permutations.

Case 1		
3 × 4	=	12
3 × 3	=	9
3 × 2	=	6
3 × 1	=	3

Let's explore these cases, one by one. The table on the left demonstrates that multiplying a positive integer by a positive integer generates a positive product (Case 1). We already know this.

Let's continue the pattern, multiplying 3 by -1, -2, -3, etc. The pattern tells us the answer to the Cases 2 and 3. Multiplying a positive integer by a negative integer generates a negative product (Case 2). Since 2 × 3 = 6 and 3 × 2 = 6, we can conjecture that this property holds for a negative factor too:

Cases 2 & 3		
3 × 0	=	0
3 × (-1)	=	-3
3 × (-2)	=	-6
3 × (-3)	=	-9
3 × (-4)	=	-12

3 × (-1) = -3 and (-1) × 3 = -3

Therefore, multiplying a negative integer by a positive integer generates a negative product (Case 3).

What about Case 4? In the table, let's multiply -4 by 3, 2, 1, 0, -1, -2, and -3. The pattern tells us the answer. Multiplying a negative integer by a negative integer generates a positive product (Case 4).

Case 4		
3 × (-4)	=	-12
2 × (-4)	=	-8
1 × (-4)	=	-4
0 × (-4)	=	0
(-1) × (-4)	=	4
(-2) × (-4)	=	8
(-3) × (-4)	=	12

ANOTHER DERIVATION

Let's reason to the same rules another way, making use of what we already know about multiplication, repeated addition of the same number, to assist us.

First, let's consider 4 × 5, a positive integer times a positive integer. We consider 5 groups of 4. We get:

4 × 5 = 4 + 4 + 4 + 4 + 4 = 20

Our product is positive (Case 1).

Second, let's consider a negative integer times a positive integer, or (-5) × 4. Considering 4 groups of -5, we get:

(-5) × 4 = (-5) + (-5) + (-5) + (-5) = -20

Our product is negative (Case 3).

Third, let's consider a positive integer times a negative integer, or 4 × (-5). Can we conceive of adding 4 to itself -5 times? We resolve this conundrum by turning the factors around, or noting that 4 × (-5) = (-5) × 4 = -20 (Case 2).

Fourth, we need to consider -5 × (-4). Before we do, is there a pattern, a dance, to the previous three demonstrations? We want to discover a rule that will help us determine the product of a negative times a negative. Put on your Sherlock Holmes thinking cap and analyze these three demonstrations with a mind to discovering a law governing the (1) sign and (2) magnitude of the product. What do you notice?

> Minus times minus equals plus. The reason for this we need not discuss.
>
> W. H. Auden, *A Certain World: A Commonplace Book* (1970).
>
> (Note: The wording is technically incorrect. We should say, "Negative times negative equals positive.")

In all cases, the magnitude of the product is the product of the absolute values of the factors. Case 1 and Case 2 show us what happens to the sign of the product when we multiply a positive or negative by a positive factor:

Table 2				
Case	**Factor**	**Factor**	**Sign of Product**	**Magnitude of Product**
1	Positive	Positive	Positive	Product of the Absolute Values of the Factors
2	Negative	Positive	Negative	Product of the Absolute Values of the Factors

Case 3 shows us what happens when we multiply a positive integer by a negative integer.

Table 3				
Case	**Factor**	**Factor**	**Sign of Product**	**Magnitude of Product**
3	Positive	Negative	Negative	Product of the Absolute Values of the Factors

When we multiply any given factor, no matter what the sign, by a negative factor the sign of the product is always the opposite of the sign of the given factor. Therefore, by this observation, we can conclude that a negative integer times a negative integer is a positive integer.

Table 4				
Case	**Factor**	**Factor**	**Sign of Product**	**Magnitude of Product**
4	Negative	Negative	Positive	Product of the Absolute Values of the Factors

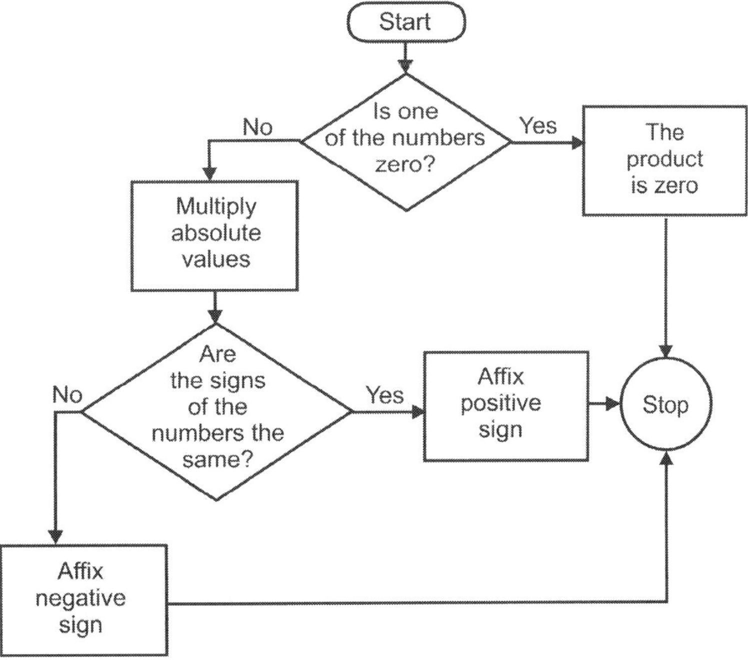

Figure 3. Flowchart for multiplication of integers.

We summarize the rules that involve factors that are negative into two statements:

1. The product of a positive integer and a negative integer is always negative.
2. The product of two negative integers is always positive.

THE GENERAL RULE

Or, stated more simply, if both factors are non-zero:

> If the signs of two factors are the same, the product is positive, else the product is negative.

It might interest you to know that a mathematician will not accept the reasoning we used to develop these rules. Mathematicians use another way to justify them. We will demonstrate the dance of that reasoning in Step 13.

We can add another multiplication property to our list. From Lesson 4.1, we know:

> Property 1. Zero multiplied by any number is zero. In symbols, $0a = 0$ for every number a.
> Property 2. One multiplied any number is that number. In symbols, $1a = a$ for every number a.

Here is the new property, a very important one in algebraic work:

> Property 3. Negative 1 multiplied any number is the opposite (Lesson 3.2) of that number. In symbols, $-1a = -a$ for every number a.

EXERCISES

1. In one sentence, define the law of multiplication of integers.

Calculate: (You may use any method for multiplication and check by casting out nines.)

2. 3 × (-4)
3. 18 – 44
4. -3 × (-16)
5. -12 × (-4)
6. 33 × 2
7. -3 × 9
8. -22 – 98
9. -80 × (-15)
10. (-18)(7)
11. (-33)(3)
12. (-28)(-4)
13. (-336)(-7)
14. (-16)(-12)
15. (-28) × 4
16. -9 × (-9)
17. 18 – (-45)
18. 188 × (-12)
19. 15 × 3
20. 22 + -(89)
21. 15 × (-3)
22. -32 × 8
23. -32 × (-8)
24. 3218 – (-2819)
25. -49 × (-7)
26. 42 + (-1000)
27. 49 × (-7)
28. -6 × 4
29. -485 – (-819)
30. 6 × (-4)

Find x: (Work these problems in your head from left to right. Remember to calculate what is in the parentheses first and that the operation of multiplication takes precedence over the operations of addition and subtraction.)

31. $20 + 12 - 3 \times (-6) = x$
32. $[16 - (-7)] + 4 \times (-1) = x$

33. $(-7) \times (12 - 6) \times (-11) = x$
34. $120 - (1 \times 12) + (-115) = x$
35. $40 \times (-3) + [(-40 - 23) \times (-5)] = x$
36. $(-3) \times [0 + (-4)] \times (-7) = x$
37. $(55 - 77) - 20 + (-8) \times (-3) = x$
38. $(-7 + 5)(-6 + 4) = x$
39. $556 \times 0 + 45 - 99 = x$
40. $1000 - [12 \times (-8)] + 20 - (-39) = x$

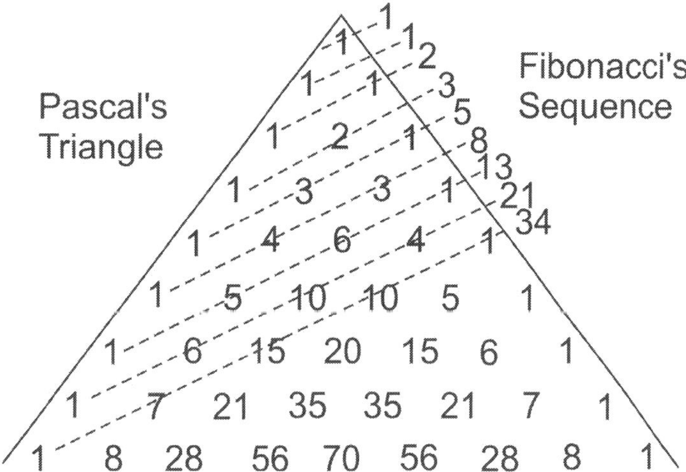

Figure 4. The Fibonacci Sequence embedded in Pascal's Triangle.

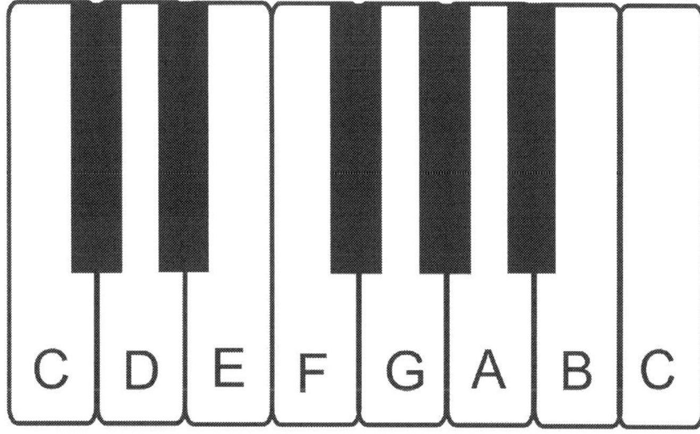

Figure 5. Fibonacci and the Piano Keyboard. The basic structure of a piano keyboard consists of an octave of thirteen notes, eight of which are white and five of which are black; and the black keys come in groups of two and three.

4.13 Two More Shortcuts

In this lesson, will explore two more shortcut methods, two important properties of numbers that you will use quite a bit in algebraic work.

Terms & Concepts Introduced
1. Binomial
2. Binomial Theorem
3. Difference of Two Squares
4. Factoring Common Factors

Shortcut #1: Another Way to Square a Number

Let's begin by investigating $8 \times 8 = 64$. Since $8 = 6 + 2$, we can restate the problem as follows:

$(6 + 2)(6 + 2) = 64$

Doing this should be familiar since we have already established that 34×27 can be rewritten as $(30 + 4)(20 + 7)$. We invoke our two-digit by two-digit paradigm (Figure 1) using the extended distributive property we learned in Lesson 4.6.

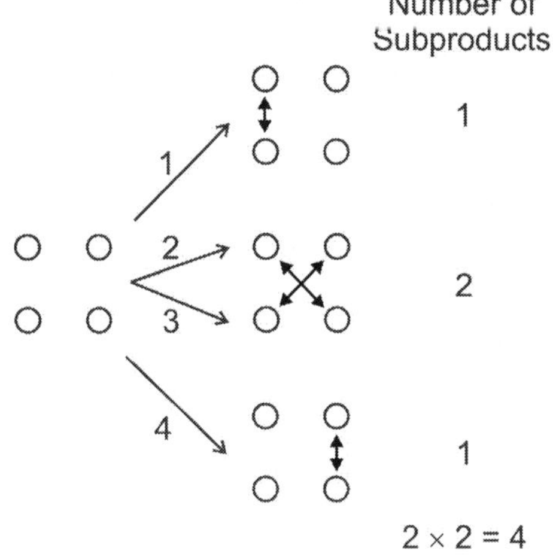

Figure 1. Two-Digit by Two-Digit Paradigm.

We set up the array as follows:

6 2

6 2

We get four subproducts as addends:

Product 1 (up/down): 6×6
Product 2 (cross): 6×2
Product 3 (cross): 2×6
Product 4 (up/down): 2×2

Notice first and last subproducts are squares: $6 \times 6 = 6^2$ and $2 \times 2 = 2^2$. The second and third subproducts add (6×2) twice or, by definition of multiplication, $(6 \times 2) + (6 \times 2) = 2(6 \times 2)$ or "twice the product of 6 and 2." Our answer is:

$6^2 + 2(6 \times 2) + 2^2 =$

$6^2 + 2(12) + 2^2 =$

$36 + 24 + 4 =$

64

Investigate Figure 2 for a visual representation of $(6 + 2)(6 + 2) = 64$. The number of dots could represent area if we want to find the area of a square garden, 8 feet by 8 feet. Make sure that you see the four subproducts: $6^2 = 36$, $6 \times 2 = 12$, $2 \times 6 = 12$, and $2^2 = 4$.

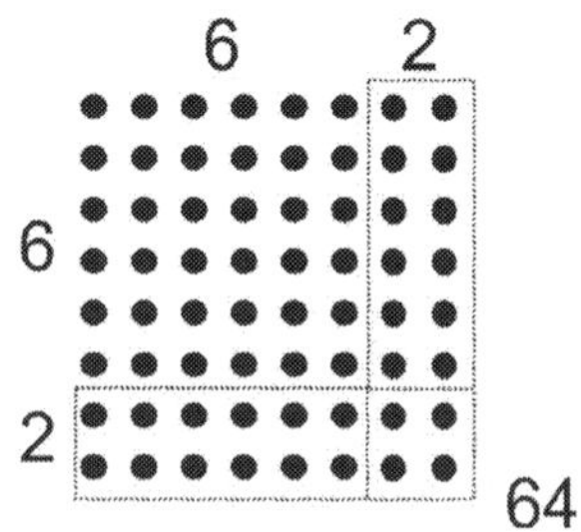

Figure 2. (6 + 2)(6 + 2)

Using this method, to find 51^2, we set $51 = 50 + 1$. Our array becomes:

50 1

50 1

$50^2 = 2500$, $2(50)(1) = 100$, $1^2 = 1$. Summing these subproducts, we get:

$$\begin{array}{r} 2500 \\ 100 \\ + \quad 1 \\ \hline 2601 \end{array}$$

We have another new shortcut tool, and you should be able to do this in your head. Try it now with:

Problem 1. 21^2
Problem 2. 32^2
Problem 3. 44^2

Answers:

Problem 1. $21^2 = 441 = 400 + 40 + 1$
 Do you see the $2(20 \times 1) = 40$ in 21^2?
Problem 2. $32^2 = 1024 = 900 + 120 + 4$
 Do you see the $2(30 \times 2) = 120$ in 32^2?
Problem 3. $44^2 = 1936 = 1600 + 320 + 16$
 Do you see the $2(40 \times 4) = 320$ in 44^2?

This problem is an elementary example of another law, one that you will learn later in your algebraic studies as the **Binomial Theorem.**[1] How would we generalize it? Using words would be too cumbersome so let's use algebraic symbols. First, we let a and b represent any positive integers. We want to find $(a + b)(a + b) = (a + b)^2$. We set up our two-by-two array:

$$\begin{array}{cc} a & b \\ a & b \end{array}$$

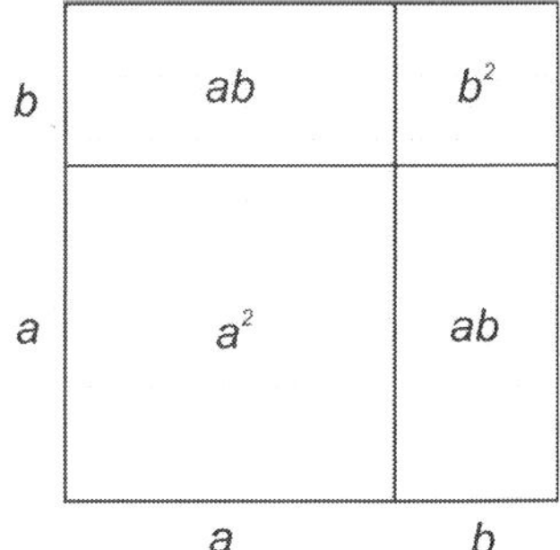

Figure 3. Square of a sum pictured geometrically.

We get four subproducts as addends, demonstrated by the area diagram (Figure 3). This is how Euclid demonstrated this relationship in Proposition 4 of Book II of *The Elements.* We calculate four products:

Product 1 (up/down): $a \times a = aa = a^2$
Product 2 (cross): ab
Product 3 (cross): $ba = ab$
Product 4 (up/down): $b \times b = bb = b^2$

Products 2 and 3, when summed become:

$ab + ab = ab(1 + 1)$ Why? Factoring or Distributive Law in reverse or factoring common factors[2]
$ab(1 + 1) = (1 + 1)ab$ Why? Switch factors (Commutative Law of Multiplication)
$(1 + 1)ab = 2ab$ Why? Sum 1 + 1

THE BINOMIAL THEOREM: SQUARE OF A SUM

In general, for any two numbers a and b, the shortcut for squaring their sum is:

---

[1] **Binomial**, from Latin, means "two names" or "two terms." In our example, the two terms are 6 + 2. As we learned in Lesson 3.7, *theorem* means "to look into" or "an insight" in the Greek.
[2] Since ab is a common factor of the two addends, we can "undistribute" it, or reverse the process of distribution. We name this reverse process **factoring common factors**.

$(a + b)^2 = a^2 + 2ab + b^2$

Sir Isaac Newton (1642-1727) proved a general case, named the Binomial Theorem, as a tool for work in Calculus, which he co-discovered with Gottfried Wilhelm Leibniz (1646-1716). Recall that a theorem is a statement that can be proved, by close and careful reasoning, by explicit and standard assumptions.

Note the coefficients of the three terms:

- a^2 (coefficient is 1 since, in general, $1k = k$ and in this case, k is a^2)
- $2ab$ (coefficient is 2)
- b^2 (coefficient is 1)

Do you see "1 2 1" in a row in Pascal's Triangle?

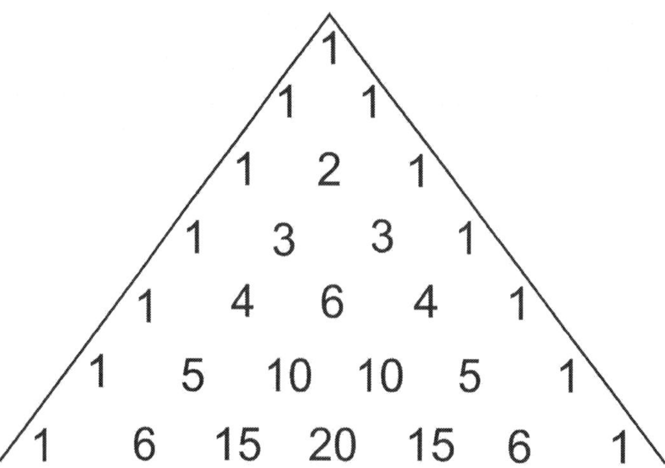

Figure 4. The Binomial Theorem embedded in Pascal's Triangle.

The subsequent rows form the coefficients of $(a + b)^3$, $(a + b)^4$, etc. For example:

- $(a + b)^3 = a^3 + 3a^2b + 3ab^2 + b^3$ (Note the 1 3 3 1 row in Pascal's Triangle and compare with the four coefficients.)
- $(a + b)^4 = a^4 + 4a^3b + 6a^2b^2 + 4ab^3 + b^4$ (Note the 1 4 6 4 1 row in Pascal's Triangle and compare with the five coefficients.)

A homework exercise in Lesson 13.15 will make use of the Binomial Theorem in this way.

THE BINOMIAL THEOREM: SQUARE OF A DIFFERENCE

Can we find 38^2 by setting $38 = 40 - 2$?

$38^2 = 38 \times 38 = (40 - 2)(40 - 2)$

Applying the Algebraic Definition of Subtraction, we get:

$(40 - 2)(40 - 2) = [40 + (-2)][40 + (-2)]$

The Algebraic Definition of Subtraction:
For any two numbers a and b,
$a - b = a + (-b)$

Setting up our array, we get:

40 -2

40 -2

We get four subproducts as addends:

Product 1 (up/down): $40 \times 40 = 1600$
Product 2 (cross): $40 \times (-2) = -80$
Product 3 (cross): $(-2) \times 40 = -80$
Product 4 (up/down): $(-2) \times (-2) = 4$

Summing, with get $1600 + (-160) + 4 = 1444 = 38^2$

Can we generalize this process using algebraic symbols? First, we change $(a - b)^2$ to $[a + (-b)]^2$ applying the Algebraic Definition of Subtraction. Next, we set up our array, demonstrated in Figure 5 where we shade the areas that have the negative sign ($-ab$). We are using the area diagram to demonstrate that the shading area is subtracted from the unshaded area to get the square of the difference of a and b.

a $-b$

a $-b$

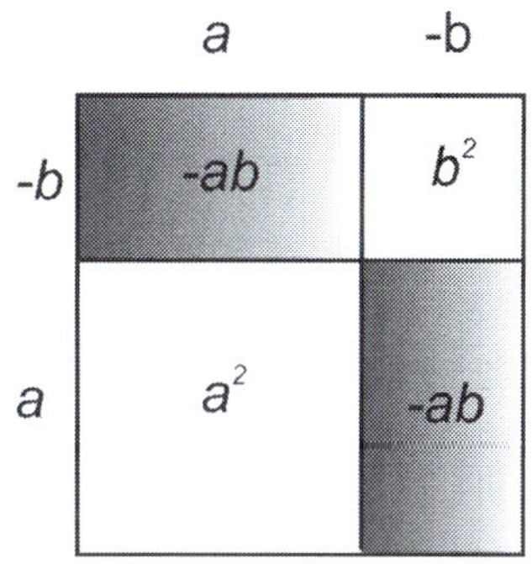

We get four subproducts as addends:

Product 1 (up/down): $aa = a^2$
Product 2 (cross): $-ab$
Product 3 (cross): $-ba = -ab$
Product 4 (up/down): $bb = b^2$

Figure 5. Square of a difference.

Products 2 and 3, when summed become:

$(-ab) + (-ab) = -ab(1 + 1)$	Why? Factoring
$-ab(1 + 1) = (1 + 1)(-ab)$	Why? Switch factors (Commutative Law of Multiplication) [Note: the coefficient of $-ab$ is -1. Or, from Property 3 of Lesson 4.13, $-ab = (-1)ab$.]
$(1 + 1)(-ab) = -2ab$	Why? Sum 1 + 1. [Note: $2(-1) = -2$] [Note: $2(-ab) = 2(-1)(ab) = -2ab$. Make sure you understand this important work.]

In general, for any two numbers a and b, the shortcut for squaring their difference is:

$$(a - b)^2 = a^2 - 2ab + b^2$$

We have two forms of the Binomial Theorem involving two terms, a and b. One form is for the square of their sum, and the other is for the square of their difference.

SHORTCUT #2: ANOTHER WAY TO MULTIPLY NUMBERS WHOSE FACTORS ARE UNIQUELY RELATED

Let's try multiplying $6 + 2$ by $6 - 2$, which is another way to write the problem $8 \times 4 = 32$. Let's see what happens when our two-by-two array to $(6 + 2)(6 - 2)$. Let's rewrite this product as follows, again invoking the Algebraic Definition of Subtraction:

$(6 + 2)(6 - 2) = (6 + 2)[6 + (-2)]$

We set up our array:

6 2

6 -2

We get four subproducts as addends:

Product 1 (up/down): $6 \times 6 = 36$
Product 2 (cross): $6 \times (-2) = -12$
Product 3 (cross): $2 \times 6 = 12$
Product 4 (up/down): $2 \times (-2) = -4$

Since $12 + (-12) = 0$ (a number added to its opposite is zero), we get $36 - 4 = 32$.

Once, early in my marriage, my wife and I were walking down the aisle of a grocery store. Looking at some prices, she asked me, "What is 17 times 23"? Within seconds, I responded, "391." Under her breath, I thought she mumbled, "Did I marry a human calculator?" No, she didn't; but I knew how to calculate this type of problem.

Here is what I did in my mind.: I noted that 17 is 3 below 20 and 23 is 3 above 20.

Therefore, we can rewrite 17×23 as follows:

$17 \times 23 = (20 - 3)(20 + 3)$

From this, we set up our array [Note: $20 - 3 = 20 + (-3)$]:

20 3

20 -3

We get four subproducts as addends:

Product 1 (up/down): $20 \times 20 = 400$
Product 2 (cross): $20 \times (-3) = -60$
Product 3 (cross): $3 \times 20 = 60$
Product 4 (up/down): $3 \times (-3) = -9$

Since $60 + (-60) = 0$, we only need to calculate $400 + (-9)$. Invoking the Algebraic Definition of Subtraction, we note:

$400 + (-9) = 400 - 9 = 391$

Therefore, 17×23 can be calculated by computing a simple difference, $400 - 9$, by this process of reasoning:

$17 \times 23 = (20 - 3)(20 + 3) = 20^2 - 3^2 = 400 - 9 = 391$

To calculate the product of 17 and 23 is as easy as calculating the difference between two squares, 20^2 and 3^2.

$20^2 - 3^2 = 400 - 9 = 391$

As a standard of order, we always calculate the operation of exponentiation first, multiply or divide next, and add or subtract last. (See Lesson 11.1.)

We can illustrate multiplying 6 + 2 by 6 – 2 with pebbles. Inspect Figure 6 carefully and you will see how $(6 + 2)(6 - 2) = 32$.

The left group of pebbles shows $6^2 = 36$. We draw a square box around four pebbles, $4 = 2^2$.

The middle group of pebbles shows the 2^2 pebbles taken away. We draw a rectangular box around eight pebbles.

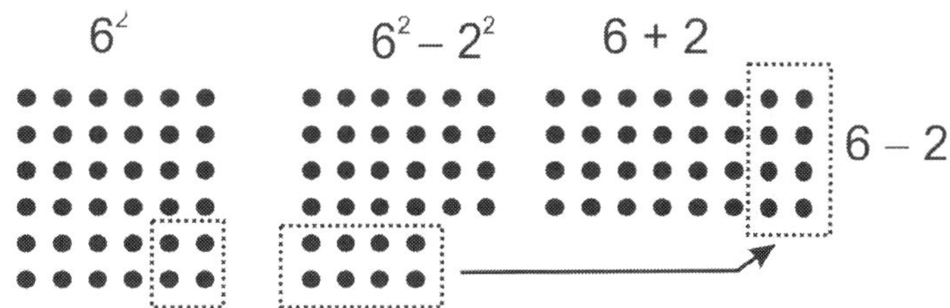

Figure 6. (6 + 2)(6 – 2)

The right group of pebbles shows us rearranging these eight pebbles. What is left is a rectangular array of pebbles, 8 by 4 or 6 + 2 by 6 – 2, totaling 32. Our result:

$6^2 - 2^2 = 36 - 4 = 32$

This law[3] is the **difference of two squares**. In general, for any two numbers a and b, the difference of their squares, $a^2 - b^2$, is $(a + b)(a - b)$.

We then set up our array, demonstrated by the area diagram at right in which the areas that have a negative sign are shaded. Here is our array:

$a \quad b$

$a \quad -b$

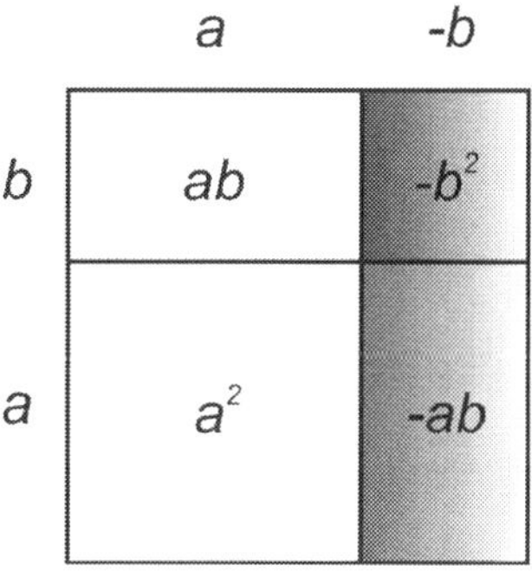

Figure 7. Difference of two squares.

We get four subproducts as addends:

Product 1 (up/down): $aa = a^2$
Product 2 (cross): $-ab$
Product 3 (cross): ab
Product 4 (up/down): $b(-b) = -b^2$

Since $ab + (-ab) = 0$, our sum becomes this:

$a^2 + (-b)^2$

THE DIFFERENCE OF TWO SQUARES

Invoking the Algebraic Definition of Subtraction, we have:

$$(a + b)(a - b) = a^2 + (-b)^2 = a^2 - b^2$$

These two laws, the Binomial Theorem and the Difference of Two Squares, are not only two important shortcut tools – knowing them gets you ready for algebraic work. Make sure that you make use of them whenever the situation warrants.

Know these three identities:
1. $(a + b)^2 = a^2 + 2ab + b^2$
2. $(a - b)^2 = a^2 - 2ab + b^2$
3. $(a + b)(a - b) = a^2 - b^2$

[3] This law is technically an identity in mathematics.

EXERCISES

Answer the following questions:

1. Define binomial.
2. State, in symbols, the two forms of the Binomial Theorem.
3. State, in symbols, the Difference of Two Squares.
4. Explain, with an example, what factoring out common factors means.

Use the Binomial Theorem to calculate the following products:

5. 15^2
6. 17^2
7. 21^2
8. 22^2
9. 25^2
10. 35^2
11. 51^2
12. 55^2
13. 59^2
14. 61^2
15. 65^2
16. 69^2
17. 72^2
18. 75^2
19. 79^2
20. 81^2
21. 85^2
22. 93^2
23. 99^2
24. 121^2
25. 158^2
26. 999^2
27. 9999^2

Figure 8. Pentagonal stamps. Source: United States Postal Service.

Use the Difference of Two Squares Law to calculate the following products:

28. 24×16
29. 25×15
30. 26×14
31. 27×13
32. 28×12
33. 29×11
34. 34×26
35. 35×25
36. 36×24
37. 37×23
38. 38×22
39. 39×21

Answer the following questions:

40. In Figure 9, consider the row of groups of window panes in the upper part of the building. Determine the optimal way to calculate the number of individual window panes to the right of the two groups of panes that are open.
41. Explain why 3 cm + 4 cm = 7 cm by factoring out the common factor cm. (Hint: "3 cm" means "3 times one centimeter" and "4 cm" means "4 times one centimeter.")
42. Explain why $x + 9x + 8x = 18x$ by factoring out the common factor. (Note: for every a, $1a = a$.)
43. Explain why $-8x - 16x = -24x$ by factoring out the common factor x.
44. Explain why $2xy + 3xy - 8xy = -3xy$ by factoring out the common factor xy.
45. Explain why $3x^2 - 6x + 8x^2 + 8 + 11x - 3 - 5x^2 = 6x^2 + 5x + 5$
46. Explain how the filled area in Figure 10 represents the difference of two squares, $a^2 - b^2$.
47. Explain how the filled area in Figure 11 represents the square of a difference, $(a - b)^2$.

Figure 9. Pybus Public Market, Wenatchee, Washington. Source: James D. Nickel.

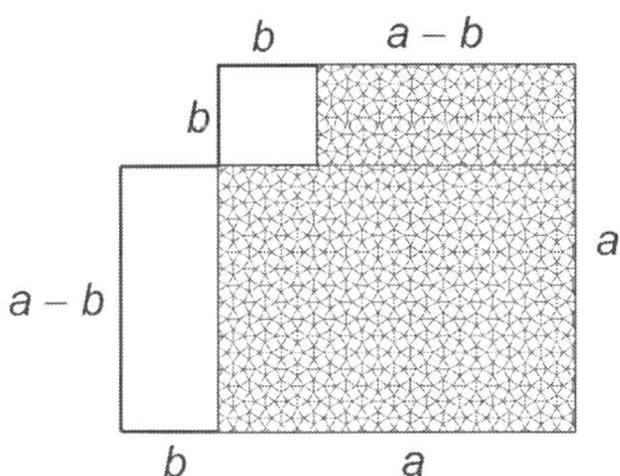

$$a^2 - b^2 = (a - b)a + (a - b)b = (a - b)(a + b)$$

Figure 10. Difference of two squares.

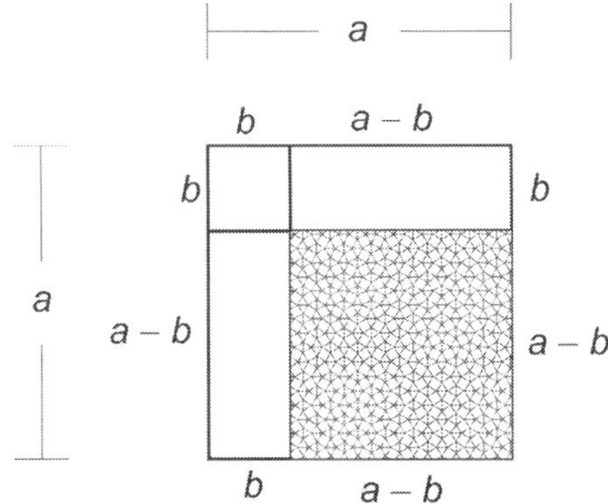

$$(a - b)^2 = (a - b)(a - b) = a^2 - ab - ab + b^2 = a^2 - 2ab + b^2$$

Figure 11. Square of a difference.

4.14 Multiplication with Fingers

With our ten fingers, we have a calculator that can quickly demonstrate these products:

6×6			
6×7	7×7		
6×8	7×8	8×8	
6×9	7×9	8×9	9×9

After we understand how finger reckoning works, we have a new shortcut, a new method of multiplication that is exceptionally quick.[1] It works for finding the products of numbers that are relatively close to each other while, at the same time, are close to multiples of 10, e.g., 9×9, 19×18, 21×23, 52×48, 99×95, etc.

Figure 1. Finger multiplication.

The Method

Let's start by looking at our hands, palms facing you: We assign the numbers 6 through 9 to four fingers, little finger to index finger. We do not assign a number to our thumbs.

The two hands on the left illustrate 6 times 6, both little fingers touching. In all the examples that follow, we count the two fingers touching and all the fingers below the point of touch and multiply that count by 10. In 6×6, two fingers are touching so we multiply $2 \times 10 = 20$. Four fingers of each hand are above the two that are touching. We multiply $4 \times 4 = 16$. We add our two results:

$20 + 16 = 36 = 6 \times 6$

Let's try 6×7. Note first that the six-finger on the left hand touches the seven-finger on the right. Two fingers are touching, and one finger is below the point of touching $(2 + 1 = 3)$ so we multiply $3 \times 10 = 30$. Four fingers are above the touch on the left hand, and three are above the touch on the right hand. We multiply $4 \times 3 = 12$. We add our two results:

$30 + 12 = 42 = 6 \times 7$

Consider 6×8. Note first that the six-finger on the left hand touches the eight-finger on the right. Two fingers are touching, and two fingers are below the point of touching $(2 + 2 = 4)$ so we multiply $4 \times 10 = 40$. Four fingers are above the touch on the left hand, and two are above the touch on the right hand. We multiply $4 \times 2 = 8$. We add our two results:

$40 + 8 = 48 = 6 \times 8$

---

[1] The "Human Calculator" Scott Flansburg published this method of "quick multiplication" in his book *Math Magic* (New York: William Morrow Company, 1993). Australian educator Bill Handley has reprised this method recently in his book *Speed Mathematics: Secrets of Lightning Mental Calculation* (New York: John Wiley & Sons, [2000, 2003] 2004). The original source of this method goes back to ancient times, probably to the Hindus of India.

Consider 6×9. Note first that the six-finger on the left hand touches the nine-finger on the right. Two fingers are touching, and three fingers are below the point of touching ($2 + 3 = 5$) so we multiply $5 \times 10 = 50$. Four fingers are above the touch on the left hand, and one is above the touch on the right hand. We multiply $4 \times 1 = 4$. We add our two results:

$50 + 4 = 54 = 6 \times 9$

Consider 7×7 (Figure 2). Note first that the seven-finger on the left hand touches the seven-finger on the right. Two fingers are touching and two fingers are below the point of touching ($2 + 2 = 4$). Therefore, we multiply $4 \times 10 = 40$. Three fingers are above the touch on each hand. We multiply $3 \times 3 = 9$. We add our two results:

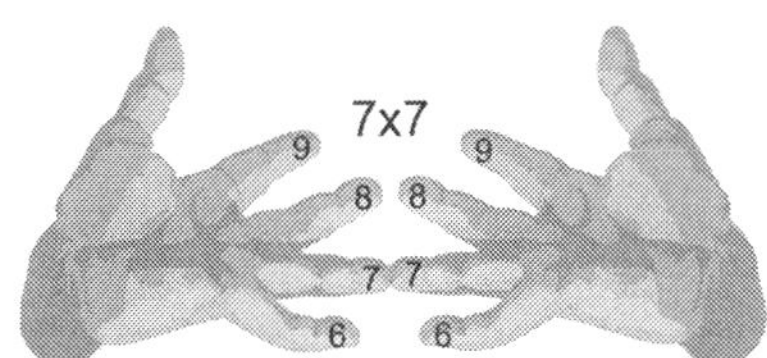

Figure 2.

$40 + 9 = 49 = 7 \times 7$

Can you do the rest (Figure 3)?

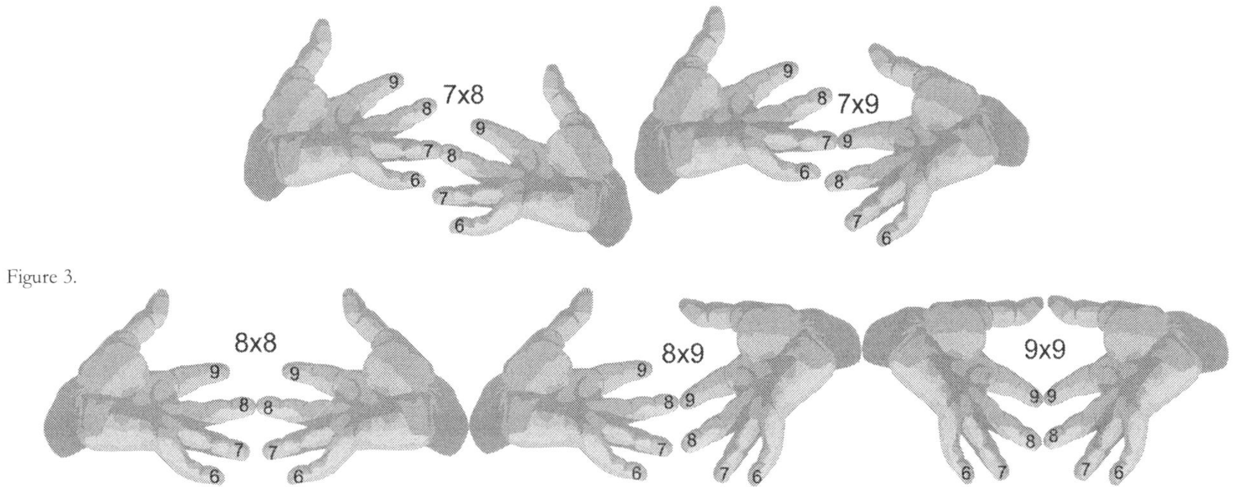

Figure 3.

THE METHOD WITHOUT FINGERS

Why multiply by 10? Let's consider how we would multiply 8×7 without using your fingers, but using the method that this finger reckoning employs. We first note that 8 and 7 are near 10 and 10, therefore, will be our base of operations. Next, we find the complements of 8 and 7: they are 2 and 3 respectively. Since 8 and 7 are both less than 10, we write each complement to the right of 8 and 7 respectively as negative integers:

8 -2

7 -3

Now, add -3 to 8 or -2 to 7:

$8 + (-3) = 5$ or $7 + (-2) = 5$

By the Algebraic Definition of Subtraction, we could also subtract 3 from 8 and 2 from 7 and get the same results:

$8 - 3 = 5$ or $7 - 2 = 5$

> The Algebraic Definition of Subtraction:
> For any two numbers a and b,
> $a - b = a + (-b)$

(Note: This addition/subtraction always generates the same answer.)

Multiply: $5 \times 10 = 50$. Finally, multiply $(-2) \times (-3) = 6$ and add 6 to 50:

$6 + 50 = 56$

In *The Ground of Artes* (1543), the English mathematician Robert Recorde (ca. 1510-1558) demonstrated how to multiply 8 by 7 by setting up the problem this way:

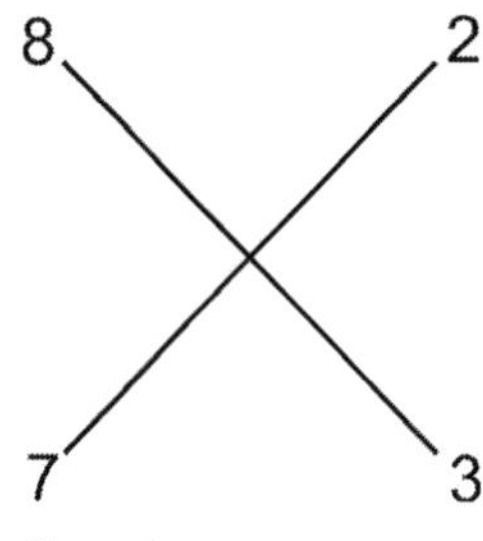

Figure 4.

He wrote 2, the complement of 8, to the right of 8 and 3, the complement of 7, to the right of 7, forming a square with four numbers at each vertex. Instead of drawing a square, he connected 8 and 3 with a line segment in indicate subtraction. In the same way, he connected 7 and 2 making a cross. He then subtracted 3 from 8, or 2 from 7, to get 5, the tens position of the product. Then, he multiplied 2 by 3 to get 6 (Note:[2] $-2 \times -3 = 6$), the ones position of the product. The answer is 56. In time, this cross shrank in size and arguably become the multiplication sign we use in arithmetic work.

Let's return to our method that uses negative integers. Let's try 6×7. First, find the complements and write them accordingly:

6 -4

7 -3

Add: $6 + (-3) = 3$ or $7 + (-4) = 3$. Or, subtract: $6 - 3 = 3$ or $7 - 4 = 3$. Multiply $3 \times 10 = 30$. Finally, multiply $(-4) \times (-3)$ and add 12 to 30:

$12 + 30 = 42$

Confirm these simple steps with your fingers and you will see that it reflects exactly what is going onyou're your notebook, do the same, without fingers, for the other eight products: 6×6, 6×8, 6×9, 7×7, 7×9, 8×8, 8×9, and 9×9.

WHY THE METHOD WORKS

Why does this finger method work? Consider 8×7. We can rewrite 8×7 as:

$[10 + (-2)][10 + (-3)]$

We set up a two-by-two array:

[2] In this lesson, we invoke the use of integers, -2 and -3 in this case. Recorde did not do this.

10 -2

10 -3

We add four subproducts:

100 + (-30) + (-20) + 6

Adding -30 to -20 we get:

100 + (-50) + 6

Invoking the Algebraic Definition of Subtraction, we get:

100 – 50 + 6

Noting that 10 is a factor of 100 and of 50, we get:

10 × 10 – 10 × 5 + 6

Factoring the 10 from the first two terms, we get:

10(10 – 5) + 6 = 10(5) + 6 = 50 + 6 = 56

Do you see the 10 – 5 = 5 generating 5 × 10 = 50 and do you see (-2) × (-3) = 6?

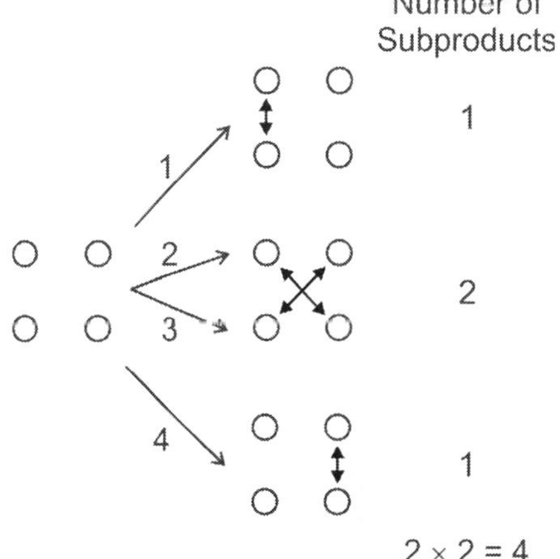

Figure 5. Two-Digit by Two-Digit Paradigm.

12 × 11

Let's continue our journey beyond our fingers by considering 12 × 11. Since these factors are close to 10, we use 10 as our base of operations. 12 and 11 are both greater than 10, by 2 and 1 respectively. We write these numbers to the right of 12 and 11, as positive integers, this way:

12 2

11 1

Next, since the numbers are above 12 and 11 consider 2 and 1 as positive. We add: 12 + 1 = 13 or 2 + 11 = 13. (Note: This addition always generates the same answer.) Multiply 13 × 10 = 130. Now multiply 2 × 1 and add 2 to 130:

2 + 130 = 132

Why? We can rewrite 12 × 11 as (10 + 2)(10 + 1). We have a two by two array:

10 2

10 1

We must add four subproducts:

100 + 10 + 20 + 2

Adding 10 to 20 we get:

100 + 30 + 2

Noting that 10 is a factor of 100 and 30, we get:

10 × 10 + 10 × 3 + 2

Factoring the 10 from the first two terms, we get:

$10(10 + 3) + 2 = 10(13) + 2 = 130 + 2 = 132$

Do you see the $10 + 3 = 13$ generating $13 \times 10 = 130$ and $2 \times 1 = 2$?
Check:

Product of factors: $3 \times 2 = 6$
Sum of the digits in the product: $1 + 3 + 2 = 6$
$6 = 6$. Check.

Verify by casting our nines each of the following answers in your notebook.

16 × 14

Let's try another one: 16 × 14. Considering 10 as our base of operations, we subtract 10 from each number and write:

16 6

14 4

Add (Note: We will always get the same answer.): $16 + 4 = 20$ and $6 + 14 = 20$. Multiply 20 by 10: $20 \times 10 = 20$. Multiply 6 by 4 and add 24 to 200:

$200 + 24 = 224 = 16 \times 14$

19 × 18

Let's now try 19 × 18. We note that both numbers are close to 20, which becomes our base of operations. Both 19 and 18 are less than 20, by 1 and 2 respectively. We write these numbers as negative integers:

19 -1

18 -2

Add (Note: We will always get the same answer): $19 + (-2) = 17$ and $18 + (-1) = 17$. Multiply 17 by 20, our base of operations: $17 \times 20 = 340$. Multiply -1 by -2 and add 2 to 340:

$2 + 340 = 342 = 19 \times 18$

52 × 47

Let's now try one where one factor is above and the other factor is below a base: 52 × 47. Since both numbers are close to 50, we choose 50 as our base of operations. 52 is 2 greater than 50 and 47 is 3 less than 50. Because of this, we write 2 to the right of 52 and the -3 to the right of 47:

52 2

47 -3

Next, we add (Note: We will get the same answer.): $52 + (-3) = 49$ or $2 + 47 = 49$. Multiply by 50. Using the Distributive Law, $49 \times 50 = (50 - 1)50 = 2500 - 50 = 2450$. We multiply, $2 \times (-3) = -6$, and then add -6 to 2450:

$2450 + (-6) = 2444$

Let's check our result:

$$\begin{array}{r} 52 \\ \times\ 47 \\ \hline {}^{4\,1} \\ 2034 \end{array}$$

Why? We can rewrite 52×47 as $(50 + 2)[50 + (-3)]$. We have a two-by-two array:

50 2

50 -3

We must add four subproducts:

$2500 + (-150) + 100 + (-6)$

Adding -150 to 100 we get:

$2500 + (-50) + (-6)$

Invoking the Algebraic Definition of Subtraction, we get:

$2500 - 50 - 6$

Noting that 50 is a factor of 2500 and 50, we get:

$50 \times 50 - 50 \times 1 - 6$

Factoring the 50 from the first two terms, we get:

$50(50 - 1) - 6 = 50(49) + - 6 = 2450 - 6 = 2444$

Do you see the $50 - 1 = 49$ generating $49 \times 50 = 2450$ and $2 \times (-3) = -6$?

56 × 48

Try another: 56×48. Choose 50 as our base of operations. 56 is 6 above 50 and 48 is 2 below. We write:

56 6

48 -2

Add: $56 + (-2) = 54$ or $6 + 48 = 54$. Multiply: $54 \times 50 = 2700$. Find the product of 6 and -2 and add it to 2700:

$2700 + (-12) = 2688$

This is fun!

99 × 99

We are now going to employ this method to solve more complicated products. Let's start with 99×99. Both numbers are close to 100 so we will choose 100 as our base of operations. We next find the difference between the factors and 100 and write:

99 -1

99 -1

Next, we add: $99 + (-1) = 98$. Since 100 is our base of operations, we multiply $98 \times 100 = 9800$. Multiply $(-1) \times (-1) = 1$. Sum the result:

$9800 + 1 = 9801 = 99 \times 99$

Isn't this amazingly fast? Confirm your result:

99

× 99 Product: 9801. Did you see how I wrote the 162 for the cross step, $81 + 81 = 162$?

1 6 8

8121

92 × 96

Let's try another: 92×96. Both are close to 100 so we subtract each from 100 and place underneath.

92 -8

96 -4

Next, we add: $92 + (-4) = 88$ or $96 + (-8) = 88$. Multiply $88 \times 100 = 8800$. Multiply $(-8) \times (-4) = 32$. Sum the result:

$8800 + 32 = 8832 = 92 \times 96 = 8832$

Check:

92

× 96 Product: 8832

7 1

8122

95 × 98

Let's try one more: 95×98. I will not elaborate the steps.

95 -5

98 -2

$93 \times 100 = 9300$

$(-5) \times (-2) = 10$

$9300 + 10 = 9310 = 95 \times 98$

102 × 101

Now consider 102×101. Subtract 100 from each number and write:

102 2

101 1

Next, we add: $102 + 1 = 103$. $101 + 2 = 103$. Now, multiply $103 \times 100 = 10{,}300$. Multiply $2 \times 1 = 2$. Sum the result:

10,300 + 2 = 10,302 = 102 × 101

Check:

```
  102
× 101
-----
10302
```

108 × 105

Let's try another: 108 × 105. Subtract 100 from each and write:

108 8

105 5

Next, add: 105 + 8 = 113 or 108 + 5 = 113. Multiply 113 × 100 = 11,300. Multiply 8 × 5 = 40. Sum the result:

11,300 + 40 = 11,340 = 108 × 105

SUMMARY

With our shortcut methods and our left-to-right paradigm, we now have a substantial arsenal of tools with which we can pry out the answer to any multiplication problem.

EXERCISES

Answer this question:

1. Using factoring, explain why the sum of 78 miles and 152 miles is 230 miles.

Showing all your work, (a) use the finger method to find the following products with 20 as your base of operations and (b) check your answer by casting out nines:

2. 11 × 11
3. 11 × 12
4. 11 × 15
5. 12 × 12
6. 16 × 16
7. 15 × 17
8. 19 × 19
9. 14 × 14
10. 17 × 18

Showing all your work, (a) use the finger method to find the following products with 30 as your base of operations, (b) with 20 as your base of operations, and (c) check your answer by casting out nines:

11. 21 × 22
12. 21 × 23
13. 21 × 24
14. 21 × 25
15. 21 × 26
16. 21 × 27

17. 21×28
18. 21×29
19. 22×22
20. 22×23

Showing all your work, (a) use the finger method to find the following products with 20 as your base of operations and (b) check your answer by casting out nines:

21. 22×25
22. 22×29
23. 23×23
24. 24×26
25. 25×25

Showing all your work, (a) use the finger method to find the following products with 30 as your base of operations and (b) check your answer by casting out nines:

26. 31×32
27. 32×33
28. 34×34
29. 36×31
30. 38×33

Showing all your work, (a) use the finger method find the following products using the base of your choosing and (b) check your answer by casting out nines:

31. 61×59
32. 72×67
33. 89×93
34. 103×98

Figure 6. The numerical and geometric symmetry of a rose window, Strasbourg Cathedral, France. Source: Public Domain.

> A power of creativity is at work in the universe, which can be viewed as a creaturely perichoresis of dynamic systems echoing the trinitarian mystery. Apart from divine intelligence, it is very hard to account for it.
>
> Clark H. Pinnock, *Flame of Love: A Theology of the Holy Spirit* (1996), p. 67.

4.15 Summary and Review

In the past lessons, we have learned the simple algorithm of streamline multiplication. The value of this method is:

- It maintains consistency (left to right paradigm).
- It provides an immediate estimation of the product.
- It streamlines the process.

It also requires mental effort to keep certain digits in your mind. This mental rigor is good for you need to regularly exercise your brain just like your muscles.

Also, we have learned how to recognize when to use these shortcuts:

1. Multiplying by multiples of 10
2. Distributive Law
3. Factor one of the factors
4. Multiplying by 11
5. Squaring numbers whose last digit is 5
6. Difference of Two Squares
7. Binomial Theorem
8. The Finger Method

These are essential tools for your multiplication toolkit. Use them wisely and efficiently. As you do, you will enjoy the challenge of finding the quickest and easiest way to calculate products.

Let's apply what we have learned to the following problem sets.

Exercises

Do these problems mentally, i.e., without pencil or paper:

1. The sum of 18 + 22 + 32 equals the product of 9 and what number?
2. The sum of 80 – 29 and 6 + 6 equals the product of 7 and what number?
3. The difference between 55 + 25 and 64 – 24 is equal to the product of what (a) two factors? (b) three factors?
4. The product of what two factors is equal to the sum of 13, 20, and 12?
5. The product of 4 and 12 is equal to 6 times what number?
6. (a) What is the sum of 5 times the product of 4 and 3 and 3 times the product of 5 and 2? (b) What is the difference?
7. What is the product of 22 + 48 – 35 and 22 – 19?
8. How much less is 50 – (4 × 9) than 28 + 32 – 45?
9. Which is greater, 7(31 – 22) or 10 × 6 + 2?
10. Adam is three times as old as Jason, and Jason is 15 years old. What is the sum of their ages?
11. Paul saved $12 a day and spent $7 a day for his repayment of a loan to a friend. How much did he save in 15 days?
12. Roger bought two pencils at $2.75 cents each, two erasers at $1.75 cents each, and one stick of gum for 10 cents. How much change should he receive from a ten-dollar bill?

Mentally (without pencil and paper), find the value of n:

13. $22 - 14 + 32 - 21 = n$
14. $2 \times (-9) - 16 = 17n$
15. $32 + (3 \times 5) = 30 + n$

16. $(5 \times 8) - [2 \times (-8)] = n$
17. $125 - (9 \times 9) = 45 - n$
18. $56 + 44 - [60 - (-25)] = n$
19. $(25 \times 25) - (25 \times 0) = n$
20. $8 \times 0 \times 13 + 44 = (-2) \times n$
21. $45 + 65 - 9n = 65$
22. $(10 \times 13) + 18 = 146 - n$
23. $78 - 19 + 22 = 9n$
24. $27 + 45 - 13n = 7$

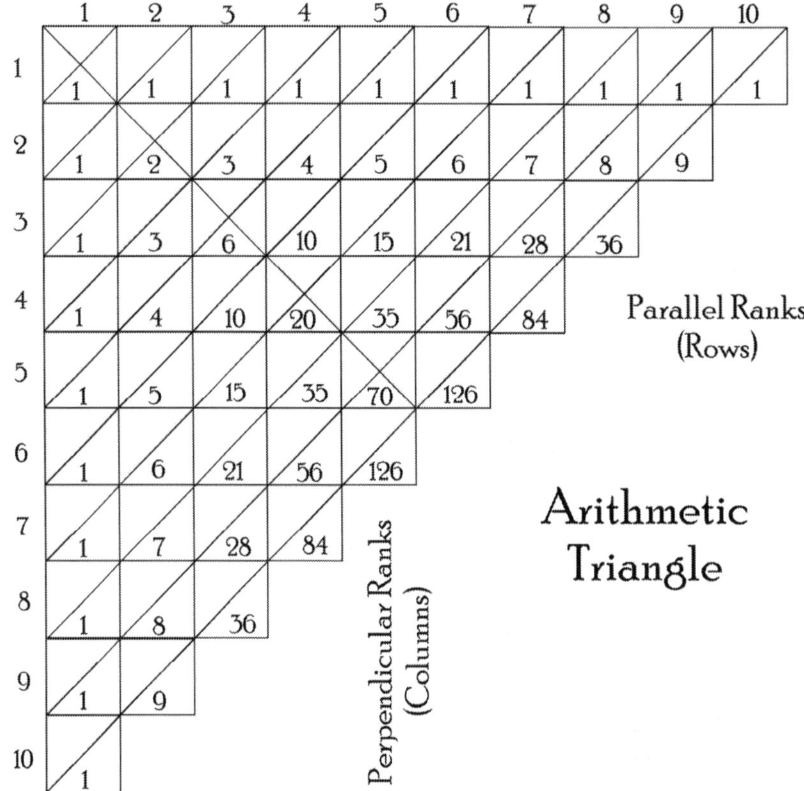

Figure 1. Pascal's original version of his Arithmetic Triangle.

In Figure 1, use the parallel and perpendicular ranks of Pascal's Triangle to answer these questions:

25. What is the relationship between the number in column four, row three, with the numbers in column four, row two and column three, row three?
26. What is the relationship between the number in column n, row $n - 1$, with the numbers in column n, row $n - 2$ and column $n - 1$, row $n - 1$?
27. What is the relationship between the number in column four, row three, with the numbers in column four, row four and column three, row four?
28. What is the relationship between the number in column n, row $n - 1$, with the numbers in column n, row n and column $n - 1$, row n?
29. What is the relationship between the number in column four, row three, with the numbers in row tow, columns one to four?
30. What is the relationship between the number in column n, row $n - 1$, with the numbers in row $n - 2$, columns one to n?

Verbally (without pencil and paper), find the product of:

31. 3 and 121
32. 3 and 525
33. 4 and 124
34. 4 and 615
35. 6 and 132
36. 5 and 724
37. 512 and 2
38. 512 and 3
39. 512 and 9

> The wondrous and structured order of the physical universe is held together, it coheres, in Christ. The stunning truth is that this spectacular and intricate framework is subject to and revealed by the laws of mathematics.
>
> James D Nickel

Simplify by factoring:

40. $7x + 3 + x - 4$
41. $x - 5y + x - y$

42. $7x^2 - 5 + x^2 + 15$
43. $x^2 + 12x - 5 + 8x^2 - 3x + 6$
44. $7x^3 - 2x^2 + 22 + (-x^2) + 3x - 21$

Answer the following questions:

45. Subtract $3x^2 - x + 7$ from $5x^2 + 9x + 6$.
    [Hint: $5x^2 + 9x + 6 - (3x^2 - x + 7) \Leftrightarrow$
    $5x^2 + 9x + 6 - 1(3x^2 - x + 7) \Leftrightarrow$
    $5x^2 + 9x + 6 - 3x^2 + x - 7$]
46. Subtract $x^2 - 4x - 3$ from $x^4 - 16x^2 - 3$.
47. Find the sum of $2x^2 + x - 5$ and $x^2 - 4x + 8$.
48. Subtract $x^2 - 4x + 8$ from $2x^2 + x - 5$.
49. Find the value of $x^2 - 4x + 8$ when $x = 2$.
50. Find the value of $x^2 - 4x + 8$ when $x = 3$.
51. Find the value of $2x^2 + x - 5$ when $x = 2$.
52. Find the value of $2x^2 + x - 5$ when $x = 3$.
53. Find the value of the sum in Question 47 when $x = 2$.
54. Find the value of the sum in Question 47 when $x = 3$.
55. Find the value of the difference in Question 48 when $x = 2$.
56. Find the value of the difference in Question 48 when $x = 3$.
57. (a) Complete this table:

	Value when x = 2	Value when x = 3
$2x^2 + x - 5$		
$x^2 - 4x + 8$		
$2x^2 + x - 5 + x^2 - 4x + 8$		
$2x^2 + x - 5 - (x^2 - 4x + 8)$		

    (b) What do you notice?

Solve the following, using any of our shortcut methods and check by casting out nines:

58. 35 × 35
59. 25 × 35
60. 37 × 99
61. 46 × 39
62. 55 × 45
63. 120 × 99
64. 203 × 197
65. 9997 × 10,002
66. 12,134,118 × 11

Use pencil and paper to solve the following problems: (Check your multiplication work by casting out nines.)

67. If Paul had an income of $6570 a month and his daily expenses averaged $175, how much can he save in a year of 365 days?
68. What number must add to 375 × 400 to make the amount 252,603?
69. What is the difference between 52,018 – (299 × 41) and [99 × (-340)] – 25,000?
70. Multiply 101 + 12(22 + 30) by [(56 – 60) × 6 – 22].
71. Multiply 736 – (13 + 38) by [(-5 × 111) – (136 – 218)].

Figure 2. 21 rows of Pascal's Triangle.

Terms/Concepts/Abbreviations

	Lesson
Absolute value	3.2
Abstract or pure numbers	2.2
Acute angle	2.7
Ad infinitum	1.3
Addend	2.4
Addition	2.3
Additive inverse	3.3
Algebra	1.1
Algebraic Definition of Subtraction	3.3
Algebraic expression	4.1
Algebraic term	4.2
Algorism	1.5
Algorithm	1.5
al-jabr	1.1
Angle	2.3
Anno Domini	1.1
Apex	2.6
Applied or denominate numbers	2.2
Area	4.2
Arithmetic	1.1
Arithmos	1.1
Associative Law of Addition	3.7
Axiom	3.7
Balance Beam Principle	3.7
Barred-gate system of counting	1.2
Base of a number system	1.3
Base of the power	4.2
BC	3.2
BCE	3.2
Binary operations	2.4
Binomial	4.13
Binomial Theorem	4.13
ca.	1.3
Calculi	4.9
Cardinal	1.4
Casting out nines	2.8
CE	3.2
Celsius	3.2

	Lesson
Centimeter	2.8
cf.	1.1
Ciphers	1.3
Circle	2.7
Circumference	2.8
Classification or unity of concept	1.4
Coefficient	4.2
Collinear points	2.7
Common denomination	2.4
Commutative Law of Addition	3.7
Complement	2.5
Complementary addition	2.8
Complementary subtraction	3.5
Concave polygon	4.6
Concrete number	2.2
Concurrent lines	2.7
Congruent figures	4.2
Contingence	1.2
Converse	3.3
Convex polygon	4.6
Coplanar points	2.7
Counterexample	3.7
Counting	2.1
Creation as a language fabric	1.2
Cuneiform	4.9
Decagon	4.6
Decimal System	1.3
Definition	1.3
Denomination of a concrete number	2.2
Diagonal	4.4
Diameter	2.8
Difference	2.4
Difference of two squares	4.13
Digit	1.3
Distributive Law of Multiplication over Addition	4.4
Distributive Law of Multiplication over Subtraction	4.4
Dodecagon	4.6
Dynamic addition	2.4
e.g.	1.2

	Lesson
Ellipsis	2.3
Elohim	1.1
Equation	2.4
Equiangular triangle	4.6
Equidistant	3.2
Equilateral triangle	4.6
Equivalent expression	3.1
etc.	1.4
Expanded notation	1.5
Exponent or degree of the power	4.2
Exponential form	4.2
Exponentiation	4.2
Expression	3.1
Factor	4.1
Factor Shortcut Law	4.11
Factoring out common factors	4.13
Fahrenheit	3.2
ff.	1.1
Fibonacci Sequence	2.3
Finite	2.3
Fixed-value system	1.3
Formula	4.2
Generalization	2.3
Geometry	1.1
Googol	4.2
Googolplex	4.2
Great circle	3.2
Heptagon	4.6
Hexagon	2.3
Hieroglyphics	4.3
i.e.	1.1
Identity	3.7
Identity Element	3.7
Identity Element for Addition	3.7
Identity Element for Subtraction	3.7
imago Dei	1.1
Infinity	2.3
Integers	3.2
Integral number	2.2

	Lesson
Intersection	2.7
Inverse operations	2.3
Latitude	3.2
Law of Change in Addition	3.7
Law of Decrease in Addition	3.7
Law of Increase in Addition	3.7
Law of Minuend Change in Subtraction	3.7
Law of Subtrahend Change in Subtraction	3.7
Left side of an equation	2.4
Like numbers	2.2
Line	2.7
Line segment	2.7
Locus	2.7
Logical equivalence	3.3
Logistica	1.1
Logos	1.1
Longitude	3.2
Magnitude	3.2
Matching	1.4
Mathematical object	2.7
Mathematics	1.1
Minuend	2.4
Multiple	4.1
Multiplicand	4.1
Multiplication	4.1
Multiplication of Integers	4.12
Multiplier	4.1
Multiplying by 11 Law	4.11
Naming as a tool of dominion	1.1
Natural numbers	2.3
Nonagon	4.6
Noncollinear points	2.7
Notation	1.3
Number	1.1
Number line	3.2
Number system or scale	1.3
Number theory	2.2
Numeral	1.3
Numeration	1.3

	Lesson
Numeric expression	3.1
Obtuse angle	2.7
Octagon	2.8
One dimension	2.7
One-to-one correspondence	1.4
Ontology	1.1
Onto-relational	1.1
Opposite	3.2
Origin on the number line	3.2
Order of units	1.3
Ordering	1.4
Ordinal	1.4
Paradigm	3.6
Parallel lines	2.4
Parallelogram	3.3
Pascal's Triangle	4.12
Pentagon	2.8
Perichoresis	1.1
Perimeter	2.8
Periods	1.3
Place Value or Positional Notation	1.3
Plane	2.7
Point	2.7
Polygon	4.6
Power	4.2
Presuppose	1.4
Prime meridian	3.2
Product	4.1
Proposition	3.7
Protractor	2.7
Providence	1.2
Quadrilateral	2.8
Quadrivium	2.2
Quantification	2.1
Radius	2.7
Ray	2.7
Rectangle	2.5
Reflex angle	2.7
Regular hexagon	2.3

	Lesson
Regular polygon	4.6
Rhetoric	1.5
Rhombus	3.3
Right angle	2.7
Right side of an equation	2.4
Roman Numeral System	4.3
Scalar	3.2
Scientific Law	2.1
Sequence	1.4
Set	2.2
Set builder notation	3.2
Sexagesimal system	4.9
Sides of a polygon	4.6
Sides of an angle	2.7
Signs of multiplication	4.1
Solid	4.2
Soroban	2.5
Space	2.7
Square	2.8
Squaring Numbers That End with 5 Law	4.11
Standard notation	1.5
Static addition	2.4
Straight angle	2.7
Stylus	4.9
Subscript notation	2.10
Substitution	2.3
Subtraction	2.3
Subtrahend	2.4
Sum	2.4
Symmetry	2.6
Tag	1.4
Tangent point	2.7
Tangent ray, line, or line segment	2.7
Terms of an equation	2.4
Theology	1.2
Theorem	3.7
Three dimensions	2.7
Triangle	2.7

Mathematical Symbols

Symbol	Lesson	Symbol	Lesson
$<$	1.4	$\angle$	2.7
$>$	1.4	$m\angle$	2.7
$=$	1.4	.	2.8
$+$	1.4	a_1	2.10
$\neq$	1.4	$\mathbb{Z}$	3.2
$\times$	1.4	W	3.2
$\div$	1.4	$\Leftrightarrow$	3.3
-	1.4	$\Rightarrow$	3.3
$\mathbb{N}$	2.3	$\in$	3.7
∞	2.3	$\notin$	3.7
$\parallel$	2.4	$\cdot$	4.1
Δ (Geometry)	2.7	$*$	4.1
$\circ$	2.7		

Greek Alphabet

Greek Letter (Capital, Lower Case)	Greek Name	English Equivalent	Greek Letter (Capital, Lower Case)	Greek Name	English Equivalent
Α, α	Alpha	a	Ν, ν	Nu	n
Β, β	Beta	b	Ξ, ξ	Xi	x
Γ, γ	Gamma	g	Ο, ο	Omicron	ŏ
Δ, δ	Delta	d	Π, π	Pi	p
Ε, ε	Epsilon	ĕ	Ρ, ρ	Rho	r
Ζ, ζ	Zeta	z	Σ, σ or ς	Sigma	s
Η, η	Eta	ē	Τ, τ	Tau	t
Θ, θ	Theta	th	Υ, υ	Upsilon	u
Ι, ι	Iota	i	Φ, ϕ or φ	Phi	ph
Κ, κ	Kappa	k	Χ, χ	Chi	ch
Λ, λ	Lambda	l	Ψ, ψ	Psi	ps
Μ, μ	Mu	m	Ω, ω	Omega	ō

General Index

Numbers in **bold** indicate where the index entry is defined.
Numbers in *italics* indicate the index entry is in a footnote.

A

B

C

D

L

M

N

O

P

Q

R

S

T

U

V

W

Y

Z

SCRIPTURE INDEX

65941293R00176

Made in the USA
Columbia, SC
15 July 2019